Lecture Notes in Electrical Engineering

Volume 55

Subhas Chandra Mukhopadhyay and
Aimé Lay-Ekuakille (Eds.)

Advances in Biomedical Sensing, Measurements, Instrumentation and Systems

Springer

Subhas Chandra Mukhopadhyay
School of Engineering and Advanced Technology (SEAT)
Massey University (Turitea Campus)
Palmerston North
New Zealand
E-mail: S.C.Mukhopadhyay@massey.ac.nz

Aimé Lay-Ekuakille
Università Salento
Dipto. Ingegneria dell'Innovazione
Piazza Tancredi,7
73100 Lecce
Italy
E-mail: aime.lay.ekuakille@unisalento.it

ISBN 978-3-642-26198-5 e-ISBN 978-3-642-05167-8

DOI 10.1007/978-3-642-05167-8

Lecture Notes in Electrical Engineering ISSN 1876-1100

Typeset: Scientific Publishing Services Pvt. Ltd., Chennai, India.

Cover Design: Steinen Broo

Printed in acid-free paper

9 8 7 6 5 4 3 2 1

springer.com

Guest Editorial

This Special Issue titled "Advances in Biomedical Sensing, Measurements, Instrumentation and Systems" in the book series of "Lecture Notes in Electrical Engineering" contains the invited papers from renowned experts working in the field. A total of 19 chapters have described the advancement in the area of biomedical sensing, instrumentation and measurement methods and system in recent times.

This Special Issue has focused on the recent advancements of the different aspects of sensing technology, i.e. information processing, adaptability, recalibration, data fusion, validation, high reliability and integration of novel and high performance sensors in the area of biomedical field. The book is not a rough collection of contributions; instead, it offers an inside view of the topics, that is, a coherent vision in the foreseen area. Advances in technological devices unveil new architectures for instrumentation and improvements in measurement techniques. Sensing technology, related to biomedical aspects, plays a key role in nowadays' applications; it promotes different advantages for: healthcare, solving difficulties for elderly persons, clinical analysis, microbiological characterizations, signal processing, smart algorithms, augmented reality for new frontiers in surgery issues, use of EEG signals for late onset epilepsy, sensing system for breathing investigation, etc.. This book intends to illustrate and to collect recent advances in biomedical measurements and sensing instrumentation, not as an encyclopedia but as cleaver support for scientists, students and researchers in other to stimulate exchange and discussions for further developments.

We do sincerely hope that the readers will find this special issue interesting and useful in their research as well as in practical engineering work in the area of biomedical sensing technology. We are very happy to be able to offer the readers such a diverse special issue, both in terms of its topical coverage and geographic representation.

Finally, we would like to whole-heartedly thank all the authors for their contribution to this special issue.

Subhas Chandra Mukhopadhyay, Guest Editor
School of Engineering and Advanced Technology (SEAT),
Massey University (Turitea Campus)
Palmerston North, New Zealand
S.C.Mukhopadhyay@massey.ac.nz

Aime Lay-Ekuakille, Guest Editor
Dipartimento d'Ingegneria dell'innovazione
University of Salento
Lecce, Italy
aime.lay.ekuakille@unisalento.it

Dr. Subhas Chandra Mukhopadhyay graduated from the Department of Electrical Engineering, Jadavpur University, Calcutta, India in 1987 with a Gold medal and received the Master of Electrical Engineering degree from Indian Institute of Science, Bangalore, India in 1989. He obtained the PhD (Eng.) degree from Jadavpur University, India in 1994 and Doctor of Engineering degree from Kanazawa University, Japan in 2000.

During 1989–90 he worked almost 2 years in the research and development department of Crompton Greaves Ltd., India. In 1990 he joined as a Lecturer in the Electrical Engineering department, Jadavpur University, India and was promoted to Senior Lecturer of the same department in 1995.

Obtaining Monbusho fellowship he went to Japan in 1995. He worked with Kanazawa University, Japan as researcher and Assistant professor till September 2000.

In September 2000 he joined as Senior Lecturer in the Institute of Information Sciences and Technology, Massey University, New Zealand where he is working currently as an Associate professor. His fields of interest include Sensors and Sensing Technology, Electromagnetics, control, electrical machines and numerical field calculation etc.

He has authored/co-authored over 200 papers in different international journals and conferences, edited eight conference proceedings. He has also edited six special issues of international journals as guest editor and four books with Springer-Verlag.

He is a Fellow of IET (UK), a senior member of IEEE (USA), an associate editor of IEEE Sensors journal and IEEE Transactions on Instrumentation and Measurements. He is in the editorial board of e-Journal on Non-Destructive Testing, Sensors and Transducers, Transactions on Systems, Signals and Devices (TSSD), Journal on the Patents on Electrical Engineering, Journal of Sensors. He is in the technical programme committee of IEEE Sensors conference, IEEE IMTC conference and IEEE DELTA conference. He was the Technical Programme Chair of ICARA 2004, ICARA 2006 and ICARA 2009. He was the General chair/co-chair of ICST 2005, ICST 2007, IEEE ROSE 2007, IEEE EPSA 2008, ICST 2008 and IEEE Sensors 2008. He is currently organizing the IEEE Sensors conference 2009 at Christchurch, New Zealand as General Chair.

Aimé Lay-Ekuakille has a Master Degree in Electronic Engineering, a Master Degree in Clinical Engineering, a Ph.D in Electronic Engineering from Polytechnic of Bari, Italy. He has been technical manager of different private companies in the field of: Industrial plants, Environment Measurements, Nuclear and Biomedical Measurements; he was director of a Health & Environment municipal Department. He has been a technical advisor of Italian government for high risk plants. From 1993 up to 2001, he was adjunct professor of Measurements and control systems in the University of Calabria, Univer sity of Basilicata and Polytechnic of Bari. He joined the Department of Innovation Engineering, University of Salento, in September 2000 in the Measurement & Instrumentation Group. Since 2003, he became the leader of the scientific group; hence, he is the co-ordinator of Measurement and Instrumentation Lab in Lecce. He has been appointed as UE Commission senior expert for FP-VI (2005–2010). He is still: chair of IEEE-sponsored SCI/SSD Conference, member of Transactions on SSD and Sensors & Transducers Journal editorial board. He is Associate Editor of the International Journal on Smart Sensing and Intelligent Systems. He is currently organizing the next ICST2010 in Lecce, Italy. He is a member of the following boards and TCs: Association of the Italian Group of Electrical and Electronic Measurements (GMEE), SPIE, IMEKO TC19 Environmental Measurements, IEEE, IEEE TC-25 Medical and Biological Measurements Subcommittee on Objective Blood Pressure Measurement, IEEE-EMBS TC on Wearable Biomedical Sensors & System and included in different IPCs of conferences. Aimé Lay-Ekuakille is a scientific co-ordinator of different international projects. He has authored and co-authored more than 95 papers published in international journals, books and conference proceedings. His main researches are on Environmental and Biomedical instrumentation & measurements and, measurements for renewable energy.

Contents

Distributed System Architecture Using a Prototype Web E-Nose

M. Branzila and C. Donciu

Tehnical University of Iasi, Faculty of Electrical Engineering, Bd. D. Mangeron 53, 700050, Iasi, Romania

Abstract. In this paper we propose a web based design of an Electronic Nose so called "Web E-Nose". The devise have the possibility to transmit information's over the Internet and can be used in remote control mode. The system collects and automatically save data about the temperature, relative humidity and specific odors. Our hardware design allows various types of sensors to be used for different applications. The remote user can perform sensors test, analyze historical data and evaluate statistical information. Database based software with neural network facility was designed to interface the built hardware and to process the electronic nose signals before being classified.

Keywords: RS232 data acquisition, Internet transmission, Web Electronic Nose, Olfactory, Gas Sensor, Hardware, Software.

1 Introduction

The recent wide diffusion of (i) easy-to-use software tools for the implementation of Graphical User Interfaces (GUIs); and (ii) communication-oriented instrumentation, often provided with Ethernet interface, in addition to the more traditional GPIB and RS-232 ones, can be particularly exploited in the field of measurement teaching. It is well known, in fact, that for a better understanding of the teaching issues in such a field, the students have to practice with real instrumentation. The computer-based simulations are often inadequate to assure a good experience in that direction. The tools mentioned above give the possibility of accessing real measurement instrumentation from a remote location, such as the students' homes [8-10]. Moreover, it could be possible to repeat the same experience many times in order to make all students able to operate the measuring instrumentation without devoting expert technicians to such activity for many days [11-15].

In measurement teaching, the great increase of students on the one hand, and the reduced number of technicians on the other, greatly requires the possibility of accessing real measurement instrumentation for remote experiments [16-17].

In electrical and electronic measurement courses, particularly, these problems become more severe as consequences of the more sophisticated and expensive apparatus now available which makes it difficult to keep the technical staff up-to-date, and the necessity for repeating the same experience many times in order to make all students able to operate the measuring instrumentation.

S.C. Mukhopadhyay and A. Lay-Ekuakille (Eds.): Adv. in Biomedical Sensing, LNEE 55, pp. 1–15.
springerlink.com

There are two basic options when it comes to atmospheric monitoring: use of portable gas detectors or use of fixed detection systems. Portable monitors are battery-operated, transportable devices worn by the person using it and generally can detect only few gases at a time. There are places where many potentially hazardous gases can be permanently present: refineries, chemical plants, gas production plants, laboratories, mines, a.s.o. In these cases, a fixed system placed in the area where leaks of potentially dangerous gases likely to occur may provide general continuous monitoring.

The objective of our research is the development of compact fixed systems for real-time monitoring of the air contaminants, suited for gas leak detection, environmental control, worker protection or other industrial applications. In the first stage, we have designed a system capable to detect only several combustible gases, using a Web E-Nose. The monitoring of more air pollutants increases the system complexity. In this paper we present a system which performs the following main functions: detection of combustible gases (methane, ethanol, isobutane, hydrogen) and concentration measurement of toxic gases (carbon monoxide, hydrogen sulfide, ammonia).

Our system provide an alarm when the concentration of detected gases in the air reaches a dangerous level: % LEL (Lower Explosive Limit) – for combustible gases, TLV/TWA (Threshold Limit Value/Time Weighted Average), IDHL (Immediately Dangerous to Life or Health) – for toxic gases.

LEL is a specific minimum concentration above which an ignition source will cause an explosion or flame front propagation. These limits are different for every gas. Typical settings for the alarm circuit are 20% LEL for low alarm, 40% LEL for high alarm and 60% LEL for high-high alarm.

TLV/TWA is the time-weighted average concentration of a substance for a normal 8-hour workday, to which nearly all workers may be repeatedly exposed.

IDHL represents the maximum concentration level of a gas from which one could escape within 30 minutes without escape-impairing symptoms or any irreversible effects (for instance, 300 ppm for hydrogen sulfide).

The EU-funded conference on "Environment, Health, Safety: a challenge for measurements", held in Paris in June 2001, recognized the need to improve the performance of environmental measurement systems and their harmonization at EU level, to foster the dialogue between the providers of measurement methods and the users of measurement results, and to prepare the base - by establishing special communication tools – for the integration of research expertise and resources of environmental monitoring across Europe. The concept presented herein aims to respond to this actual challenge by combining the latest software trends with the newest hardware concepts in environmental monitoring, towards providing reliable measurement results and representative environmental indicators, evaluating trends and quantifying the achieved results in order to manage the potential environmental risk in compliance with European legislation and local particularities.[1-4]

For ages, the human nose has been an important tool in assessing the quality of many products, food products being good examples. While all others parts of production processes, including these of the food industry, were getting more and more automated, there was still no "objective" means for using the "subjective" information confined in the smell of products. This changed in 1982, when Persaud and Dodd introduced the concept of an electronic nose. They proposed a system, comprising an array of essentially non-selective sensors and an appropriate pattern recognition system, often called "e-nose".

The task of an electronic nose is to identify an odorant sample and perhaps to estimate its concentration. The E-Nose consists of two main components: an array of gas sensors, and a pattern-recognition algorithm. Electronic odour sensing systems can include a combination of hardware components such as sensors, electronics, pumps, fans, air conditioners and flow controllers, and software for hardware observation and data processing. The gas sensors most commonly used in electronic noses are based on metal oxide semiconductor and conducting polymer techniques. Metal oxide sensors were first produced in Japan in the 1960s for use in gas alarms and depend on an alteration in conductance caused by contact with the odour and the reaction that result.

The proposed Web E-Nose consists of three main components: an array of gas sensors, a pattern-recognition algorithm and an Ethernet module with a static IP.

2 System Architecture-Overview

An adaptive architecture based on web server application is proposed, in order to increase the performance of the server that hosts a dedicated (environmental monitoring) Web site, and customize the respective site in a manner that emphasizes the interests of the clients. The most virtual laboratories normally provide access either to one remote application, or accept only one user at a time. The system presented below provides a multitask connection, by accessing different detectors, working with different clients, and offering different variants for dedicated remote jobs, including technical tests of terminals, direct measurements of environment parameters, remote expertise's, technical demonstrations or vocational training and education [6-7].

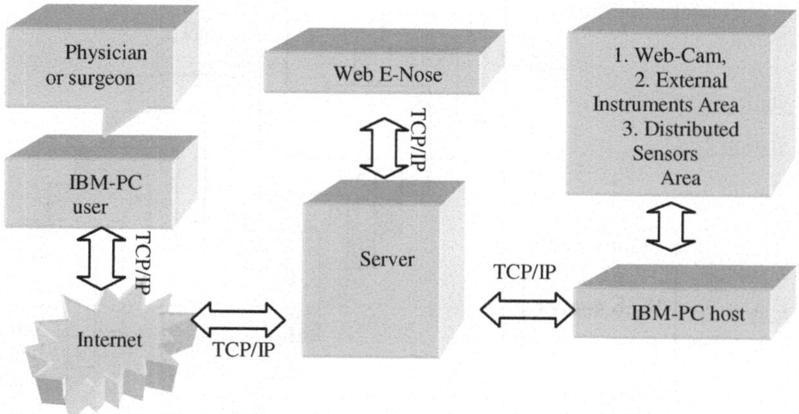

Fig. 1. System architecture-overview

An educational measurement remote system is proposed. The architecture is composed by to part: user and measurement provide as shown in figure 1. The student is able to watch for all the transformation that he command from front panel. Two cases are possible for remote teaching and education. In the first case, the professor from server room, after he set the students connected In this way, the students from their home study points can receive and follow the lessons. Number of students connected in the same time is unlimited.

In the second case the server is set to all user master. The students are able to perform the connection via modem and provider until server, in order to training and practice the programs.

The architecture is based on the virtual instrumentation concept and is composed of three main parts: gas detection block (including sensors and conditioning circuits), data acquisition part and data processing part. A virtual instrument performs the communication control and the data processing functions. In order to realize the continual measurement of gas concentration, artificial neural network structures are implemented in the software component.

The instrumentation control and communication software has been designed under LabView graphical programming language. In particular, the PC-server – via TCP/IP protocol and the client-server - via CGI (Common Gateway Interface) technology, have the important role of developing the PC-instruments communication. CGI simply defines an interface protocol by which the server communicates with the applications.

A dedicated software package supports the CGI applications in form of virtual instruments, used to develop interactive applications for Web-enabled experimental set-ups (that may be geographically distributed stations or expensive instruments, distributed areas of specialized sensors for water, air, soil etc. and Web E-Nose pollution monitor).

The main web page is located in server that allows the access to every station, using a connection link. In this machine a web server is run. The LabVIEW server represents the back up for the individual stations, as shown in figure 2.

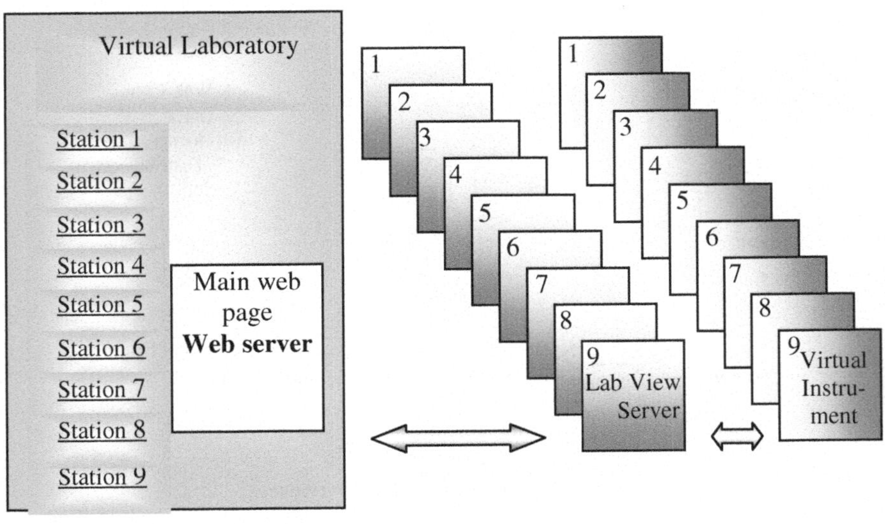

Fig. 2. Web pages configuration

The paper proposes an adaptive Web server application that tries to increase the performance of the Web server that hosts a Web site, as seen from the point of view of clients.

The adaptiveness is based on the customization of the Web site in a manner that emphasizes the interests of the clients.

Our server with dynamic allocation of client number is auto restarting. The usage of Web sites can vary widely during a day making difficult to predict the exact number of requests during any given time. Web server performance has a large impact on the popularity of a given Web site and therefore extensive effort has been done to improve it, especially for the sites that service a high number of requests. To customize a Web site means to offer different versions of the site to different users. We divide users into classes, each class receiving a version of the site optimized to their requirements. The classification of the users is based on a wide variety of parameters including both capabilities of clients and server performance measures, as the clients perceive them. Some of the parameters used for the classification are: download time of Web pages, client's bandwidth, the client's connection type, Web server's load, the type of the browser's plug-in components. The parameters' values are stored in a centralized database situated on the server side. The classification information is computed at every access done by the user and therefore the class the user belongs to can be dynamically changed.

3 Gas Detection

The sensing element is a metal oxide semiconductor mainly composed of SnO2. This element is heated at a suitable operating temperature by a built-in heater. Exposure of the sensor to a vapor produces a large change in its electrical resistance. In fresh air the sensor resistance is high. When a combustible gas such as propane, methane etc. comes in contact with the sensor surface, the sensor resistance decreases in accordance with the present gas concentration. Semiconductor gas sensors based on SnO2 are widely used as safety monitors for detecting most combustible and pollution gases. However, most of the commercial gas sensors are not selective enough to detect a single chemical species in a gaseous mixture. It is desirable that a single sensor should be able to selectively detect several kinds of gases. Previously, multiple gas identification efforts with a single sensor had assumed a particular temperature modulation, either sinusoidal with a fixed frequency or cyclic heating such as linear ramps. However, trials to implement a practical usage of such systems seem to have been lacking. The main problem to be overcome is still the non-linearity of the sensor response. Recently, it has been shown that microscopic fluctuation of the resistance of the chemical sensor contains more information about the measured chemical than the mean value of the resistance. The effect of various vapors on the power spectra of the resistance fluctuations has been studied for conducting polymer thin-film resistors.

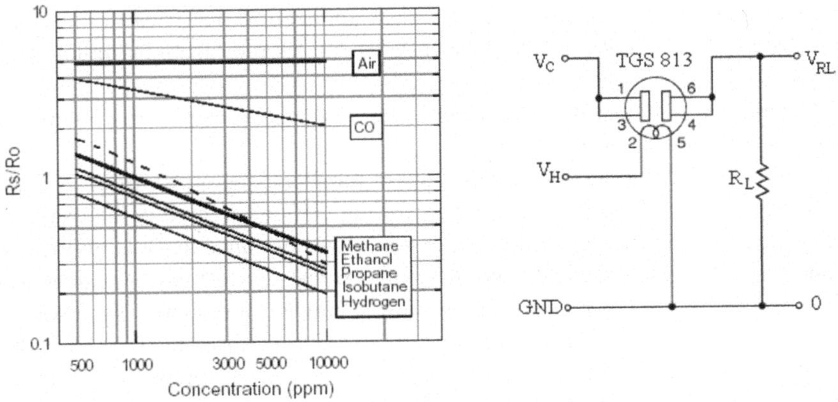

Fig. 3. Sensitivity characteristics and detection circuit

Figure 3 shows the basic detection circuit. The change in the sensor resistance is obtained as the change of the output voltage across the load resistor (R_L) in series whit the sensor resistance (R_S). The constant 5V output of the data acquisition board is available for the heater of the sensor (V_H) and for the detecting circuit (V_C). The relationship between R_S and V_{RL} is expressed by the following equation.

$$R_s = \frac{V_c - V_{R_L}}{V_{R_L}} \cdot R_L \tag{1}$$

The interaction of the chemical with the surface and bulk of the sensor induces spontaneous fluctuations. Recently, new methods have been proposed for chemical sensing that utilizes the analysis of the stochastic component of the sensor signal in Taguchi type sensors. It has been shown that even a single sensor may be sufficient for realizing a powerful electronic nose. However, there are no studies of the power spectrum in different types of commercial gas sensors under different gas atmospheres. This paper studies the stochastic signal in commercial semiconductor gas sensors measured under different atmospheres.

A unique gas detection block is used for both architectures of the system. It contains an array of five sensors and the corresponding detection circuits.

To detect hydrogen sulfide (H_2S), ammonia (NH_3) and combustible gases we use Taguchi type gas sensors produced by Figaro Co. The detection principle of TGS sensors is based on chemical adsorption and desorption of gases on the sensor surface. The sensing element is a tin dioxide (SnO_2) semiconductor that is heated at a suitable operating temperature by a built-in heater. In the presence of a detectable gas, the sensor conductivity increases depending on the gas concentration in the air. A simple electrical circuit converts the change in the sensor resistance to an output voltage, which corresponds to the gas concentration.

TGS 813 sensor has a good sensitivity to a wide range of combustible gases for concentrations from several ppm to over 10,000 ppm. Because of poor sensor selectivity, it is used only to detect the presence of some flammable gases in the environment (methane, ethanol, isobutane, and hydrogen).

TGS 825 and TGS 826 sensors have good sensitivity and selectivity to H_2S and NH_3, respectively. The relationship of sensor resistance to gas concentration is non-linear within the practical range of gas concentration (from several ppm to 100 ppm). In the data processing part, two artificial neural networks approximate the sensitivity characteristics of these sensors for the continual measurement of H_2S and NH_3 concentration.

4 The Prototype Data Acquisition Board

The data acquisition system is a low cost board realized around the chip LM12H458 that is an integrated DAS and offers a self-calibrating 12-bit a sign A/D converter with choice of single ended, fully differential, or mixed inputs, with on-chip differential reference, 8-input analog multiplexer, sample-and-hold, an impressive, flexible programmable logic system and a choice of speed/power combinations. The programmable logic has the circuitry to perform a number of tasks on its own, freeing the host processor for other tasks. This logic includes:

1. An instruction RAM that allows the DAS to function on its own (after being programmed by the host processor) with programmable acquisition time, input selection, 8-bit or 12-bit conversion mode.
2. Limit registers for comparison of the inputs against high and low limits in the "watchdog" mode.
3. A 32-word FIFO register to store conversion results until read by the host.
4. Interrupt control logic with interrupt generation for 8 different conditions.
5. A 16-bit timer register.
6. Circuitry to synchronize signal acquisition with external events.
7. A parallel microprocessor/microcontroller interface with selectable 8-bit or 16-bit data access.

The board can be used to develop both software and hardware. Since the parallel port is limited to 8-bit bidirectional data transfers, the BW pin is tied high for 8-bit access. Multiplexed address/data bus architecture between the DAS and LPT is used.

The circuit operates on a single +5V supply derived from the external supply using an LM7805 regulator or from USB port. This greatly attenuates noise that may be present on the computer's power supply lines.

Digital and analog supply pins are connected together to the same supply voltage but they need separate, multiple bypass capacitors. Multiple capacitors on the supply pins and the reference inputs ensure a low impedance bypass path over a wide frequency range.

All digital interface control signals (/RD, /WR, ALE, /INT, /CS), data lines (DB0–DB7), address lines (A0–A4) connections are made through the motherboard LPT connector using a standard LPT cable. All analog signals applied to, or received by, the input multiplexer (IN0–IN7), VREF+, VREF–, VREFOUT, and the SYNC signal input/ output are applied through a connector on the rear side of the board.

The voltage applied to VREF– is GND and VREF+ is selected using a jumper. This jumper selects between the LM12H458 internal reference output, VREFOUT, and the voltage applied to the corresponding pin applies it to the LM12H458 VREF+ input.

The board can provide 4 current inputs by manually acting to the jumpers (JP). The conversion of the unified currents into voltages is accomplished by precise resistors (R) calibrated for its own input.

A SYNK push button is available on the DAQB. With signal SYNC configured as an input, it is possible to synchronize the start of a conversion to an external event. This is useful in applications such as digital signal processing (DSP) where the exact timing of conversions is important.

Because the LM12H458 is so versatile, working with them may appear to be an overwhelming task. However, gaining a basic understanding of the device will prove to be fairly easy and using it to be as easy as programming the host microprocessor or microcontroller.

The main features of our DAQB are: 4 full-differential channels, 12 + sign ADC resolution, 100 ksamples acquisition rate, 20 ksamples transfer rate, 1 LSB linearity, 0,5 LSB accuracy, auto-zero and full calibration procedures, ± 5V input voltage span, 30 mW power dissipation.

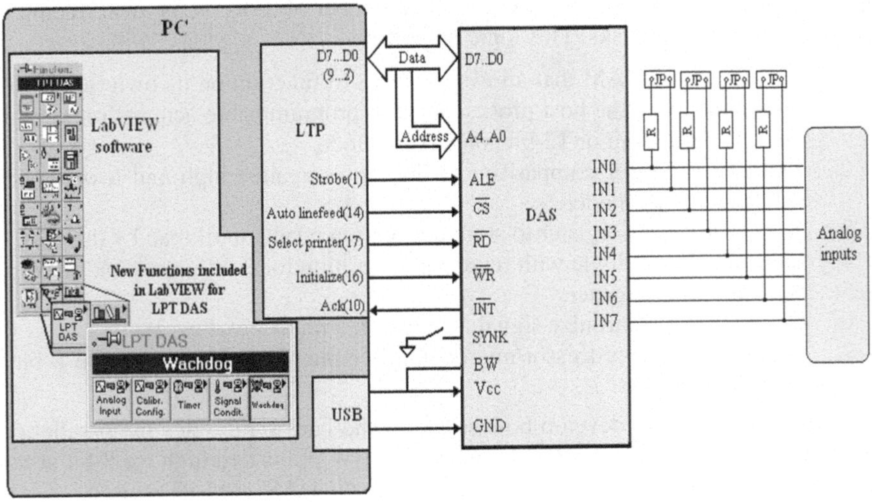

Fig. 4. The Virtual library and the architecture of a prototype data acquisition system

DAQB presented have the capability to perform tasks that diminish the host processor work and is capable to communicate with the host computer by using a set of drivers associated in LabVIEW software[20]. The novelty of our work mostly consists in the drivers and functions associated that are gathered into a library easily accessed by LabVIEW and assure the flexibility and the portability of the system. One of the performances consist in the fact that you can plug-in the DAQB to the running host computer externally.

5 Web E-Nose System

We developed a simple and original WebE-Nose prototype to test pattern recognition techniques that are necessary for building remote electronic nose systems.

Gas sensors tend to have very broad selectivity, responding to many different substances. This is a disadvantage in most applications, but in the electronic nose, it is an advantage. Although every sensor in an array may respond to a given chemical, these responses will usually be different [18-19].

Sensor array "sniffs" the vapors from a sample and provides a set of measurements. The pattern-recognizer compares the pattern of the measurements to stored patterns for known materials [5].

The implemented Web E-Nose system consists in three main components (figure 5):

1. a gas sensors array,
2. the pattern recognition algorithm, and
3. Ethernet with IP static module.

The initial experiment was performed with a number of low selectivity gas sensors – calibrated to identify a threshold value of the most important polluting gases occurring in the atmosphere, combined with SHT11 humidity and temperature sensor, allowing immediate temperature and humidity compensation. The sensor array was also trained to recognize, by different sets of measurements, the hazard patterns for different polluting factors acting in the monitored area, as well as identify accidental patterns of polluting factors with external causes. The Ethernet module, having a static IP, give the possibility to share, over the World Wide Web, information's about "remote polluters and potential effects, hazard level etc." with a clear identification of the instrument and area. Finally, the result of Web E-Nose expertise may be visualized either as a code image of a given combination of volatile compounds, or may offer a review of the concentrations of individual molecule species detected in a complex environment.

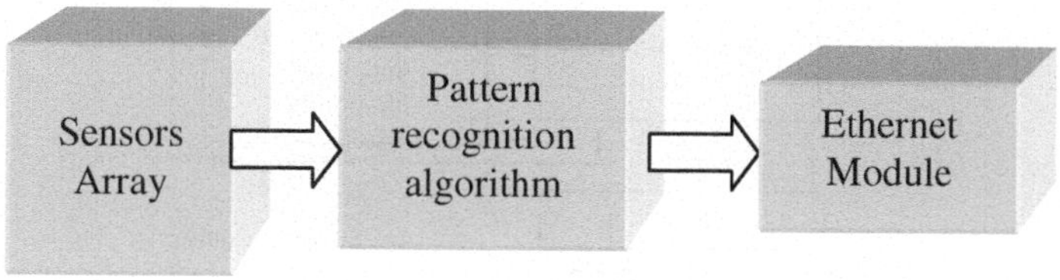

Fig. 5. Main components of the Web E-Nose

The response of the sensor array is numerically converted using a prototype data acquisition system SADI (integrated data acquisition system). This response is registered by microcontroller as "case pattern", compared and classified with the ones predefined within the training library. The microcontroller, playing the role of Web E-Nose "brain", communicates with SADI or with IP-Static module server by a serial interface.

Hence, the most important function of the Web E-Nose system consists in detecting and evaluating toxic gases or mixtures at minimum threshold quantities, especially those odourless to human senses. The information, acquired by the gas sensor arrays and rough calibrated by SHT11 temperature and humidity sensor, is subject of further processing for pattern recognition and transmission to the decision block by RS232 protocol to the Ethernet server.

The Web E-Nose system has five sequential stages: pre-processing, feature extraction, classification, decision making and decision transmission to the network. The decision making, based on pattern recognition, is assisted by a neural network with both training and extraction functions.

It goes without saying that the Web E-Nose system was not projected to substitute human capability of detecting hazardous situations by "smelling". In addition, the exquisite sensitivity of the dog's nose for sniffing out odours associated to drugs or other hazardous vapours has not yet been matched by currently designed E-nose.

But the system is well suited for repetitive and accurate measurements, and provided not to be affected by saturation, a common disadvantage of natural smelling senses.

Our human nose is elegant, sensitive, and self-repairing, but the Web-E-Nose sensors do not fatigue or get the "flu". Further, the Web-E-Nose can be sent to detect toxic and otherwise hazardous situations that humans may wish to avoid. Sensors can detect toxic CO, which is odorless to humans. And humans are not well suited for repetitive or boring tasks that are better left to machines. No wonder the E-Nose is sometimes referred to as a "sniffer".

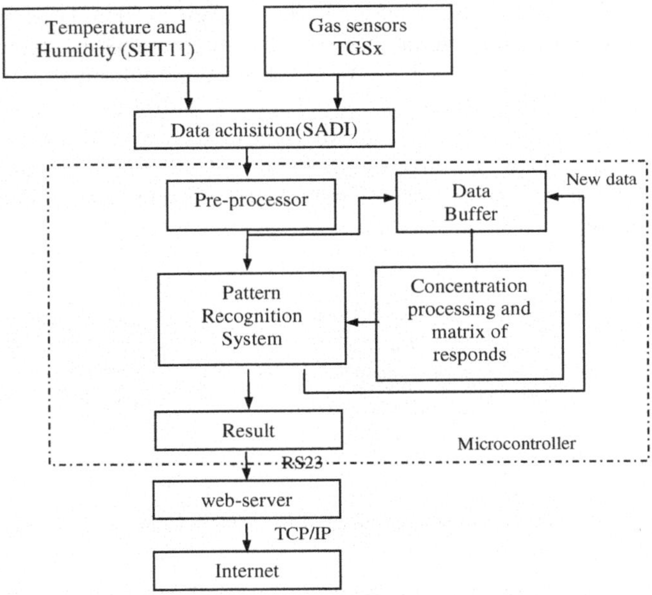

Fig. 6. Web E-Nose communication architecture

However, the human nose is still preferred for many situations like the selection of a fine wine or to determine the off-odor of recycled plastics. In addition, the exquisite sensitivity of the dog's nose for sniffing out drugs or contraband at an airport is legendary already. These skills have not yet been matched by any currently designed E-Nose.

6 Virtual Instrument

The virtual instrument is the main part of the monitoring system and, in the same time, the human interface, providing parameters control. The main functions of the VI are:

1) SADI programming and communication control using Nr.Scan function from the specific functions palette.
2) data processing after data reading from SADI analog input channels
3) environmental temperature calculus starting from the analog input voltage(CH2) V_T.

$$T = V_T \cdot 1000 - 273{,}15 \quad [^{\circ}C] \tag{2}$$

4) Sensors resistance calculus starting from CH_0 and CH_1 read voltages (V_{RL1} and V_{RL2}):

$$Rs = \frac{V_C \times R_L}{V_{RL}} - R_L \quad [\Omega] \tag{3}$$

5) Mean pollutant concentration calculus for a user presetting time interval (ppm/30min, ppm/8h, ppm/24h) for pollution level testing.
6) Pollution agent concentration limits exceeding verify for knowing the immediate effects fort health, lighting and voice user warning
7) Decrease the environmental temperature influence using a compensation subVI
8) Data base saving.

Admitting that the temperature and humidity have a great influence to Taguchi sensor resistance we have to make a compensation of the effect very utile when the system is used for outside. Knowing the R_S/R_0=f(T) dependency characteristic of the sensors and the temperature from AD590 temperature sensor the VI realize a temperature compensation by parts.

At 65% relative humidity, a characteristic linearization is made on the next intervals: -10ºC ÷ 20ºC, 20ºC ÷ 30ºC, 30ºC ÷ 40ºC. A slope determination for the three straight lines is done and for each temperature values is established the compensation factor. The main sub-VI's are:

1. *Nr.Scan* do the samples acquisition from an analog input channel.
2. *Vrl to RpeR0.vi* do the determination of Rs/R_0
3. *Compens_term.vi* realizes the compensation of temperature influence on sensor resistance (TGS). The VI inputs are: current temperature (ºC) and measured value of Rs/R_0. The VI output is R_S/R_0 value after thermo-compensation. The compensation of temperature influence is realized by equation implementation of linear variation R_S/R_0=f(T) on temperature interval previously mentioned.
4. *R to Concentratie.vi* determine the methane concentration based on sensor measured resistance using the next equation:

$$G_s = S_0 \cdot C^b \tag{4}$$

where G_s=1/R_S is the sensor conductivity at certain methane concentration C.

S_0, b – constants determinate for two concentration (C1=1000 ppm, C2=3000 ppm) when we know the value of sensor resistance. At VI input is applied the sensor resistance (R_S) after the thermal effect compensation obtaining to the output the calculated value of concentration.

5. *Tens to grdC.vi* give the temperature dependence on input voltage of analogical channel 2 (1m/K).

Intelligent system achievement witch is dedicated for particular application is not easy. It presume a selection of chemical sensors area witch provide a large information quantity and complex algorithms development for signal processing [1-2].

The developed environmental monitoring systems (EMS) that use a prototype data acquisition board perform different tasks like: multi-sensors/multi-point measurement, continuum real-time monitoring, across limits warnings, save data etc.

Air quality parameters can be monitories, from interested areas like public places, enterprises etc. The desktop PC and LabVIEW software have the fallowing functions:

- DAQB control,
- Data processing and results display,
- Data storage and data administration
- User and warning beneficiar/utilizator,
- Analysis and decision etc.

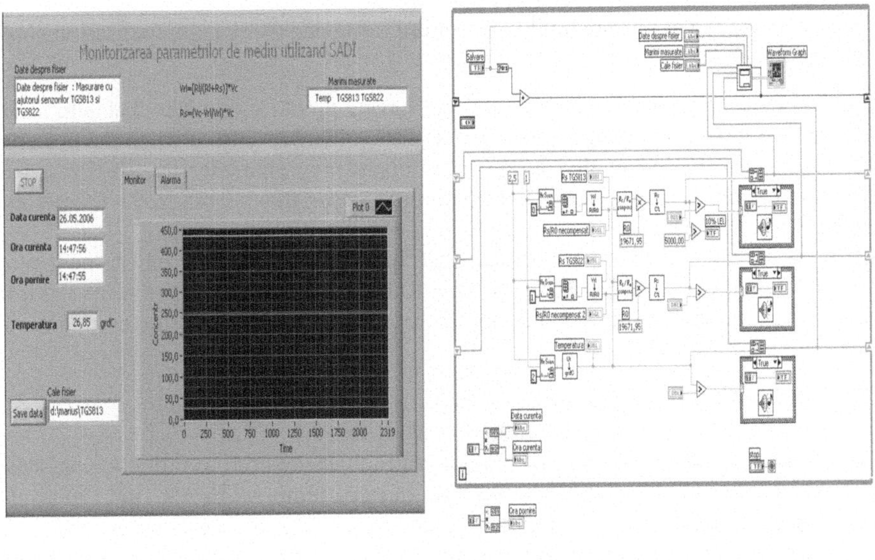

Fig. 7. Front panel and diagram of VI using a prototype SADI

7 Virtual Laboratory-System Architecture

The Web concept itself is changing the way the measurements are made available and the results are distributed/communicated. Many different options are occurring as regards reports publishing, data sharing, and remotely controlling the applications.

The low-cost availability of new communication tools based on Internet is opening more and more horizons to remote teaching. Interactive on-line tutorials based on World Wide Web (WWW) sites now can be followed directly on the web site. The hardware of monitoring systems (sensors, conditioning circuits, acquisition and communication) must usually be complemented with processing blocks to perform different tasks associated to one-dimensional or multi-dimensional data that flow on the system measurement channels.

The main objective of this work is to realize a complex monitoring model based on specialized sensors that are connected in a unit system. It can be very useful in the new society information to create a Virtual Laboratory for a remote teaching [3-4].

An adaptive architecture based on web server application is proposed, in order to increase the performance of the server that hosts a dedicated Web site, and customize the Web site in a manner that emphasizes the interests of the clients. The most virtual laboratories normally provide access either to one remote application, or accept only one user at a time. The system presented below provide a multitask connection, with possible variants for remote education. In this way, two parts compose the architecture of the system:

- client user that uses a client computer and
- measurement provider who disposes the server with the web site of the virtual laboratory.

The users will be able to perform the lab work, controlling the applications and accessing the virtual library. Number of users connected in the same time is unlimited.

The LabVIEW environment was incorporated in centre concept towards creating a unique and powerful distributed application, combining together different measurement nodes and multiple users into a unique measurement controlling system, in order to integrate and revolutionize the fundamental architecture of actual PC-based measurement solutions. All communication software is designed under LabVIEW graphical programming language.

In the figure 8 is presented the main web page of application.

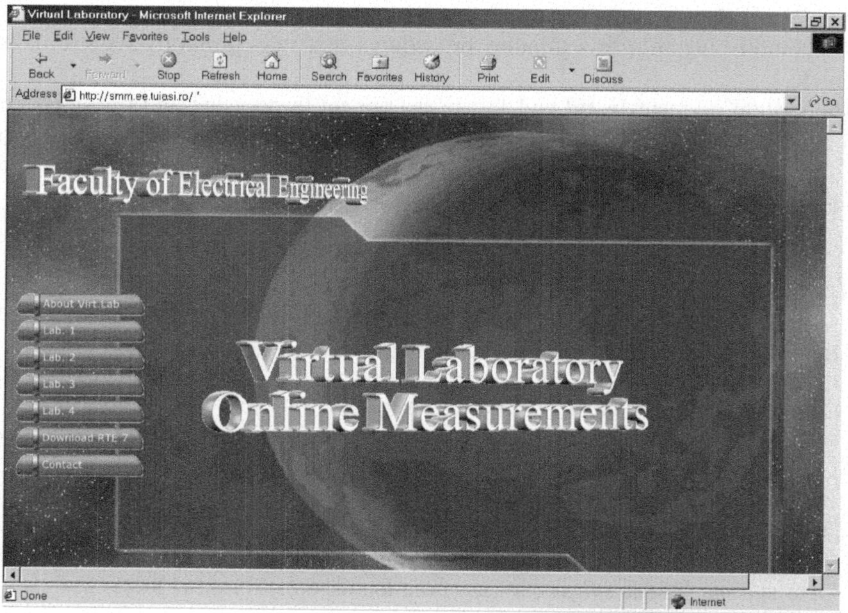

Fig. 8. The main web page of Virtual Laboratory

The main web page is located in server that allows the access to every station, using a connection link. In this machine a web server is run. The LabVIEW server represents the back up for the individual stations.

8 Conclusion

The Internet continues to become more integrated into our daily lives. This is particularly true for scientists and engineers, because designers of development systems view the Internet as a cost-effective worldwide standard for distributing data.

The paper presents the architecture of a versatile, flexible, cost efficient, high-speed measurement centre, based on remote instrumentation, and having as final purposes the monitoring of the air quality and the advertising of the air pollution. On the other hand the E-Medicine becomes a very interesting domain for physicians and bio-engineers. That for the purposed system can be very useful tool for them.

In many locations a basic infrastructure to evaluate the E-Medicine already exists, but a unitary concept of an E-Medicine centre can be used to deliver services of comparable or higher quality, at a clear lower cost and a higher speed and reliability.

The Web-E-Nose system was tested, and provided to be well suited for repetitive and accurate measurements, without being affected by saturation. But the successful implementation of such Web E-Nose concepts for air pollution evaluation at larger scales will require a careful examination of all costs, either direct or indirect, and should demonstrate its societal benefit over time.

The remote and distributed measurement system developed as environmental centre may be also particularized as virtual laboratory for on-line environmental monitoring, helping the formation of well trained specialists in the domain.

The Web E-Nose is a tool that may be used for safety, quality, or process monitoring, accomplishing in a few minutes procedures that may presently require days to complete.

The system performs good and fast measurement, processing and transmission of the odors. It is very useful in the new society information to create a Virtual Laboratory for a remote teaching or to get information about gas mixtures or odors from a remote site.

References

1. Branzila, M.: Intelligent system for monitoring of exhaust gas from hybrid vehicle. In: 4th International IEEE Conference on Intelligent Systems, IS 2008, Varna, Bulgaria, September 6-8, vol. 1, pp. 5-2–5-5 (2008) ISBN: 978-1-4244-1739-1
2. Branzila, M.: New DAQB and associated virtual library included in LabVIEW for environmental parameters monitoring. In: IEEE Conference on VECIMS 2008, Istanbul, July 14-16, pp. 121–124 (2008) ISBN: 978-1-4244-1927-2, INSPEC Accession Number: 10156672
3. Branzila, M.: Virtual environmental measurement center based on remote instrumentation. Environmental Engineering and Management Journal 6(6), 517–520 (2007)
4. Branzila, M.: Virtual meteorological center. International Journal of Online Engineering 3(4), 45–480 (2007)

5. Branzila, M.: Design and Analysis of a proposed Web Electronic Nose (WebE-Nose), IAŞI – ROMÂNIA Buletinul Institutului Politehnic din Iasi, Tomul LII (LIV), Fasc. 5, pp. 971–976 (2006) ISSN 1223-8139
6. Donciu, C.: Urban traffic pollution reduction using an intelligent video semaphoring system. Environmental Engineering and Management Journal 6(6), 563–566 (2007)
7. Donciu, C., Costea, O., Temneanu, M., Damian, R., Branzila, M.: New prototype architecture for automated irrigation based on power line communications, pp. 505–516; Grid Enabled Remote Instrumentation, p. 600. Springer, Heidelberg, ISBN 978-0-387-09662-9.1
8. Arpaia, P., Daponte, P.: A distributed measurement laboratory on geographic network. In: Proc. of 8th IMEKO Int. Symp. on New Measurement and Calibration Methods of Electrical Quantities and Instruments, Budapest, Hungary, pp. 294–297 (1996)
9. Arpaia, P., Daponte, P.: A measurement laboratory on geographic network for remote test experiments. IEEE Trans. on Instrumentation and Measurement 49(5), 992–997 (2000)
10. Arpaia, P., Daponte, P.: Ethernet application perspectives for distributed measurement systems. In: Proc. of XIV IMEKO World Congress, Tampere, Finland, June 1997, vol. IVA, pp. 13–18 (1997)
11. Dyer, S.A., Dyer, R.A.: Emphasizing the interdependence of topics in required undergraduate electrical engineering courses: a case study. In: Proc. of IEEE IMTC 1997, Ottawa, Canada, May 1997, pp. 1320–1325 (1977)
12. Hancock, N.H.: I&M in the 21st century engineering curriculum-a direction for change. In: Proc. of IEEE IMTC 1997, Ottawa, Canada, May 1997, pp. 1326–1331 (1977)
13. Carlosena, A., Cabeza, R.: A course on instrumentation: the signal processing approach. In: Proc. of IEEE IMTC 1997, Ottawa, Canada, May 1997, pp. 1326–1331 (1977)
14. Arpaia, P., Daponte, P.: A distribuited laboratory based on object-oriented systems. Measurement 19(3/4), 207–215 (1996)
15. Bertocco, M., Cappellazzo, S., Carullo, A., Parvis, M., Vallan, A.: Virtual environment for fast development of distributed measurement applications. In: Proc. of IEEE International Workshop on Virtual and Intelligent Measurement Systems, VIMS 2001, Budapest, Hungary, May 2001, pp. 57–60 (2001)
16. Donciu, C., Cretu, M.: Communication in Virtual Instrumentation. In: Management of technological changes, Iasi, Romania, pp. 69–74 (2001)
17. Donciu, C., Rapuano, S.: A Remote Inter-University System For Measurement Teaching. In: 12th IMEKO TC4 International Symposium, Zagreb, Croatia, pp. 420–424 (2002)
18. Keller, P.E.: Three Neural Network Based Sensor Systems for Environmental Monitoring. In: IEEE Electro 1994 Conference Proceedings, pp. 377–382. IEEE Press, Piscataway (1994)
19. Keller, P.E.: Electronic Noses and Their Applications. In: Proceedings of the World Congress on Neural Networks1996, pp. 928–931. Lawrence Erlbaum Associates Inc., Mahwah (1996)
20. Schreiner, C., Branzila, M.: Air quality and pollution mapping system, using remote measurements and GPS technology. Global NEST Journal 8(3), 315–323 (2006)

Magnetic Fluids for Bio-medical Application

Bruno Andò, Salvatore Baglio, and Angela Beninato

DIEES, Engineering Faculty, University of Catania, V.le A. Doria, 6
Catania, Italy
bruno.ando@diees.unict.it

Abstract. The use of magnetic fluids in the bio-medical field is becoming quite appealing due to the amusing properties and potentialities of these materials as bio-compatible markers to localize target molecules and to transport functional entities. Moreover, the use of magnetic fluids in integrated devices is very promising to implement lab-on-chip systems. Although this strong efforts towards bio-applications, magnetic fluids reveals interesting properties and peculiarities for the realization of inertial sensors and actuators with very performing performances and characteristic.

This chapter is dedicated to a review of the use of magnetic fluid in the field of bio-applications with a specific focus on the implementation of actuators aimed to control small amount of fluid.

Keywords: Magnetic fluids, bio-application, pumps, versatility, reliability.

1 Introduction

Magnetic fluids are stable suspensions of magnetic particles in a carrier liquid coated with a dispersant [21]. The coating provides an elastic shield which avoids particle friction and at the same time allows to implement bio-assay strategy.

The rheological properties of these fluids, such as viscosity, depend on particle density, particle size and magnitude of eventual external magnetic fields. Actually, the size of the particles makes the difference between two types of fluids: magneto-rheological fluids and ferrofluids.

In magneto-rheological fluids the magnetic particles have a diameter in the order of microns. The rheological behaviour of these fluids is strongly controllable by an eternal magnetic field up to experience the medium solidification for high value of the magnetic stimulus.

Ferrofluids are colloidal suspensions of magnetic beads with a diameter ranging in the order of nanometers. Conversely to magneto-rheological fluids, ferrofluids maintain their liquid status also in the presence of a strong magnetic field. Usually, a low magnetic field is required to bring the material magnetization to the saturation state while high magnetization strength implies high magnetic pressure exerted by the fluid. The latter property is strategic for the implementation of transducers adopting ferrofluids as functional or inertial masses.

A typical response of a ferrofluid to an external magnetic field is observable in Figure 1. Magnetic fluids show also interesting patterns, coming from ferrohydrodynamic instabilities, such as lines, labyrinths and various structures which could be

S.C. Mukhopadhyay and A. Lay-Ekuakille (Eds.): Adv. in Biomedical Sensing, LNEE 55, pp. 16–28.
springerlink.com

exploited to produce actuation [25]. Actually the flat surface of a horizontal ferrofluid layer subjected to a normal magnetic field could become unstable, thus showing the Rosensweig instability [9]. At a certain intensity of the field, peaks appear at the fluid surface, which typically form a static hexagonal pattern at the final stage of the pattern forming process.

Parametric waves on the surface of a ferrofluid can be excited using a vertical or a horizontal alternating magnetic field [7, 16]. This phenomenon is called the magnetic Faraday instability and typical patterns are shown in Figure 2.

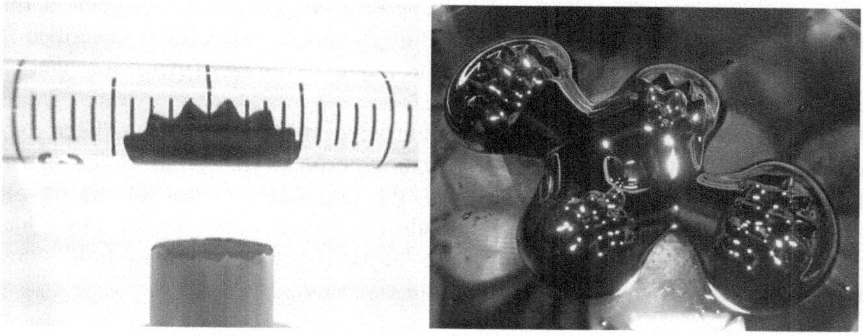

Fig. 1. Examples of ferrofluids subjected to magnetic fields

Practical interest in magnetic fluids derives from the possibility to implement valuable and efficient conversion of elastic energy into mechanical energy [10]. Sealing for several industrial processes, loudspeakers, inertial dampers, angular position sensors, computer disk drives are examples of applications where this kind of materials has been widely adopted [24].

The use of magnetic fluids in transducers is now widely diffused due to valuable properties of these fluids compared to traditional materials [3]. Actually, fast response, shock resistance and their intrinsic feature to be shapeless allow to develop suitable sensors and actuators. Moreover, the possibility to control some material characteristics (such as viscosity) by an external magnetic field permits the development of smart devices with tuneable operating ranges for an extended set of applications. Various examples of devices exploiting non-conventional materials such as ferrofluids to implement sensors and transducers are available in the literature.

The possibility to use these materials in combination with Micro-Electro-Mechanical-Systems (MEMS) to develop efficient sensors and actuators [23, 29] is pushing the interest of the scientific community toward the development of new and efficient magnetic fluids and specific coatings. Miniaturized permanent magnets, ferromagnetic patterned layers, and integrated coils are traditionally implemented in MEMS technology. Anyway, such realizations host small volumes of magnetic material which can provide a negligible magnetic force which is unsuitable for actuation purposes. The use of MEMS cavity filled with ferrofluids could be a valuable solution to obtain high volumes of magnetic material of appropriate density. Concerning the implementation of actuation systems, the possibility to easily drive ferrofluids through micro-channels of various geometry represents a convenient solution. Nanogenerators and nanomotors [10], micropipettes [11], alternating micropumps and valves [12], rotating micropumps [13] and electromagnetic micropumps [14] are examples of ferrofluidic integrated transducers. In [11] a micropipette exploiting the magnetic force generated by external electromagnets is proposed. The mechanism implemented allows to suck and release fluid through the same gate. To perform the pumping operation external magnets moved by a DC motor are used; moreover, an interesting solution is proposed to realize the valve by exploiting a dedicated geometry of the channel. In [12] another example of alternating micropump is given. A rotating micropump implementing fluid sampling through two ferrofluidic is discussed in [13]. In this case the caps are created and managed by two permanent magnets actuated through a DC motor. In [14] a pump handling magnetic particles is described. The device exploits an array of electromagnets to move a volume of magnetic nanoparticles from an inner tank to a destination tank.

The behaviour of ferrofluid in a travelling wave magnetic field has been experienced and can be suitable exploited for actuation purposes [8, 27-28]. Below a critical magnetic field strength the fluid moves in the opposite direction of the travelling wave while above the critical field the ferrofluid moves in the same direction. Quantities affecting the critical magnetic field are frequency of the magnetic field, concentration of the magnetic particles, and the fluid viscosity.

(a) (b)

Fig. 2. Ferrofluidic patterns: (a) Peaks due to the Rosensweig instability; (b) Waves due to the Faraday instability

2 Ferrofluids and Bio-applications

Apart from above mentioned traditional uses of magnetic fluids which are widely available in literature, ferrofluids are nowadays assuming a fundamental role in biomedical applications for diagnostic and therapy [22]. These materials find interesting applications in magnetic bio-assay tasks, such as magnetic separation, drug delivery, hyperthermia treatments, magnetic resonance imaging (MRI) and magnetic labelling [22]. In fact, due to the different and controllable sizes of their particles (from few nanometres up to micrometres), they can interact with the biological entity of interest like a cell (10–100 μm), a virus (20–450 nm), a protein (5–50 nm) or a gene (2 nm wide and 10–100 nm long), thus offering attractive possibilities to develop efficient solutions in the bio-medical field. Examples of such techniques can be found in [15, 18, 19, 20, 26, 30].

Magnetic particles coated with biocompatible molecules act as markers to identify bio-entities, as schematized in Figure 3. The main advantage of this form of labelling as compared to other techniques is related to the simple mechanisms to identify, to localize and to transport magnetic labelled entities. Actually, all these mechanisms are based on the use of magnetic fields which are also intrinsically penetrable into human tissue. Magnetic labelling is used for both entities localization and separation.

Localization tasks require remote sensing usually performed by AC susceptometry or SQUID magnetometry which show high sensitivity while require expensive cryogenics and instrumentations. Alternative solutions use giant magnetoresistance sensor (GMR) [17] which are not intrinsically sensitive as SQUIDS but performs very well to sense magnetic fields generated by magnetic beads. Fluxgate magnetometers represent an alternative low cost solution to sense weak magnetic fields or field perturbations [4]. Recently, Residence Times Difference (RTD) Fluxgate have been proposed as competitive devices to the traditional second harmonic architectures. The working principle is based on the different value of Residence Times as function of the presence/absence of magnetic beads. On absence of any target field, the hysteresis loop is symmetric and two identical Residence Times are obtained. The presence of magnetic beads leads to a skewing of the hysteresis loop with a direct effect on the Residence Times, which are no longer the same.

Magnetic separation is becoming widely adopted in biomedical and biological applications due to the very promising performances of this method to select and to separate small amount of target cells [19, 20, 22, 26]. A typical target of magnetic separation is the detachment of biological entities from their environment in order to increase the entities concentration in the sample to be analyzed. It is a two-step process: the first step consists of labelling the molecules to be detected with magnetic beads; this is obtained by coating the magnetic particle surface with specific biocompatible molecules to allow the binding with the target entity. The second step consists of separating the labelled molecules from their native by blocking the magnetic particles via a magnetic field. As an example the fluid containing the magnetic labelled entities can flow through a region subjected to a magnetic field.

The same principle is used to remove unwanted biomaterials from a fluid, capturing the magnetic particles with a magnet or with a magnetic field gradient and letting flow the remaining fluid.

A promising use of magnetic fluid in biomedicine is drug delivery which has peculiar advantages as compared to traditional techniques [15, 18, 22, 30]. In fact, today therapeutic drugs act on the entire body (e.g. by attacking both tumour tissue and healthy ones) while with selective drug delivery only specific locations of the body are attacked. This approach allows to reduce the required drug dosage, to increase drugs specificity as well as prolonging the use of these effective agents. The magnetic particles bounded with the drugs are injected through the circulatory system while an external high-gradient magnetic field is used to deliver tasks.

Another possibility for tissues (e.g. cancer) treatments is the hyperthermia [22]. Such technique consists of embedding magnetic particles into the target tissue and applying a suitable magnetic field to cause particles heating. This heat radiates into the surrounding tumour tissue thus to destroy, after a specific length of time, the cancer cells. While other hyperthermia methods produces unacceptable heating of human healthy tissue, magnetic hyperthermia overcome these drawbacks by a selective heating of target cells.

MRI [22] is a non-invasive technique that allows the characterization of morphology and physiology in vivo through the use of magnetic scanning. The human body is mainly composed of water molecules containing protons which under powerful magnetic field align with the direction of the field. A second radio frequency electromagnetic field is then briefly turned on causing the protons to absorb some of its energy. When this field is turned off the protons release this energy at a radio frequency which can be detected by a scanner. Relaxation times of protons in diseased tissue, such as tumours, are different than in healthy tissue and this information is used for the sake of inspection. Magnetic nanoparticles can be conveniently used in MRI techniques as contrast agent. In particular, dextran coated iron oxide nanoparticles are commonly used being biocompatible and easily excreted via the liver after the treatment.

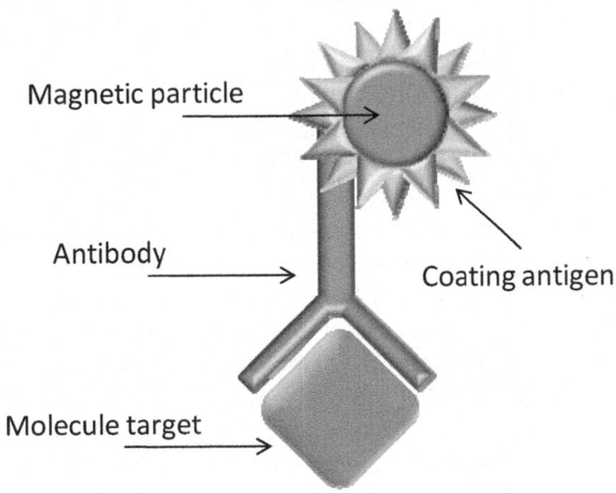

Fig. 3. A magnetic particle is coated with a suitable biocompatible molecule (antigen). The coating allows to link the target bio-entity with the magnetic particle as an example through a specific antibody.

The need for handling small amount of fluids is another issue in the field of biomedical devices. Although conventional micropipette can sample small liquid volumes with a precision ranging in 3-5%, the development of novel solutions showing high reliability (against shocks), bio-compatibility, adaptability (configurable specifications) and flexibility (easy implementation into preexistent pipes) are highly demanded. Moreover, the use of low cost transducers is becoming mandatory especially in the contexts of point-of-care.

The research community is hence focusing towards the development of innovative and cheap solutions to control liquids in channels and micro-channels for biomedical applications (diagnostic, drug delivery, genetic sequencing, Lab-On-a-Chip, flow control of bio-fluids). Figure 4 shows an example of a Lab-On Chip system.

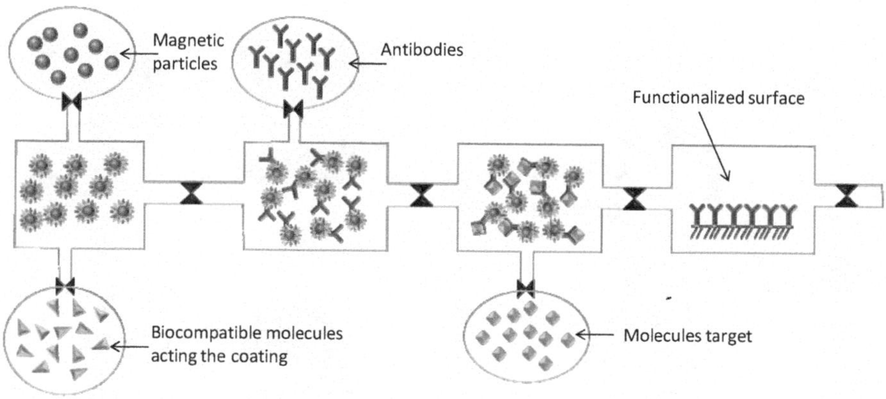

Fig. 4. Basic schematization of a Lab-On-Chip system regulated by valves. In the first chamber magnetic particles bind with the bio-molecules implementing the coating; in the second chamber the link with the antibody and in the third chamber the link with the molecules target are achieved; finally, the realized binding is sent to a chamber with a functionalized surface which, as an example, implements the sensing element of the readout system adopted to estimate the amount of labeled target entities.

In particular, efforts are dedicated to the use of both new materials and innovative actuation mechanisms. The idea of using small volumes of ferrofluids as active masses to implement the pumping action is emerging [5, 6]. In [6], the authors investigated a ferrofluidic pump realized by a glass channel filled with deionized water where three ferrofluidic drops have been injected to implement two valves and one plunger: the FP3 device. The main advantages of such solution reside in the absence of mechanical moving parts, thus avoiding stress and increasing the life time and the reliability of the device. Moreover, such approach allows for the implementation of a pump in a section of a preexistent pipe without actions damaging or modifying the original structure of the channel: this feature is valuable for applications where the pump must be installed occasionally in specific locations (e.g. vascular surgery application), thus increasing the flexibility and the applicability of the methodology proposed. In [5], a novel pumping architecture using one drop of ferrofluid is presented. This architecture exhibits the same advantages of the FP3 pump while uses a single drop of ferrofluid and an array of electromagnets to be clamped on the pipe. This solution boosts the advantages of the FP3 device in terms of dimension shrinking and implementations into preexistent pipes.

The possibility to exploit these strategies for flow control into a pre-existing channel also in the field of bio-medical application is outlined in Figure 5.

Although several amusing ferrofluidic transducers have been recently developed [1-6], the highlighted interest in novel pumping mechanisms led to the choice of focusing the next section on the Ferrofluidic Pump FP3 developed at the D.I.E.E.S. of the University of Catania – Italy [6]. In particular, a detailed description of the device is given along with some indications on its design and estimated performances.

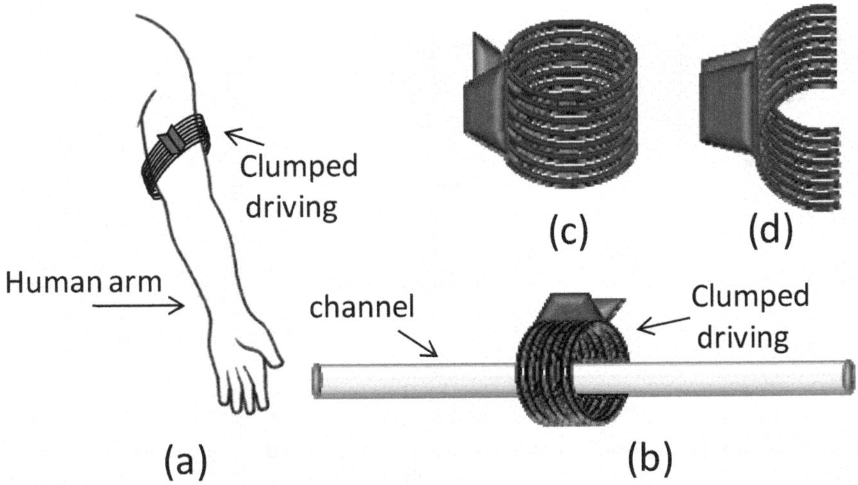

Fig. 5. Implementation of pumping mechanisms into pre-existing channels: a) human arteries; b) generic channel; c) the electromagnetic driving before clamping; d) the electromagnetic driving ready to work

3 A Novel Pumping Mechanisms with Ferrofluids

The FP3 pump uses three masses of ferrofluid (one to realize the plunger and two for the valves) injected into a glass channel to implement a traditional pumping sequence. A set of external electromagnets are used to control the mechanical action of the ferrofluid volumes. The main idea is to move a small amount of liquid through the glass channel by controlling the activation sequence of the plunger and the valves. As already announced main claims of this solution are the device robustness against physical shocks, the electric tunability of its specifications and the versatility to different real uses.

Figure 6 schematizes the architecture of the pump, which uses a glass channel filled with the liquid to be handled, with an inner diameter, *d*, of 4 mm and a length of 100 mm.

For the realization of the valves and the plunger, the EFH1 ferrofluid by Ferrotec was used. In particular, a 0.04 ml volume of ferrofluid was injected into each specific valve location. The valve action is controlled by a driving system made of two electromagnets. The adopted ferrite cores for the electromagnets have a section of 3 mm x 10 mm and a length of 30 mm; the number of coil layers is 15 for the top actuator and 10 for the bottom actuator while the number of coils per layer is ~128.

Only two status are allowed for the valve operation: OPEN and CLOSE. The magnetic fields generated by the electromagnets must properly shape the ferrofluidic mass to implement a suitable valve action. In particular, the bottom electromagnet was placed with the symmetry axis parallel to the fluid flow in order to optimize the flattened state of the ferrofluidic valve in the OPEN mode, while the top electromagnet is perpendicular to the fluid flow to assure the channel sealing during the CLOSE mode.

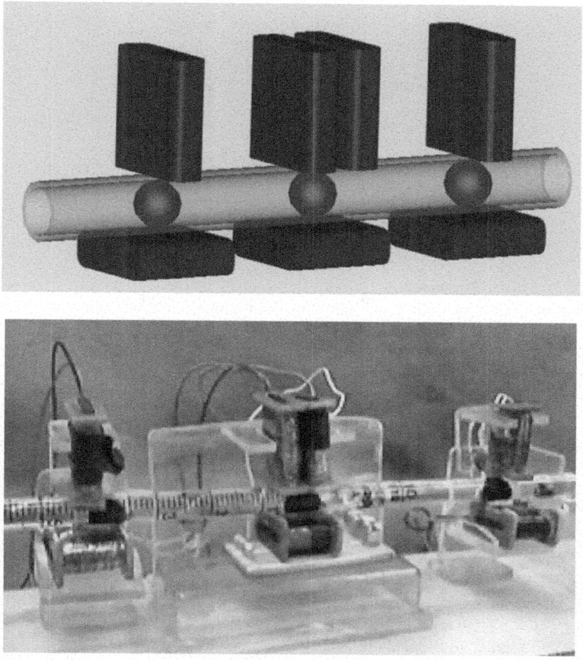

Fig. 6. (© [2006] IEEE) A schematization and the real view of the FP3 pump

Figure 7 shows the schematization and real views of the valves and the actuation strategy adopted.

The plunger idea and its functioning are shown in Figure 8. To implement the plunger a 0.05 ml of ferrofluid was injected into the channel section while the electromagnets configuration was chosen to assure a symmetric plunger operation. In particular, to guarantee the best pushing performance the upper coils must be in counter polarity (one each other) while the bottom coil polarities must be in accordance with the closest pole of the top coils. The allowable states for the plunger are LEFT, RIGHT perform the liquid pushing and the OPEN state allowing the redistribution of the fluid handled during the returning phase of the pumping cycle.

Figure 9 schematizes the actuation timings to perform the pumping sequence. Actuators, valves and plunger, are controlled via a digital strategy implemented through a PC based system connected to a control board. An user interface allows to set the duration of the whole pumping sequence and to monitor the state of each electromagnet.

The performances of the FP3 pump has been characterized in terms of valves behavior and flow rate.

In particular, concerning the sealing feature of the valve several tests have been performed to estimate the drop pressure as a function of the current value applied to the driving electromagnets. To such aim the FP3 pump was arranged in a pressure

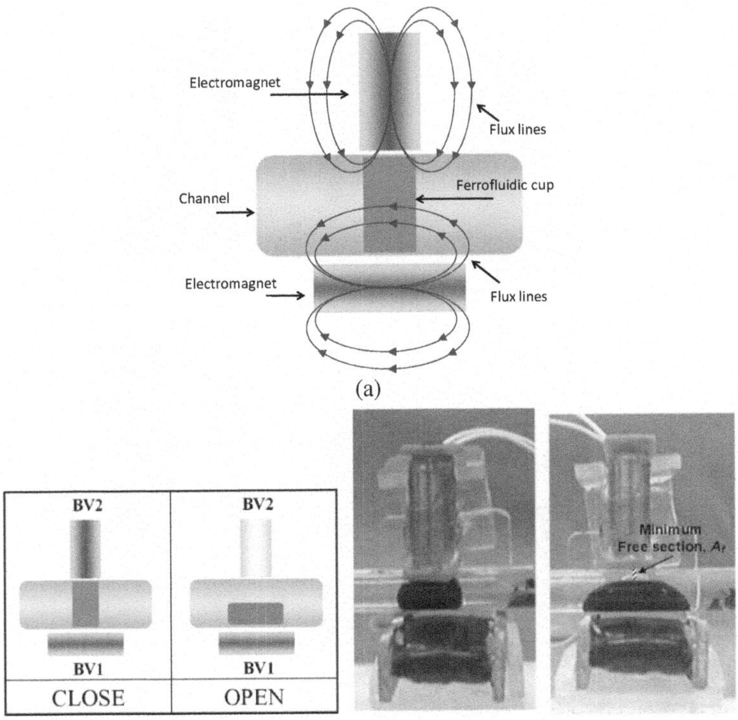

(b) (c)

Fig. 7. (© [2009] IEEE) Implementation of the valve and its mechanical action. (a) Magnetic field distribution; (b) Schematization of the valve operation; (c) Real views. The CLOSE mode (on the left) is forced by activating both the electromagnets, while the OPEN mode (on the right) is controlled by turning off the top electromagnet.

reference system connected to the glass pipe. For each driving current, the valve (set to the CLOSE state) was forced with increasing pressure until the cap collapsed and the corresponding pressure value was assumed as the drop pressure. Figure 10 shows the valve drop pressure as a function of the driving current of the top electromagnet, while the bottom electromagnet current was set to 200 mA. The results presented are the mean value of ten tests while dotted lines represent the uncertainty in the estimation of the drop pressure.

The valve drop pressure estimated for another device with a channel diameter of 8 mm is also reported for the sake of comparison. As it can be observed the thinner device performs better than the other thus confirming the suitability of shrinking the channel section.

A successful interpolation of data in Figure 10 can be obtained by a quadratic form which find basics in the following expression of the tolerable drop pressure, *DP*, of a ferrofluidic mass inside a channel and subjected to a magnetic field [6]:

$$\Delta P = \mu(M_1 H_1 - M_2 H_2) \tag{1}$$

where μ is the ferrofluid permeability, M_i and H_i (i=1,2) are the magnetization and magnetic field strength at the ferrofluid volume respectively, and the indexes 1 and 2 are used to indicate the high pressure surface and the low pressure surface of the ferrofluid cap.

Supposing a first-order dependence of the magnetic field on the driving current and a linear relationship between M_i and H_i, Eq. (1) leads to a quadratic relationship between the drop pressure and the current driving the top electromagnet. It must be considered that from a practical point of view, the driving current is definitively a more convenient parameter to be measured and to be controlled as respect to the magnetic field strength. The matching between experimental data and the predicted behavior supports the theoretical assumption above discussed.

Another important feature to be considered is the efficiency of the valve in the OPEN state estimated in terms of free section of the pipe allowing fluid flow. To such aim dedicated image processing tools were used and for the FP3 device under investigation an efficiency of 45%±5% was obtained.

For the estimation of the pump flow rate the prototype was plugged into two graded tanks, a source tank and a destination tank, and the amount of liquid pumped over several cycles was measured. Results obtained revealed a liquid volume handled

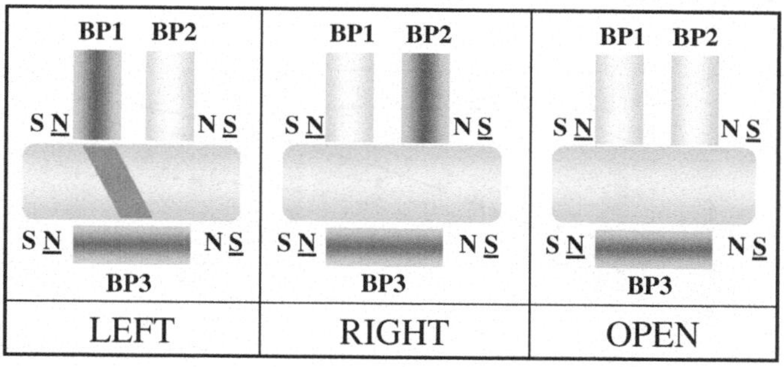

(a)

(b)

Fig. 8. (© [2009] IEEE) (a) Plunger operation. The states allowed for the plunger are: LEFT (coils BP1 and BP3 ON and coil BP2 OFF), RIGHT (coils BP2 and BP3 ON and coil BP1 OFF), and OPEN (coil BP3 ON and coils BP1 and BP2 OFF). (b) Real views of the plunger prototype.

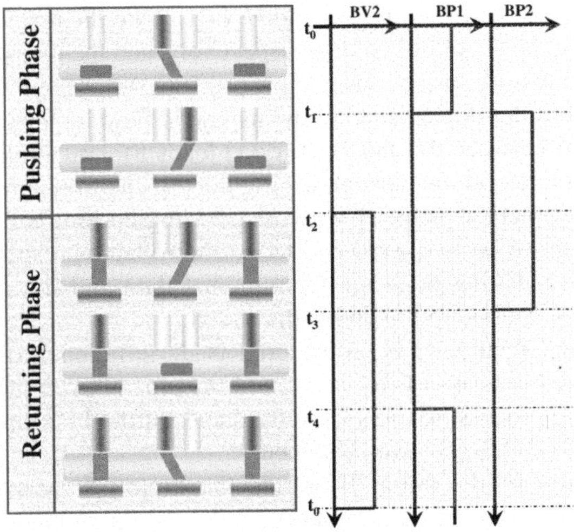

Fig. 9. (© [2009] IEEE) The whole actuation sequence and the timing signal for the top actuation coils of the valves and the plunger. The bottom actuators are always turned on.

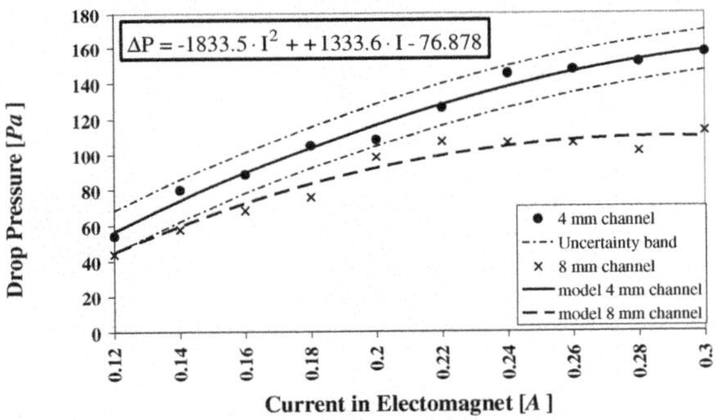

Fig. 10. (© [2009] IEEE) Drop pressure of the valve as a function of the excitation current of the top electromagnet. On top the relationship between the valve drop pressure and the driving current.

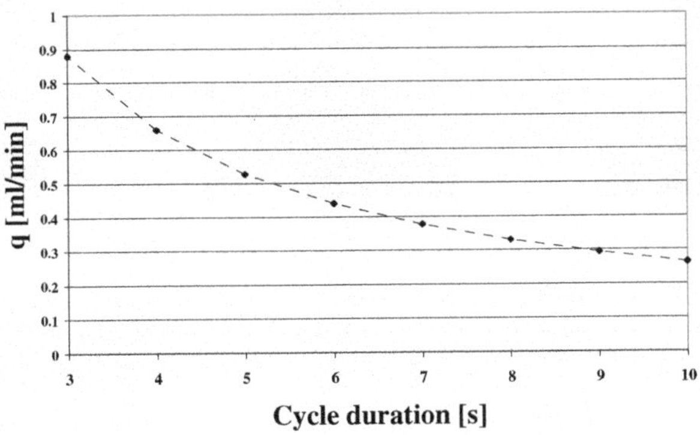

Fig. 11. (© [2009] IEEE) The pump flow rate as a function of the cycle duration

per cycle equal to 0.04 ml±0.01 ml. Since for the considered device the minimum cycle duration assuring a reliable behavior of the whole pumping system is 3 s, a maximum flow rate of 0.9 ml/min±0.1 ml/min was estimated. The flow rate trend as a function of the pumping cycle duration is sketched in Figure 11.

4 Conclusion

Magnetic fluids with their properties and peculiarities allow to implement valuable solutions for bio-medical applications. The interesting aspect of these materials is their versatility which makes them suitable as bio-compatible labels as well as functional masses in inertial transducers.

The latter field is leading to the design of innovative solutions which are really convenient for several application contexts from fluid control in pre-existing channels to fluid management in bio-channels in a future vision. Moreover, the intrinsic nature of magnetic fluids and the possibility to control their physical properties through the use of external magnetic fields allow to implement reliable devices with tuneable specifications.

Acknowledgements

Authors wish to thank Dr. Alberto Ascia for the precious support in developing the research activity on ferrofluidic pumps presented through this chapter.

References

1. Andò, B., Ascia, A., Baglio, S., Savalli, N.: A Novel Ferrofluidic Inclinometer. IEEE Transactions on Instrumentation and Measurement 56(4), 1114–1123 (2006)
2. Andò, B., Ascia, A., Baglio, S., Franco, G., Savalli, N.: A Novel Ferrofluidic Gyroscope. In: Proceeding of Eurosensors 2006, pp. 1–4 (2006)
3. Andò, B., Ascia, A., Baglio, B., Pitrone, N.: Magnetic fluids and their use in transducers. IEEE Magazine on Instrumentation and Measurements 9(6), 44–47 (2006)
4. Andò, B., Ascia, A., Baglio, S., Bulsara, A.R., In, V.: RTD Fluxgate performance in magnetic label-based bioassay: preliminary result. In: 28th Annual International Conference of the EMBC 2006, USA, pp. 5060–5063. IEEE, Los Alamitos (2006)
5. Andò, B., Ascia, A., Baglio, S., Pitrone, N.: Development of a pump with a single ferrofluidic mass. In: Proceeding of Eurosensors XXII, pp. 581–584 (2008)
6. Andò, B., Ascia, A., Baglio, S., Pitrone, N.: Ferrofluidic Pumps: a valuable implementation without moving parts. Accepted for the publication on IEEE Transactions on Instrumentation and Measurement
7. Browaeys, J., Bacri, J.-C., Flament, C., Neveu, S., Perzynski, R.: Surface waves in ferrofluids under vertical magnetic field. The European Physical Journal B 9(2), 335–341 (1999)
8. Calugaru, G.H., Cotae, C., Badescu, R., Badescu, V., Luca, E.: A new aspect of the movement of ferrofluids in a rotating magnetic field. Revue Roumaine de Physique 21, 439–440 (1976)
9. Cowley, M.D., Rosensweig, R.E.: The interfacial Stability of a Ferromagnetic Fluid. The Journal of Fluid Mechanics 30(4), 67–688 (1967)
10. Gazeau, F., Baravian, C., Bacri, J.C., Perzynski, R., Shliomis, M.I.: Energy conversion in ferrofluids: Magnetic nanoparticles as motors and generators. Physical Review E 56(1), 614–618 (1997)

11. Greivell, N.E., et al.: The design of ferrofluid magnetic pipette. IEEE Transactions on Biomedical Engineering 44(3), 129–135 (1997)
12. Hartshorne, H., Backhouse, C.J., Lee, W.E.: Ferrofluid-based microchip pump and valve. Sensors and Actuators B 99(2-3), 592–600 (2004)
13. Hatch, A., Kamholz, A.E., Holman, G., Yager, P., Böhringer, K.F.: A Ferrofluidic Magnetic Micropump. Journal of Microelectromechanical Systems 10(2), 215–221 (2001)
14. Joung, J., Shen, J., Grodzinski, P.: Micropumps Based on Alternating High-Gradient Magnetic Fields. IEEE Transactions on Magnetics 36(4), 2012–2014 (2000)
15. Kularatne, B.Y., Lorigan, P., Browne, S., Suvarna, S.K., Smith, M.O., Lawry, J.: Monitoring tumour cells in the peripheral blood of small cell lung cancer patients. Cytometry 50(3), 160–167 (2002)
16. Mekhonoshin, V.V., Lange, A.: Faraday instability on viscous ferrofluids in a horizontal magnetic field: Oblique rolls of arbitrary orientation. Physical review E 65(1-6), 061509.1–061509.7 (2002)
17. Miller, M.M., Sheehan, P.E., Edelstein, R.L., Tamanaha, C.R., Zhong, L., Bounnak, S., Whitman, L.J., Colton, R.J.: A DNA array sensor utilizing magnetic microbeads and magnetoelectronic detection. Journal of Magnetism and Magnetic Materials 225(1-2), 138–144 (2001)
18. Molday, R.S., MacKenzie, D.: Immunospecific ferromagnetic iron–dextran reagents for the labeling and magnetic separation of cells. Journal of Immunological Methods 52(3), 353–367 (1982)
19. Morisada, S., Miyata, N., Iwahori, K.: Immunomagnetic separation of scum-forming bacteria using polyclonal antibody that recognizes mycolic acids. Journal of Microbiological Methods 51(2), 141–148 (2002)
20. Mura, C.V., Becker, M.I., Orellana, A., Wolff, D.: Immunopurification of Golgi vesicles by magnetic sorting. Journal of Microbiological Methods 260(1-2), 263–271 (2002)
21. Odenbach, S.: Ferrofluids: Magnetically controllable fluids and their applications. Lecture Notes in Physics, 253 pages. Springer, Heidelberg (2002)
22. Pankhurst, Q.A., Connolly, J., Jones, S.K., Dobson, J.: Applications of magnetic nanoparticles in biomedicine. Journal of Physics D: Applied Physics 36, 167–181 (2003)
23. Perez-Castillejos, R., Plaza, J.A., Esteve, J., Losantos, P., Acero, M.C., Cane, C., Serra-Mestres, F.: The use of ferrofluids in micromechanics. Sensors and Actuators 84(1-2), 176–180 (2000)
24. Raj, K., et al.: Commercial applications of ferrofluids. Journal of Magnetism and Magnetic Materials 85(1-3), 233–245 (1990)
25. Rosensweig, R.E.: Ferrohydrodynamics. Cambridge University Press, Cambridge (1985)
26. Tibbe, A., de Grooth, B., Greve, J., Liberti, P., Dolan, G., Terstappen, L.: Optical tracking and detection of immunomagnetically selected and aligned cells. Nature Biotechnology 17, 1210–1213 (1999)
27. Zahn, M., Pioch, L.L.: Magnetizable fluid behaviour with effective positive, zero, or negative dynamic viscosity. Indian Journal of Engineering & Materials Sciences 5(6), 400–410 (1998)
28. Zahn, M., Pioch, L.L.: Ferrofluid flows in AC and travelling wave magnetic fields with effective positive, zero or negative dynamic viscosity. Journal of Magnetism and Magnetic Materials 201(7), 144–148 (1999)
29. Zahn, M.: Magnetic fluid and nanoparticle applications to nanotechnology. Journal of Nanoparticle Research 3, 73–78 (2001)
30. Zigeuner, R.E., Riesenberg, R., Pohla, H., Hofstetter, A., Oberneder, R.: Isolation of circulating cancer cells from whole blood by immunomagnetic cell enrichment and unenriched immunocytochemistry in vitro. The Journal of Urology 169(2), 701–705 (2003)

Design of the New Prognosis Wearable System-Prototype for Health Monitoring of People at Risk

Alexandros Pantelopoulos and Nikolaos Bourbakis

Assistive Technologies Research Center, Wright State University,
Dayton Ohio 45435, USA
{pantelopoulos.2,nikolaos.bourbakis}@wright.edu

Abstract. The paper presents the design framework of the Prognosis wearable system, which aims at realizing continuous and ubiquitous health monitoring of people at risk along with providing embedded decision support to enhance disease management and/or prevention. The system's functional scheme is built on top of a formal language for describing and fusing health symptoms that are extracted from the acquired measurements of a variety of wearable sensors, which may be distributed over a patient's body. Moreover, by incorporating an automated intelligent and interactive dialogue system, additional health-status feedback can be obtained from the user in terms of described symptoms, captured using a voice recognition module, which can further enhance the autonomous decisional capabilities of the system. A simulation framework built in Java is also presented in detail and is based on a previously presented Stochastic Petri Net functional model of the system, which was used to model the event-based operation of Prognosis and to capture the concurrency issues that arise in such a system design. Finally, the illustrated paradigmatic application scenarios clearly highlight the benefits of using an interactive and intelligent health monitoring system for remote care of patients in critical medical conditions.

Keywords: wearable systems, formal language, health monitoring, signal processing, SPN, simulation, voice recognition, embedded decision support.

1 Introduction

Health monitoring of out-of-hospital patients is of increasing significance as it has been recognized that the global population is not only growing but also ageing [1]. In addition to that, there has been a corresponding increase in chronic disease cases, such as congestive heart failure, chronic obstructive pulmonary disease, diabetes etc. To address these issues as well as increasing health care costs and at the same time to increase the quality of health care provided to other types of high risk patients such as postoperative rehabilitation patients, a lot of research groups worldwide have focused on realizing low-cost wearable health monitoring systems (WHMS) that can monitor a patient's physiological parameters in a continuous, ubiquitous and unobtrusive manner [2], [3].

In this work we will present a detailed functional description of the Prognosis WHMS design framework. The presented architectural scheme, which is modelled using Stochastic Petri Nets, is focused on realizing a system that is capable of

S.C. Mukhopadhyay and A. Lay-Ekuakille (Eds.): Adv. in Biomedical Sensing, LNEE 55, pp. 29–42.
springerlink.com

i) acquiring a variety of physiological measurements via several on-body bio-sensors, ii) processing and fusing those measurements in order to derive an estimation of the patient's overall health condition and iii) interacting with the patient in terms of an automated dialogue system in order to extract significant diagnostic information about any possible non-measurable symptoms.

2 Review of Research in WHMS

Wearable systems employed for health monitoring purposes constitute a widely researched topic and through the last decade there have been numerous prototypes reported in the corresponding literature mainly propelled by advances in miniaturized sensors, wireless communication techniques, low-power integrated circuits and advanced signal processing schemes. The most significant ones of these prototype systems will be briefly described in this section.

The AMON [4] project developed a wrist-worn device, which is capable of measuring several bio-signals such as blood pressure, skin temperature, oxygen saturation, ECG and activity status of the user. That system was targeting high risk/respiratory patients and was capable of classifying the estimated health condition of the user as being one of normal, deviant, in risk or in high risk based on manually predefined limit values for each signal.

MyHeart [5], WEALTHY [6] and MERMOTH [7] are further examples of EU founded projects, which in collaboration with various industrial partners developed what is referred to as smart-clothing systems. These are wearable garments that employ smart sensing fabrics and interactive textiles to facilitate multi-parameter physiological monitoring of high-risk patients and thus to enable early health risk detection and prevention.

LiveNet [8], a system developed in MIT, was an early wearable prototype capable of real-time data capture, streaming and context classification, aiming at medical cases such as automated Parkinson symptom detection, epilepsy seizure detection and long-term behaviour modelling. Researchers in [9] describe Smart Vest, a wearable physiological monitoring system constituted by a wearable vest, which uses various sensors integrated on the fabric to simultaneously collect a wide variety of bio-signals in a non-invasive and unobtrusive manner. These sensors are connected to central processing unit, which is capable of correlating the acquired measurements to derive an overall picture of the wearer's health.

The medical embedded device for individualized care (MEDIC) [10] is based on a general architecture for wearable sensor systems that can be customized according to individual patient's needs. The proposed design framework includes embedded artificial intelligence in the form of an inference engine based on naive Bayes classifiers for detection of patient conditions and for managing system's resources in a dynamic manner to possibly improve diagnostic certainty. Furthermore, MEDIC has the capacity for remote configuration and control due to the fact that it can be reconfigured via remote commands sent to the system.

The authors in [11] describe a heart attack self-test application implemented on a Personal Health Monitor system which includes a conventional mobile phone and a Bluetooth enabled ECG sensor. The mobile phone is capable of analyzing the streaming data from the sensor in real-time and also transmitting them to a heart specialist. In the demonstrated application, a simple user interface on the phone is used to acquire feedback from the user about his symptoms and in case the patient's health condition is found to be at risk based on his answers, the emergency services will be contacted.

Table 1.

Project Title/Institution	Hardware Description	Communication Modules	Measured Signals	Applications
AMON (EU IST FP5 program)	Wrist-worn device	GSM link	ECG, blood pressure, temperature, SaO2, activity	High-risk cardiac-respiratory patients
MyHeart (EU IST FP6 program)	PDA, Textile & electronic sensors on clothes + heart belt	conductive yarns, GSM, Bluetooth	ECG, respiration, other vital signs, activity	Prevention and early diagnosis of CVD
WEALTHY (EU IST FP5 program)	Textile & electronic sensors on jacket	conductive yarns, GPRS, Bluetooth	ECG, EMG, respiration, temperature, activity	Rehabilitation patients, elderly, chronic diseases
MERMOTH (EU IST FP6 program)	PDA, knitted dry electrodes	conductive yarns, RF link	ECG, respiration, temperature, activity	General health monitoring
LiveNet, (MIT)	PDA, microcontroller board	wires, 2.4GHz radio, GPRS	activity, ECG, EMG, GSR, temperature, respiration, Sa02, blood pressure	Parkinson symptom & epilepsy seizures detection, behav. modelling
MEDIC (Un. of California)	PDA & comm. available microelectr. components and sensors	Bluetooth, GPRS, Wi-Fi	ECG, activity, posture	Abnormal gait diagnosis & other diagnostic operations
Smart Vest (National Pr. On Smart Materials, India)	Vest, microcontroller	woven wires, 2.4 GHz ISM RF	ECG, PPG, blood pressure, temperature, GSR	General remote health monitoring
Personal Health Monitor (Un. Of Tech. Sydney)	Cell phone & comm. available BT bio-sensors	Bluetooth, GPRS	ECG, activity, blood pressure	Heart-attack self-test for CVD patients
HeartToGo (Un. Of Pittsburgh)	Cell phone & comm. available BT bio-sensors	Bluetooth, GPRS	ECG, activity	Individualized remote CVD detection
Lifeshirt (Vivometrics)	Sensors embedded in vest, PDA	Bluetooth & wires	ECG, respiration, activity	All-day remote health monitoring
SmartShirt (Sensatex)	Shirt with conductive fiber sensors, PDA	conductive yarns, Bluetooth or Zigbee	ECG, blood pressure, temperature, respiration	Remote monitoring of vital signs

HeartToGo [12] is another cell-phone-based wearable platform, which is capable of continuously monitoring the user's ECG signal via a wireless ECG sensor, analyzing the electrocardiogram in real time and possibly detecting any abnormal patterns pertaining to cardiovascular disease. The novelty of the proposed system is found in its ability to adapt to the individual user's physiological conditions through the use of artificial neural network – based machine learning schemes, which can possibly result in more accurate classification of ECG patterns. In addition to the previously mentioned research efforts, some noteworthy examples of commercially available wearable systems for health monitoring are LifeShirt [13] from VivoMetrics, which is a washable light-weight vest with bio-sensors embedded on the garment itself, and SmartShirt [14] from Sensatex, a T-shirt based wearable system using conductive fiber sensors to measure a variety of vital signs and signals. Finally, there has been a great number of research projects on Body Area Networks (BANs) during the last 10 years, which are systems based on wireless sensor networks and Zigbee-based motes. For a comprehensive review of such systems the reader should refer to [1].

The systems discussed above are a sample of the ongoing research on wearable health monitoring systems and an overview of these systems' components, capabilities and applications is presented in Table 1. It should be noted that this list is not exhaustive and it is not the purpose of this paper to provide a comprehensive review of WHMS. However, these research efforts illustrate the feasibility of developing low-cost portable systems that will be able to monitor the health status of the user in a continuous, unobtrusive and ubiquitous manner. However most of the WHMS reported in the literature are limited to serving as remote physiological data loggers. By enabling wearable systems to collect a wider variety of physiological measurements and also by employing them with embedded decision support capabilities, these systems may possibly become able to perform multi-parametric data analysis and correlation and thus derive advanced diagnostic statistics about the individual user.

3 A General WHMS Architecture

In this section we present a general architectural model of a wearable health monitoring system as depicted in Fig. 1. There is a variety of bio-signals produced by the human body, including vital signs such as blood pressure, pulse rate, body temperature and respiration rate as well as other physiological signals like electrocardiogram, electromyogram and oxygen saturation. These signals can be captured via wearable bio-sensors, which constitute the front-end components of a WHMS. These sensors can be either embedded in clothing as smart textiles [15] or they can be integrated on other types of wearable devices, such as wrist devices, ear-lobe sensors, finger sensors, arm bands, chest belts, waist belts etc. In the latter case, the distributed bio-sensors constitute a Body Area Network [16], which can be either formed through Bluetooth enabled devices or through Zigbee motes. In that case some light-weight operations such as filtering, amplifying and analog-to-digital conversion or even basic feature extraction may be performed on-site and then the processed data can be transmitted to the system's central node.

The central node of the WHMS can be some type of portable platform, such as a personal digital assistant (PDA), smart-phone or a microcontroller board. The functions that the central node is responsible for are: 1) handling the communication with the on-body distributed biosensors, which includes collecting physiological measurements, communication synchronization, sending control signals for setting up a sensor's parameters such as sample rate and accuracy as well as receiving sensor status data, 2) applying additional digital signal processing on the acquired signals, e.g. for feature extraction, 3) verifying the received data, 4) comparing the extracted features or values from each signal with the thresholds, limits or patterns located in the local signal database, which may contain patient-specific information about abnormal states, in order to possibly detect health risks, 4) generating alarm signals for the user, 5) displaying the collected measurements on the GUI in real-time and finally 6) transmitting the extracted medical information about the user to a remote medical station, e.g. to a medical center or to a physician's cell phone, either in real-time or in terms of report forms when requested.

The functional description of the WHMS given above pertains to the majority of the up-to-now developed wearable system prototypes or commercially available products in the field of remote health monitoring. However as it can be seen from Fig. 1, we propose to include an additional component in the system design in order for the system to be able to get additional feedback from the user about non-measurable symptoms. Examples of such symptoms include cough, malaise, chest pain, headache etc. Such a functionality can be enabled via an automated dialogue system between the WHMS and the user, which will be described in greater detail in the following sections. Therefore a speech recognition module would be required in order for the vocal feedback given by the user to be comprehensible by the central node. The acquisition of such otherwise non-measurable symptoms can greatly enhance the system's decisional power [17] as it can possibly provide greater insight and context about what is referred to as the clinical presentation of a certain health condition.

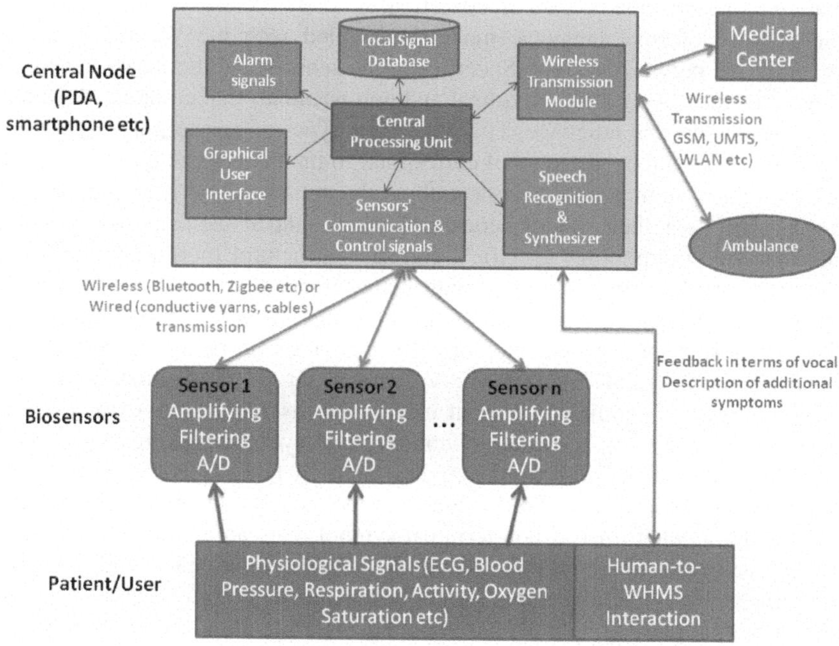

Fig. 1. General WHMS architecture

Finally it should be noted that in any case, wearable systems for health monitoring are not meant and also will never replace doctors, e.g. they cannot be considered as substitutes of a doctor's expertise, instead their goal should be "to enhance and support the human who is ultimately responsible for the clinical diagnosis" [18]. In that context, the WHMS should be capable of transmitting either raw data or extracted signal patterns and values along with patient feedback and any possibly generated alarms to the medical center or even also to a dispatched ambulance. As the medical center maintains a database with long-term detailed medical history of the patient, the received physiological measurements and the detetcted patient symptoms can be put into a wider context and also be evaluated by a supervising physician in order to derive a more accurate estimation of the patient's health. However it has been made obvious, that a WHMS with the above described functional capabilities can greatly enhance, simplify and speed-up such ambulatory diagnostic operations.

4 The Prognosis Formal Language

The Prognosis language, as presented in [19], is a theoretical model, around which the wearable monitoring and early prognosis system has been designed. The basic hypothesis of this model is that the various body or physiological signals produced by the human body are composed of "symptoms of health" whose presence under certain conditions may indicate the presence of a specific health risk. The aim of the Prognosis formal language is to provide an efficient and compact representation of the multiple combinations of extracted physiological measurements in order to aid in the association of "pathological" symptoms and patterns with the detection or estimation of a corresponding health condition.

The proposed formal language model is coupled with the WHMS architecture described in the previous section. Specifically the sensors and the measurements they provide constitute the symbols (terminal and non-terminal) that compose the alphabet of the Prognosis formal language. The three basic types of non-terminal symbols that can be derived from the start symbol correspond to the three basic types of signals or information the system is able to collect about the patient, namely: a) signals that are "value-specific", e.g. their instantaneous value carries the actual diagnostic content, b) signals that are "morphology-specific", e.g. their structural morphology and timing are the elements that carry important diagnostic information and c) voice recordings that may reveal the presence of a specific health symptom as described by the user.

The Prognosis language can be formally described as being generated from a grammar: $G = (V_N, V_T, S_T, PR)$, where: V_N is the set of terminal-symbols, which includes all types of parameters present in the analysis and feature extraction of the monitored bio-signals, V_T is the set of non-terminal symptoms corresponding to the actual values that the signal's parameters may actually take, S_T is the start symbol of the language and PR is a set of production rules the most fundamental of which being: $S_T \rightarrow S_T|S_TA|AS_T$, where A is a non-terminal symbol that can be any one of the three basic types described in the previous paragraph. Another significant production rule is that the type-specific non-terminal symbols may lead to the generation of either healthy or pathological symptoms, based on their interpretation according to the local signal database and/or the pattern recognition mechanism employed to classify samples of the type b) signals. A healthy non-terminal will lead to the production of the empty string or word, since it is conveying no important disease-related information. As a result the types of words that get generated from the Prognosis grammar have the general form:

$$w_i(t) = S_{pi(t)} S_{pj(t)} \dots S_{pk(t)}, \tag{1}$$

where the index t is used to indicate the common time window, in which the symptoms were detected. Although in general the words in a formal language are not necessarily associated with a specific semantic content, in the case of Prognosis, all words that can be derived from the language are assigned such a semantic role, which could correspond to either one or generally to more than one estimated health conditions.

An important thing to note here is that, as in the case of natural languages, a word associated with a certain, perhaps sometimes abstract as well meaning when processed on its own, may actually take on a different or more specific meaning when seen at the context of its neighbours. The underlying reason or principle supporting this statement is that pathologic symptoms may indicate various conditions when examined only at a certain time instant. When these symptoms take place concurrently with additional symptoms and these are also taken into account, e.g. forming a word of the language, additional context or meaning is provided regarding their occurrence and nature. By moving one more step further and considering these words in an even wider context, e.g. in that of some form of sentences or simply sequences of strings, the time relation between adjacent symptoms and thus also the medical history of the patient can be effectively taken into account. However, before we end this introductory discussion to the Prognosis language we should stress the following point to avoid any confusion to the reader: the word context was used in this paragraph with no relation to the language being context-free or context-sensitive (we still consider the productions rules to be context-free). The word context was simply used to refer to the semantics a Prognosis word may acquire according to its spatial and temporal environment.

5 A Simulation Framework for the Prognosis WHMS

In this section we will describe the simulation framework we have developed according to the SPN-based model of WHMS, which has been presented in [20]. The purpose of the derived simulation is to illustrate the feasibility as well as the actual operation of the Prognosis WHMS. In addition to that it can be used to study the dependencies between the various components of the system and to highlight the synchronization and concurrency issues that arise in such a system design. The presented framework has been implemented in Java, since it can facilitate a modular and hierarchical design approach, in correspondence to the SPN model. Furthermore Java provides the required synchronization and concurrency primitives for simulating the various tasks that can be present in a WHMS model as explained below.

As described in section 3, a wearable health-monitoring system may comprise a variety of bio-sensors. In the WHMS simulator, the types of sensors that can be selected to be included in the system's simulation include an ECG monitor, a thermometer, a blood pressure monitor, a respiration rate sensor, a pulse oximeter and a speech recognizer/synthesizer module to capture additional feedback from the user in terms of verbally-described symptoms. However in order to simplify the simulator's design and to address the issue that bio-signals in the simulation are created in a semi-random manner we have assumed that the ECG monitor includes an automated classifier, thus its output corresponds to per beat classification results. Under this assumption, the central node can simply poll the distributed biosensors (or the ring buffers dedicated to each type of signal) for data in a round-robin manner and at a constant rate and therefore explicit synchronization is enforced in the way the data are collected by the main process.

A different scenario would be to have dedicated threads (or even dedicated hardware in a real system design) to collect the measurements from each sensor and asynchronously signalling the event of new data being collected. Obviously such an approach is more challenging, demanding and complex especially when dealing with distributed biosensors each one of which may have a different sampling or data delivery rate, for example 250 Hz for ECG and 75 Hz for PPG sampling. In order to indirectly take this scenario into consideration as well, we have assumed the WHMS is equipped with an automated ECG classifier, which is responsible for collecting ECG signal samples and then producing per beat and rhythm classification results. Such a classifier can be implemented via a light-weight and computationally-inexpensive algorithm for ECG classification, an issue addressed in [11], [12] and which is also currently being researched by our group along the guidelines presented in [20].

As a result, in the WHMS simulator all the bio-sensors (or bio-signals) are represented by individual classes – threads, which constantly generate new data in a semi-random manner and deposit them in dedicated ring buffers. It is up to the main process to read new data in the buffers before they get overwritten, but this issue is dealt with in this case by explicitly defining identical data delivery rates for all types of sensors. Furthermore signal samples are not generated in a totally random way, but in a rather biased manner based on previous data and also based on recent values of other bio-signals, e.g. low oxygen saturation is more possible to coincide with high respiration rate etc.

Once a new measurement is acquired, it is first checked for validity. In the case that there is a redundant number of sensors used to monitor each bio-signal, erroneous values can be relatively easy detected by considering the correlation of the measured data between sensors of the same type. However the fact that a single sensor is employed for each bio-signal makes this task very challenging. A simple approach to this issue is to make use of a linear predictor for each type of signal and then calculate the difference between the predicted value and the actual value that was measured by the sensor. Then if that difference is found to be greater than a predefined threshold the corresponding measurement can be considered erroneous. The drawback of this method is that it requires calibration for each sensor in order to estimate the predictor's order and parameters as well as the threshold value to be used.

After validating each acquired measurement (or even pattern), a decision is made about whether the new data are to be considered as "healthy" or as "pathologic". This is done according to the threshold values saved in the local signal database and which may be adjusted as time progresses. Signal values or states may be generally classified as being in several states of importance or risk, for example as normal, moderate, slightly abnormal and very risky, as it happens in the case of blood pressure readings where the corresponding scales are hypotension, normal, prehypertension, stage 1 hypertension and stage 2 hypertension. In general, values not being in a normal state will be kept and contribute to the generation of strings to be parsed by

Prognosis. Even moderately irregular values should be examined, as there could be an underlying trend or pattern giving signs of a future pathologic health condition which could possibly be early detected.

Furthermore any possible additional symptoms obtained via verbal feedback from the user are fused with the acquired physiological data to create a more comprehensive estimation of the user's health status. The way this is done is explained in the next section in detail. Finally the system re-evaluates the condition of the user at any given time that any new "pathological" or slightly abnormal signal state is detected and according to the derived results, it may choose to either alert the user or to request further information from him/her or even to send an alarm message to a remote medical center or to the supervising physician's cell-phone.

Fig. 2 depicts the main screen of the WHMS simulator. For every bio-signal the following information are shown: the type of the sensor that gathers the data, the type of the bio-signal, the status of the sensor (active, inactive, malfunctioned, out of battery), the instantaneous value (or pattern) and a graph of the most recent values (for the ECG this has been omitted in this work due to the reasons described earlier). In addition to that, there is a text box where any possible extracted health symptoms are printed out, for example tachycardia or hypotension or fever etc. An alarm-detection box is used to print out any recognized health conditions, such as hypothermia or left ventricular failure for example, represented in the local signal database in terms of specific symptom strings or words. Finally a health-threat-level display bar is used, which illustrates the risk level of the user's health as perceived by the wearable system and which will simply change state to a higher level of risk whenever one or more of the monitored body-signals moves into a more "hazardous" value range.

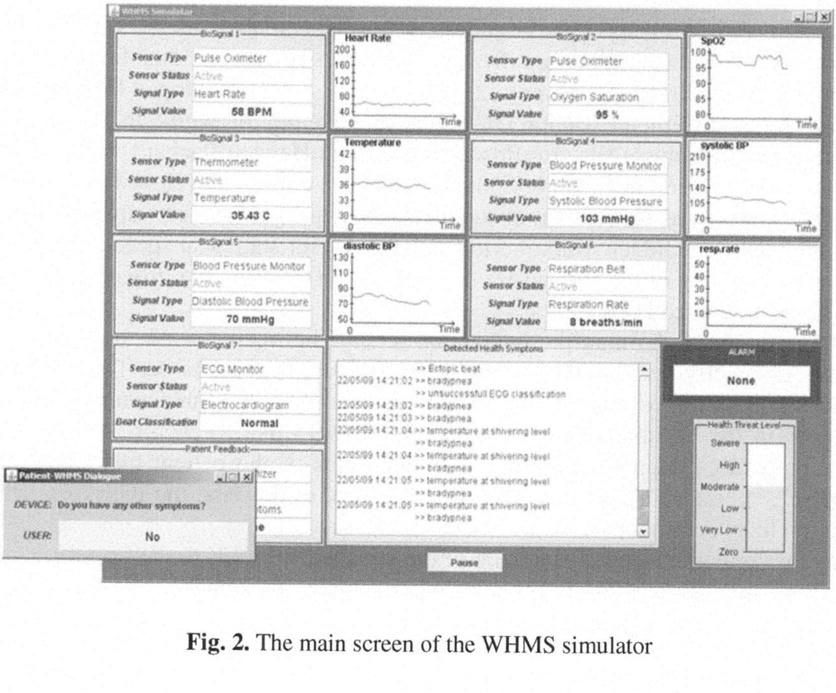

Fig. 2. The main screen of the WHMS simulator

6 Human Device Interaction

As mentioned in the previous section, we want the user of the Prognosis wearable platform to be able to interact with the system in a meaningful and helpful manner. This can be done in terms of verbal feedback from the user to provide important diagnostic information regarding non-measurable health symptoms. However in order for the system to be capable of capturing such information from a voice recording, the interaction scheme between the user and the system must be well-defined. In addition to that, we need to take two more criteria into consideration, namely the fact that the feedback mechanism must not be allowed to grow very high in complexity since it is going to be implemented on a hardware-constrained mobile platform, e.g. on a cell-phone or a PDA or even on a dedicated processing board. Secondly the interaction process must be user-friendly to ensure usability and thus also user acceptance. In order to achieve those goals it is necessary for the system to be able to respond to the user's feedback, e.g. to request more information, to ask the user to repeat something he said, to inform him about an event etc.

In accordance to the previously described goals and requirements, we have created a prototype of an automated dialogue system between the user and the wearable platform. We have employed a widely used statistical speech recognition tool, which enables the recognition of a word, phrase or sentence uttered by the user by matching it with a finite set of possibilities according to a predefined grammar. In the following the interaction scheme is described in greater detail.

6.1 Prototyping the Human Device Interaction

There is a variety of speech recognition software available, most of which being proprietary. The most important freely available software in this field are HTK developed at Cambridge University [21] and Sphinx-4 [22] developed from the CMU Sphinx Group at Carnegie Mellon University in collaboration with Sun Microsystems Laboratories, Mitsubishi Electric Research Labs and Hewlett Packard. HTK's source code has been written in C, while Sphinx-4 has been entirely developed in Java. Both tools are considered to be state-of-the-art speech recognition software and both of them are based on Hidden Markov Models (HMMs) for statistical speech recognition. For the prototyping of the Human-Device Interaction (HDI) in the Prognosis wearable system we have used HTK.

The first step in defining the interaction scheme between the user and the system consists of specifying what the vocabulary to be used is going to be. The vocabulary includes all possible words that the system needs to be able to understand. Examples of words in the vocabulary could be simple answers like "Yes" or "No", a set of nouns or phrases describing the possible symptoms of the patient like "headache", "malaise", "chest pain" etc and a set of adjectives to describe the type of a certain symptom like "continuous", "acute" etc.

Another important step is matching of the vocabulary words to their phonetics, e.g. describing how each word is pronounced. Examples of some of the words in the dictionary are given bellow:

YES	y eh s
NO	n ow
HEADACHE	hh eh d ey k
ACUTE	ah k y uw t
CHEST PAIN	ch eh s t . p ey n
PERSISTENT	p er s ih s t ah n t
COUGH	k aa f
DIZZINESS	d ih z iy n ah s

After defining the vocabulary and the grammar of the interaction scheme, an HMM or a set of HMMs needs to be trained for every word or phrase and thus for their corresponding phonemes, in the grammar by using a number of examples for that spoken word or phrase. Training of the HMM refers to the process of estimating the parameters of each HMM and thus constructing acoustic models for words and phrases. In the case of continuous speech recognition each word will be defined as a sequence of phoneme-based HMMs.

Finally to recognize some unknown word or phrase the likelihood of each model generating that word or in the more complex case the likelihoods of all the paths in the HMM interconnected network will be calculated and the most likely model or path will be identified.

It should be noted that the performance of the HTK in terms of accuracy and speed is strongly dependent on the size of the employed vocabulary. Furthermore by trying to limit the grammar to one-word phrases and thus requiring mostly isolated word recognition the performance of the speech recognizer can be further increased (or not allowed to deteriorate). In accordance to these guidelines, we have initially decided to include only one-word phrases in the vocabulary for the HDI in Prognosis, since there is also a considerable amount of symptoms and symptom descriptions that we need to guarantee that the system will be able to comprehend and also since the end goal is to integrate the speech recognition modality on a portable and thus resource constrained platform.

6.2 An Example of HDI via Voice

In this section we present a possible scenario to illustrate the operation of the Prognosis wearable system and especially the interaction between the user and the device via the voice module. Consider the case where the user of the wearable system is an elderly individual with a personal medical history of signs or tendency towards cardiovascular disease. We can imagine this person going about his daily activities either in his own home or even at a care center, while he is wearing the wearable health-monitoring system, for example a chest belt measuring ECG, body temperature and respiration rate, a pulse oximeter and a cell-phone placed on his waist with a hands-free or Bluetooth earpiece.

Now consider the following scenario: The patient starts to feel a slight discomfort in the chest along with a feeling of malaise. His heart rate may also increase and his heartbeats may become more irregular. Furthermore, his respiration effort may also become stronger and quicker and there might be a slight decrease, perhaps in a longer period of time, of his blood oxygen saturation level. The patient decides he should sit down but does not consider his condition alarming or hazardous enough to call for help. However the moment the system detects all the above events, e.g. the raise in the levels of some of the bio-sensors' measurements along with a possible arrhythmia, it decides to take action and initiates a brief dialogue with the user in order to obtain more information about his condition and then to act accordingly. The first thing the system may ask the user for, is if he needs immediate help or not. Supposing the user answers no to the previous question, the system may continue to inquire about any other possible symptoms that the patient may be experiencing. Then the patient mentions the term chest pain. In that case a small series of chest pain related questions may follow, which can lead to the establishment of such a health-condition context so that the system may indeed choose to send a notification to the medical station or to the supervising physician. Eventually appropriate help and care may be provided on time to the patient in order to possibly avoid the deterioration of his condition.

The previously described scenario is illustrated in the following in terms of the events that occur and the actual dialogue that may take place:

User: <Feels a slight chest discomfort and some of his vitals start to go out of range>

System: *Some of your vitals are out of regular range. Do you need immediate help?*

User: *No*

System: *Do you have any other symptoms?*

User: *Yes*

System: *Please describe your symptoms*

User: *I have slight chest discomfort.*

System: <the system does not acknowledge the user's phrase>
Please state your symptom again in a brief manner

User: *chest pain*

System: <Recognizes the symptom>
Is the symptom intermittent or continuous?

User: *continuous*

System: *Is the symptom instantaneous or persistent?*

User: *persistent*

System: *Is the pain radiating to your arms?*

User: *Yes*

System: *Is the pain aggravated by movement or breathing?*

User: *No*

System: *Are you experiencing any other symptoms, such as hemoptysis or sputum?*

User: *No*

System: *Describe your pain level on a scale from one until ten*

User: *five*

System: <decides that according to the user's answers and his recent vital signs the user could be in an alarming health condition>
< notifies the medical center by sending an alarm message via GPRS>
I have notified the medical center about your current condition

7 Conclusion

In this paper we have presented our approach towards the modelling and the design of a novel wearable health monitoring system able to support embedded decision making in order to increase the efficiency of remote health monitoring of people at risk. A formal language model is used to represent combinations of various pathological health signs and to aid in the detection of health alarming or threatening conditions. Moreover, we have described the development of a simulator for a wearable system built on top of the Prognosis formal language concept and explained the mechanism according to which symptoms from a variety of on-body distributed bio-sensors can be extracted and then fused together in order to produce an overall estimation of the user's health condition.

Finally we have expanded the functionality of the system by defining a scheme for human-device interaction in terms of an automated dialogue system enabled via speech recognition. This extra feature of the system can lead to the extraction of highly significant otherwise non-measurable health information, which can in turn increase the autonomous decisional capabilities of the portable platform and potentially aid in the early detection of health threatening episodes.

A provided hypothetical scenario illustrates the applicability of the Prognosis WHMS and sets the mark for the ongoing implementation of the Prognosis prototype. The presented analysis on the system's AI, signal processing capabilities and overall architecture provide the guidelines along that path.

References

1. Hao, Y., Foster, R.: Wireless body sensor networks for health-monitoring applications. Physiological Measurement 29(11), R27–R56 (2008)
2. Pantelopoulos, A., Bourbakis, N.: A Survey on Wearable Biosensor Systems for Health Monitoring. In: 30th Intl. IEEE EMBS Conf., Vancouver, BC, pp. 4887–4890 (2008)
3. Gatzoulis, L., Iakovidis, I.: Wearable and Portable eHealth Systems, Technological Issues and Opportunities for Personalized Care. In: IEEE Engineering in Medicine and Biology Magazine (September/October 2007)
4. Anliker, U., Ward, J.A., Lukowicz, P., Tröster, G., Dolveck, F., Baer, M., Keita, F., Schenker, E.B., Catarsi, F., Coluccini, L., Belardinelli, A., Shklarski, D., Alon, M., Hirt, E., Scmid, R., Vuskovic, M.: AMON: A Wearable Multiparameter Medical Monitoring and Alert System. IEEE Transactions on Information Technology in Biomedicine 8(4), 415–427 (2005)
5. Habetha, J.: The MyHeart Project – Fighting Cardiovascular Diseases by Prevention and Early Diagnosis. In: 28th Annual Internatonal Conference, IEEE Engineering in Medicine and Biology Society (EMBS), NY City, USA, August 30 – September 3 (2006)
6. Pacelli, M., Loriga, G., Taccini, N., Paradiso, R.: Sensing Fabrics for Monitoring Physiological and Biomechanical Variables: E-textile solutions. In: Proceedings of the 3rd IEEE-EMBS International Summer School and Symposium on Medical Devices and Biosensors, September 4-6. MIT, Boston (2006)
7. Weber, J.L., Porotte, F.: MEdical Remote MOnitoring with clothes. PHealth, Luzerne (January 31, 2006)

8. Sung, M., Marci, C., Pentland, A.: Wearable feedback systems for rehabilitation. Journal of NeuroEngineering and Rehabilitation (June 29, 2005)
9. Pandian, P.S., Mohanavelu, K., Safeer, K.P., Kotresh, T.M., Shakunthala, D.T., Gopal, P., Padaki, V.C.: Smart Vest: Wearable multi-parameter remote physiological monitoring system. Medical Engineering & Physics 30, 466–477 (2008)
10. Wu, W.H., Bui, A.A.T., Batalin, M.A., Au, L.K., Binney, J.D., Kaiser, W.J.: MEDIC: Medical embedded device for individualized care. Artificial Intelligence in Medicine 42, 137–152 (2008)
11. Leijdekkers, P., Gay, V.: A self-test to detect a heart attack using a mobile phone and wearable sensors. In: Proc 21st Intl. IEEE Symposium on Computer-Based Medical Systems, pp. 93–98 (2008)
12. Jin, Z., Oresko, J., Huang, S., Cheng, A.C.: HeartToGo: A Personalized Medicine Technology for Cardiovascular Disease Prevention and Detection. In: IEEE/NIH Life Science Systems and Applications Workshop (LiSSA 2009), pp. 80–83 (2009)
13. Heilman, K.J., Porges, S.W.: Accuracy of the Lifeshirt® (Vivometrics) in the detection of cardiac rhythms. Biological Psychology 75, 300–305 (2007)
14. Sensatex, Inc. Development of the Sensatex SmartShirt. pHealth (2006)
15. Paradiso, R., Loriga, G., Taccini, N.: A Wearable Health Care System Based on Knitted Integral Sensors. IEEE Transactions on Information Technology in Biomedicine 9(3) (September 2005)
16. Milenkovic, A., Otto, C., Jovanov, E.: Wireless sensor networks for personal health monitoring: Issues and an implementation. Computer Communications 29, 2521–2533 (2006)
17. McPhee, S.J., Papadakis, M.A., Tierney Jr., L.M.: CURRENT Medical Diagnosis & Treatment, 47th edn. McGraw-Hill, New York (2008)
18. Berner, E.S. (ed.): Clinical Decision Support Systems, Theory and Practice, 2nd edn. Health Informatics Series. Springer Science+Business Media, LLC, Heidelberg (2007)
19. Pantelopoulos, A., Bourbakis, N.: A Formal Language Approach for Multi-Sensor Wearable Health-Monitoring Systems. In: 8th Intl. IEEE BIBE Conf., Athens, October 8-10, pp. 1–8 (2008)
20. Pantelopoulos, A., Bourbakis, N.: SPN-based Simulation of a Wearable Health Monitoring System. Submitted to IEEE EMBS 2009 (2009)
21. HTK, `http://htk.eng.cam.ac.uk/` (accessed at 27/05/2009)
22. CMU Sphinx Group, `http://cmusphinx.sourceforge.net/html/cmusphinx.php` (accessed at 27/05/2009)

Ultra Wide Band in Medical Applications

S. D'Amico[1], M. De Matteis[1], O. Rousseaux[2], K. Philips[2], B. Gyselinck[2], D. Neirynck[2], and A. Baschirotto[3]

[1] Università del Salento, Lecce, Italy
[2] IMEC-NL, Eindhoven, The Netherlands
[3] Università di Milano-Bicocca, Milano, Italy

Abstract. Ultra Wide Band (UWB) technology has been developed in the last years in the framework of short-range low data rate communications. Due to the wide channels bandwidth and low power characteristics it provides a very different approach to wireless technologies compared to conventional narrow band systems. This makes it interesting in medicine area with many potential applications. In this chapter, the discussion is focused on the application of this technology in medical monitoring, and Wireless Body Area Networks.

UWB signals are moreover naturally suited for accurate estimation of the distance between two radios even in severe multipath or non-line-of-sight scenarios, virtually enabling indoor position sensing with an accuracy of 50 cm or better. This features can be exploited to realize a patient motion monitoring system in hospital.

Moreover, one of the key advantages of UWB in this context is its potential for the development of ultra low power consumption radio chipsets, resulting in significantly increased battery lifetimes for sensor nodes spread in the environment or even autonomous operations from waste energy. The recently developed IEEE 802.15.4a standard is capitalizing on these two advantages of UWB, and it is expected to enable the combination of accurate position sensing and ultra low power communications in one single system.

Keywords: Wireless Body Area Network (WBAN), Ultra-Wide Band (UWB), UWB radar.

1 Introduction

UWB systems are wireless systems which transmit signals across a much wider frequency spectrum than conventional wireless systems. The US Federal Communications Commission (FCC) and ITU-R (International Telecommunication Union-Radiocommunication sector) define UWB in terms of a transmission from an antenna for which the emitted signal bandwidth exceeds the lesser of 500 MHz or 20% of the center frequency. The UWB systems were originally used for radar, sensing, military communications and some niche applications. In February 2002, FCC regulate that the frequency for the UWB technique is from 3.1GHz to 10.6GHz in America. However, in Europe, the frequencies include two parts: from 3.4 GHz to 4.8 GHz and 6 GHz to 8.5 GHz. The power radiation requirement of UWB is strict and it would not disturb the existing equipments because UWB's spectrum looks like background noise. The UWB spectrum mask defined by FCC is shown in Fig. 1.

S.C. Mukhopadhyay and A. Lay-Ekuakille (Eds.): Adv. in Biomedical Sensing, LNEE 55, pp. 43–60.
springerlink.com

This new regulation has sparked great research and development interest from both academia and industry in the UWB systems and their potential applications such as high date-rate consumer electronics, and low power and low complexity wireless sensor networks.

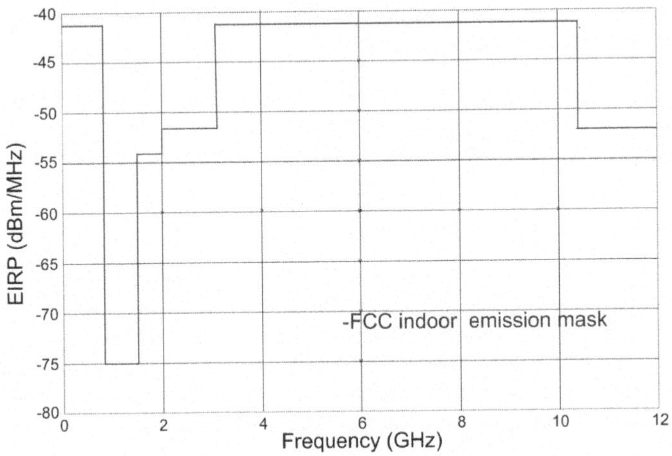

Fig. 1. UWB Spectrum Mask as defined by FCC

Typically, there are two different approaches to generate RF signals with ultra wide bandwidth: the multi-band CDMA or OFDM approach for high-data-rate (20Mbs or greater) applications, and the impulse radio (IR) approach for low-data-rate applications. The IR UWB approach transmits pulses or pulsed waveforms with very short duration, which is usually on the order of nanoseconds or even shorter. The IR UWB systems have the advantages of low complexity, low power consumption, and good time-domain resolution allowing for location and tracking applications. With all these advantages, the IR-UWB approach has been selected by the IEEE 802.15.4a standard as the alternative PHY for the IEEE 802.15.4 standard in providing communications and high precision ranging/location capability (50cm accuracy), high aggregate throughput, and ultra low power; as well as adding scalability to higher data rates, longer range, and lower power consumption and cost.

2 Medical UWB Applications

Applying Ultra-WideBand (UWB) technology in medical applications is an emerging research trend in recent years. Although many researchers have been working on UWB technology for years, concern about UWB radar arose in 1993 when it was reported that a Lawrence Livermore National Laboratory (LLNL) engineer, Thomas McEwan, discovered an original implementation of UWB radar. While working on a new high speed, low-cost sampler for pulse laser research, McEwan developed a system named Micropower Impulse Radar (MIR). The patents on MIR technology [1]-[2] describe a spectrum of applications coming from the low cost MIR technology: from plastic bodied mine detection, to remote vital signs monitoring, to the "3-D radar camera." In 1994,

MIT began an educational project for the "Radar Stethoscope". In 1996, the biomedical use of UWB radars is better described with photo and sample tracings, and in the same year, the US Patent [2] was awarded. Since then, UWB is often deemed as a possible alternative to remote sensing and imaging. Compared with X-ray imaging, UWB radar probes use non-ionizing electromagnetic waves which proved to be harmless to human body. Moreover, the UWB radar has very low average power level and is very power efficient. Thus is suitable to be a potentially cost effective way of human body imaging, especially in real time imaging. By 1999, many works have begun for UWB medical applications in cardiology, obstetrics, breath pathways and arteries. Another feature of UWB is the high precision ranging at centimeter level based on the ultra-short pulse characteristic. High precision of ranging also means strong multi-path resolving capability. The conventional wireless technique used continuous wave and the standing time is much longer than multi-path transmission time. The UWB pulse is much shorter, thus it has very strong temporal and space resolving capability (for 1 nano second pulse, the multi-path resolving power equals to 30cm), which is suitable for the localization and detection in the medical applications.

A key feature of UWB is the low electromagnetic radiation due to the low radio power pulse less than -41.3dBm/MHz in indoor environment. The low radiation has little influence on the environment, which is suitable for hospital applications. Furthermore, the low radiation is safe for human body, even in the short distance, which makes it possible to apply UWB to the clairvoyant equipments.

Since UWB uses very short pulses for radio transmission and very careful design of signal and architecture, the transmitter could be designed simple and allows extreme low energy consumption, which also enables the usage of long-life battery-operated devices. These features are quite similar with the Wireless Sensor Networks (WSN) nodes which must work under extreme condition and require very strict power control mechanism and high power efficiency. The inherent noise-like behavior of UWB systems make it highly possible to deploy medical sensor with UWB since the signal is hard to detect and also excel in jamming resistance. Wireless Body Area Networks (WBANs) for surveillance of human body can be deployed due to this feature.

In short, these bunch of features of UWB that are discussed in this section make it very suitable for medical areas. In the following sections, we will discuss some tipical UWB medical applications which benefit these features, especially for medical monitoring and Wireless Body Area Network (WBAN) implementation.

3 UWB as Radar in Medical Monitoring

Because of the highly intense pulses used in UWB technology, it is possible to use UWB radar in medical field for remote monitoring and measuring the patients' motion in short distance. This monitoring function could be applied in intensive care units, emergency rooms, home health care, pediatric clinics (to alert for the Sudden Infant Death Syndrome, SIDS), rescue operations (to look for some heart beating under ruins, or soil, or snow).

When two UWB devices communicate with each other, the distance between the two devices can also be determined, which is known as ranging [3]. From these ranges, the actual location of the nodes can be estimated. This is referred to as positioning.

Looking at a received signal strength indicator (RSSI) is the simplest way to perform ranging. However, because of the short pulse duration, UWB receivers need fine timing circuits. These can be used to determine the actual time of flight of the signal. Assuming the receiver can distinguish between pulses 1 ns apart, the distances can be detected with accuracy down to 30 cm, much better than what can be achieved using simple RSSI.

In time-of-arrival (TOA) based two-way ranging (TWR), a device that wants to find out the distance to another will start a timer at the moment it sends a ranging packet to the other device. That node will send a reply that includes the time it required to process the ranging packet. Once this reply reaches the originating device, it can stop its timer. The distance between the nodes follows simply from the difference between the two timer values (see Fig. 2).

TWR is ideal when the nodes are not synchronised. When the actual sensor nodes are synchronised, one-way ranging can be used. The transmitting device simply includes a time-stamp of when it sends the packet and the receiver(s) can determine the distance from the packet's arrival time. In a network with synchronised anchor nodes on known locations, time-difference-of-arrival (TDOA) can be used. In one embodiment, the node with unknown location sends out a signal. From the time difference between the arrivals of that signal at the anchors, the position of the unknown node can be determined.

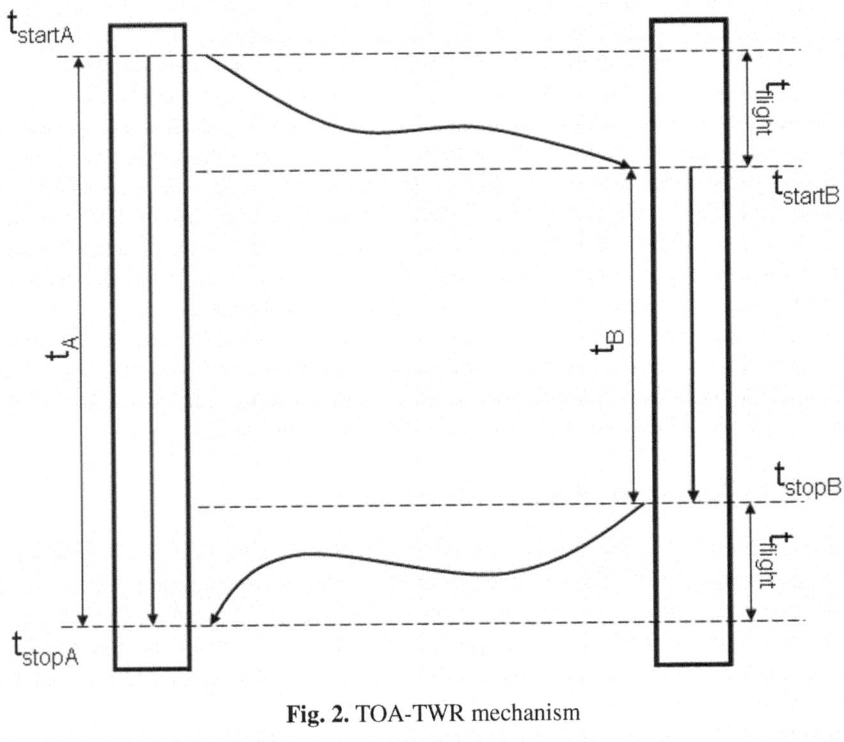

Fig. 2. TOA-TWR mechanism

In practice, a number of factors complicate the performance of ranging and positioning. Most importantly, multipath propagation causes the receiver to pick up multiple copies of the transmitted signals at slightly different time instances. In order to determine the actual range, the so-called leading edge of the power delay profile needs to be extracted. However, because of simplicity, some receivers determine the arrival time of the strongest multipath component. This is referred to as peak detection ranging. In non-line-of-sight situation in particular, leading edge detection is superior to peak detection.

In the recent IEEE802.15.4a standard, a number of provisions have been made in order to support ranging in all its flavours [3]. In the PHY header, a 'ranging bit' can be set to indicate that the current packet is a ranging packet. If this bit is set, the receiver is supposed to calculate the arrival time of the start of the preamble. The ranging information is communicated between nodes in the form of a time stamp report. This report includes information about the timer start value, the timer stop value, the clock tracking difference and interval, as well as a figure of merit. In TWR, the timer start and stop value should correspond to the time the start of the preamble arrives and leaves the receiver antenna respectively. Preferably, the devices should be aware of internal propagation delays in the analogue and digital circuits before the timer. Since both nodes are unsynchronised, any difference between the clocks should be tracked. The receiver needs to count the number of clock cycles it needs to add or drop from its own clock to remain synchronised. The number of clock cycle adjustments and the time during which these were necessary should be communicated using the clock tracking difference and interval. Finally, the figure of merit allows the UWB device to inform the others about its estimate of the reliability of its reported timings. These can be used, by maximum-likelihood positioning algorithms for example, to increase the accuracy of the position estimates.

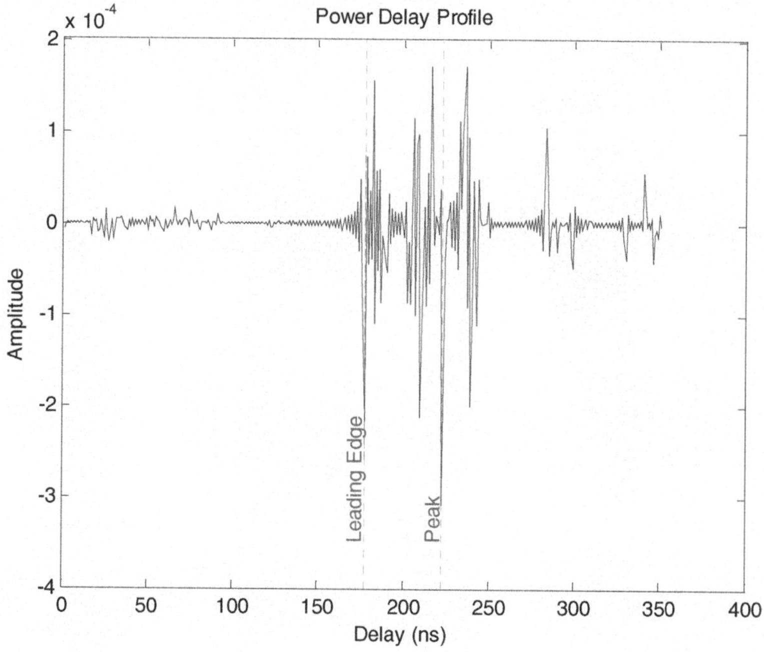

Fig. 3. Leading edge versus peak detection ranging

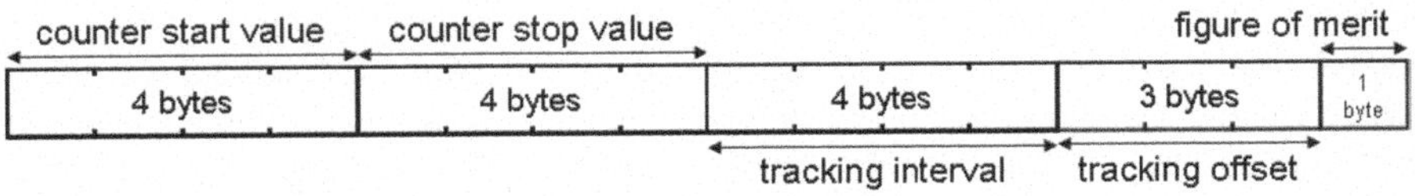

Fig. 4. 802.15.4a time stamp report

This capability of UWB devices to accurately inform the network of their location is extremely useful. For example, in a medical environment, essential equipment can be tagged and tracked such that its location is immediately known at the time of an emergency [4]. Patients and staff can also be equipped with tags such that their location is known at all times. In one application [5], this can be used to automate the billing process simply by taking into account how often patient and staff or equipment are located together. UWB has also been considered for the positioning of in-body nodes [6] but due to the non-constant speed at which the electromagnetic waves travel in different tissues, the accuracy in-body is quite low.

The energy requirement for UWB devices is small and is suitable for the sensing for a long time. Zigbee and Bluetooth devices are less suitable for medical applications because their energy requirement is higher and data transmitting speed is low.

4 Impulse Radio Ultra Wide Band Receivers for Wireless Body Area Networks

In Wireless Body Area Network (WBAN) sensors are designed to operate around the human body, in order to improve the quality of life. A typical WBAN consists of a

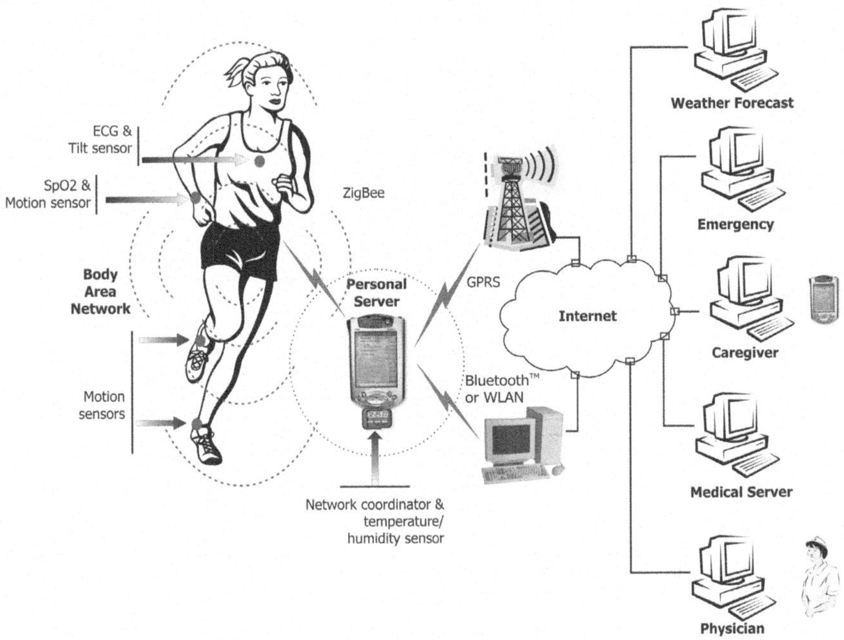

Fig. 5. Typical WBAN example

number of inexpensive, lightweight, and miniature sensor platforms, each featuring one or more physiological sensors, e.g. motion sensors, electrocardiograms (ECGs), electromyograms (EMGs), and electro-encephalograms (EEGs). Typical example of this type of application is ambulatory monitoring of user's activity [7][8]. The sensors could be located on the body as tiny intelligent patches, integrated into clothing, or implanted below the skin or muscles (see Fig. 5).

The main requirements of a WBAN are:

- a low transmit power per node to minimize interference and cope with health concern;
- an efficient support of low data rate transmission;
- an ultra-low power consumption for the radio due to the lack of space for an external power source;

The immediate consequence of these requirements is the use of a broadband signaling scheme as the Impulse Radio Ultra Wide Band (IR-UWB) technology. Some works have demonstrated the potential of impulse radio ultra-wideband communication as a solution to overcome the energy gap [9]–[14] of sensor networks. Moreover, the IEEE 802.15.4a standardization committee [15] has recently been formed to propose a physical layer for low-power sensor network communication using IR-UWB as a key technology.

However, IR-UWB brings its own challenges. Although low power solutions have been demonstrated for transmitters [11], the design of the receiver remains very challenging. The large bandwidth requirements of the analog front-end and baseband circuits, the high sampling rates in the digital conversion, the high timing precision and signal synchronization are still impediments for the realization of a low-power UWB receiver. In this paragraph, the feasibility of a low-power UWB receiver through implementation is demonstrated. Supported by a careful architectural trade-off analysis where various topologies are explored, the optimal architecture is selected for hardware implementation. A single-chip receiver realized in a standard 0.18μm CMOS technology [16] is then described without any external components (besides an external 20 MHz crystal oscillator) so as to satisfy the size and cost requirements of sensor network applications.

4.1 Receiver Topology Selection

Over the last decade, a wide range of receiver topologies has been proposed for IR-UWB communication. These topologies, ranging from purely digital to almost entirely analog solutions, differ significantly in terms of power consumption and receiver performance. In the context of low-power, low-data-rate wireless sensor networks, a good metric to compare different architectures is the *energy per useful bit* [18]. It corresponds to the energy consumed to receive a single bit of data with a certain bit error rate (BER), including channel estimation and synchronization overhead. This metric is used to select the optimal topology by carefully weighting the power consumption and receiver BER performance. In this way, both communication theory and implementation requirements are jointly considered. Three different state-of-the-art UWB receiver topologies are analyzed: the direct conversion (DC), the quadrature analog correlation (QAC), and the transmitted reference (TR).

The DC UWB receiver was introduced in [10] for the 0–1 GHz frequency band and later refined in [19] for the 3–10.6 GHz band. In this topology, the incoming pulses are, after a quadrature down-conversion to baseband, directly sampled by two parallel ADCs and processed in the digital domain. This results in a very flexible receiver, capable of doing wide parallel processing, hence, minimizing the acquisition time. The price to pay is the need for high-speed ADCs, which contribute significantly to the power budget. Some authors suggest to reduce the power consumption by limiting the ADC resolution to 1 bit [10], [20]. This, however, degrades the receiver performance, since narrowband interferers can easily saturate the front-end. The high-speed ADCs can be avoided by moving the matched filtering of the incoming pulses with the pulse template from the digital to the analog domain [9]. This operation reduces the required sample frequency from Nyquist rate to pulse rate, a reduction by a factor 10 to 100. The lower sampling speed enables the use of a higher resolution ADC for increased robustness against interferers, with almost no power penalty. The tradeoff for this reduction in power consumption is a decrease in receiver performance, since channel compensation is limited in the analog domain [21]. Moreover, the acquisition time will increase due to less parallel processing capability. The quadrature analog correlating (QAC) receiver architecture, discussed in this paper, is based on this topology. It correlates the incoming pulses with windowed sines in the analog domain. Simplified channel compensation is done by repetitively opening and closing the window with a resolution of 2 ns [21]. A third architecture worth considering is the transmitted reference (TR) receiver [22]. Every transmitted pulse is replaced by two pulses: a reference pulse and a data pulse. Instead of generating a pulse correlation template in the receiver, the reference pulse is delayed in the receiver to be correlated with the data pulse. TR receivers do not need any channel estimation and inherently capture all multipath components of the received signal. On the other hand, they suffer a large performance degradation due to the very noisy correlation template. Moreover, the implementation of the analog delay line, necessary to delay the reference pulse, is not straightforward [23]. These three different receiver classes (the DC, QAC, and TR receiver) are compared in terms of energy per useful bit. For the DC receiver, both a topology with 4-bit ADC and one with 1-bit ADC are considered.

4.1.1 BER Performance

The channel estimation and data reception of all receivers are simulated in Matlab. To make a fair comparison, the following parameters were fixed for all simulations. The used pulse form is a raised cosine pulse with a roll-off factor of 1 and a 10 dB bandwidth of 500 MHz, centered around 4 GHz. The pulse repetition frequency is set to 20 MHz, with a spreading factor of 10 pulses per bit. These values are selected targeting 2 Mb/s communication over a short-distance non-line-of-sight (NLOS) channel. As a result, the CM2 channel model, proposed by the IEEE 802.15.3 Channel Modeling Sub-committee [27], is applied in the model. Since UWB communication is typically limited by interference noise, the receiver performance is plotted against the signal-to-interference ratio (SIR). The interferer is modeled by a sine with a random in-band frequency and random phase.

Each communication packet consists of 500 data bits (bits), preceded by a preamble (bits). This preamble is used for synchronization (10 bits) and channel estimation (10 bits). The number of bits used for channel estimation differs for the various receiver types, due to the difference in parallel processing capabilities and in channel compensation techniques used. All preamble lengths are optimized to minimize the total amount of energy in a packet. The FD receiver estimates the received pulse form by averaging the pulse energy over 20 bits and uses this as its correlation template. The QAC receiver profits from a reduced channel estimation and compensation [21], but is on the other hand capable of less parallel processing, resulting in a preamble length of 100 bits. The TR reference receiver finally does not need any channel estimation, since a reference pulse is transmitted together with the data pulse. The simulation results are given in Fig. 6. The simulations show that decreasing the ADC resolution to 1 bit reduces the receiver BER performance drastically. The interferers cause the ADC to clip, resulting in a serious loss of information. The QAC receiver has less performance degradation due to clipping, since it benefits from its slower 4-bit ADC. The loss of this receiver is mainly due to the imperfect channel compensation. The TR receiver performs significantly worse than the other alternatives because of the noise cross-correlation term.

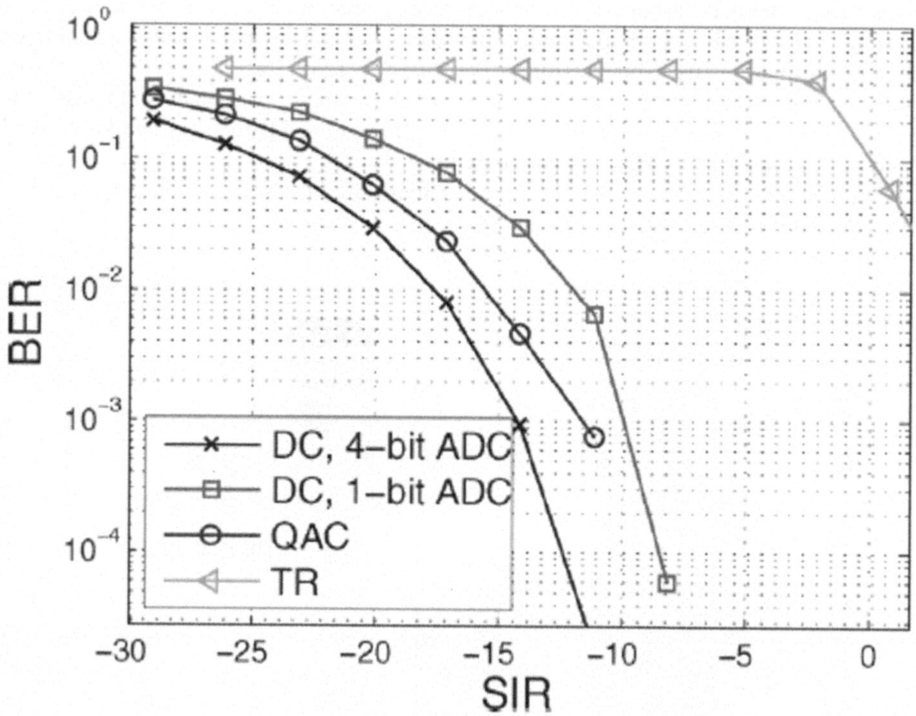

Fig. 6. Simulated BER of the various topologies in function of signal-to-interference ratio (SIR) for CM2 channels

4.1.2 Power Consumption

A back-of-the-envelope estimation of the power consumption of the different alternatives is made based on state-of-the-art components and UWB front-ends reported in literature. The power estimation is done in 0.18μm CMOS technology.

The DC topology used for power estimation is described in [19]. The topology for the QAC receiver is the one implemented in this paper. Finally, the TR receiver is estimated based on the topology presented in [23].

All receiver topologies under study in this section need a wideband low-noise amplifier (LNA) and two wideband mixers. Ref. [24] reports an LNA operating over the 3–8 GHz band, together with two 500 MHz downconversion mixers. This circuit achieves 21 dB gain, a noise figure (NF) 6 dB, while blocking the 5 GHz WLAN interferers by 6 dBm. A power consumption of 18 mW is reported. Also a 4 GHz PLL, with phase noise below 110 dBc/Hz/1 MHz [25], is present in all receiver alternatives. The PLL of Pellerano [26] fulfills these requirements, consuming 13.5 mW. The power consumption of the ADCs is estimated based on an energy consumption of 0.8 pJ/conversion.

Several implementation of the digital fraction of the fully digital receiver have been reported in literature. In [20], a very low-power, 6.7 mW digital baseband for a 1-bit FD receiver is shown. It benefits from using a one-taps correlator, which can only capture the strongest path in the channel. A one-taps receiver is, however, insufficient for NLOS environments, where the strongest path only captures a small fraction of the total energy.

Ref. [31] describes a 75 mW digital back-end for a 4-bit FD receiver capable of channel estimation. Its 1-bit version is described in [25]. Based on [20], the power consumption of this architecture can be estimated to be approximately 35 mW. Finally, the power consumed by the analog delay line of the TR receiver can be derived from [23], scaled to 500 MHz signal bandwidth.

Fig. 7 shows for each alternative the distribution of the power consumption over the different building blocks.

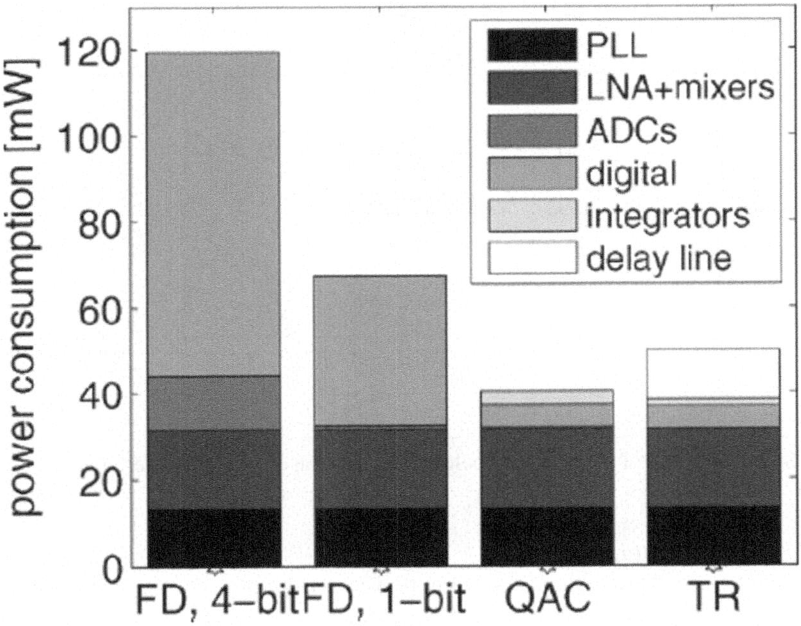

Fig. 7. Power consumption for the different topologies

Table 1. Topology Comparison in term of power, performance, and energy

	4-b FD	1-b FD	QAC	TR
N_D	500	500	500	500
N_P	30	30	90	10
$F_{ADC}[MHz]$	500	500	25	25
L_{perf}	0dB	4.3dB	2.6dB	20dB
P_{cons}	119mW	67mW	40mW	50mW
nJ/bit	56	83	27	893

4.1.3 Optimal Topology Selection

To select the optimal receiver topology, information concerning performance and power consumption must be combined. This is done by computing the energy consumed to receive one useful data bit, including preamble (synchronization + channel estimation) overhead. The metric can be expressed as:

$$\frac{energy}{bit} = \left(\frac{NP + ND}{ND} \cdot N_S \cdot T_p \cdot L_{perf}\right) \tag{1}$$

with the number of bits in the preamble, the number of data bits in a packet, the number of pulses per bit, the pulse period, and the performance loss in relation to the 4-bit DC receiver, hence, gives the number of pulses per bit that have to be received by a certain receiver type to obtain the same bit error rate as the 4-bit DC receiver using pulses per bit.

The used parameters and results are summarized in Table 1. The table shows that the QAC receiver is the most energy-efficient alternative, being at least two times more energy efficient than the other topologies. It offers the optimal tradeoff between receiver performance and power consumption. This topology selection is conducted based on a 0.18µm CMOS technology. Scaling to smaller technologies will cause a reduction in power consumption and energy per useful bit (EPUB) for all topologies. When leakage is not a dominant source of power consumption, digital power will decrease faster than analog power. The FD receiver will, hence, benefit from this scaling and become more and more attractive. Already in 90 nm technology, however, leakage has started to be a problem, slowing down digital scaling. It is therefore unclear whether FD will outperform the QAC receiver in future technologies.

4.2 Quadrature Analog Correlation Receiver

In the QAC receiver, the correlation of the incoming pulses with the pulse template is done in the analog domain to save power. The correlation operation is implemented by mixing the incoming waveform modeled as $f(t) = \tilde{f}(t) \cdot cos(\omega_{RX} t + \theta)$ with both a local oscillator (LO) and a baseband template $\tilde{g}(t)$, followed by an integration. This operation corresponds to a correlation with a template signal $g(t) = \tilde{g}(t) \cdot cos(\omega_{RX} t + \theta)$, where $\tilde{f}(t)$ is the downconverted pulse and $f(t)$ and $\tilde{g}(t)$ are real valued signals. The resulting correlation energy can be calculated from the cross-correlation function $\rho_{fg}(t)$. Assuming that the LO frequency is centered on the pulse center frequency (ω_{RX}=ω_{LO}), this cross-correlation can be written as (2)

$$\rho_{fg}(t)=\frac{1}{2}\rho_{\tilde{f}\tilde{g}}(t)\cos(\omega_{LO}t+\theta) \tag{2}$$

where $\rho_{\tilde{f}\tilde{g}}(t)$ is the cross-correlation function of the equivalent baseband signals. Any phase or frequency error between the receiver LO and the pulse carrier introduces an error that, in first order, only affects the oscillating term . Therefore, combining the energy from two branches in quadrature, results in a correlation vector with an amplitude that is only determined by the envelope of the pulse.

The signal-to-noise ratio (SNR) after correlation is optimized by maximizing:

$$SNR=\frac{\left|\rho_{\tilde{f}\tilde{g}}(0)\right|^2}{\frac{N_0}{2}\cdot\int_{-\infty}^{\infty}\left|G(\omega)\right|^2 d\omega} \tag{3}$$

To lower the receiver complexity and power consumption, the ideal template $\tilde{g}(t)$ is, however, replaced by a rectangular window, introducing a minor loss of less than 1 dB (in line-ofsight, LOS, channels). The correlation with this rectangular template can easily be implemented by integrating during a time window defined by the start of the integration and the sampling instant. The resulting architecture of the QAC receiver is depicted in Fig. 8.

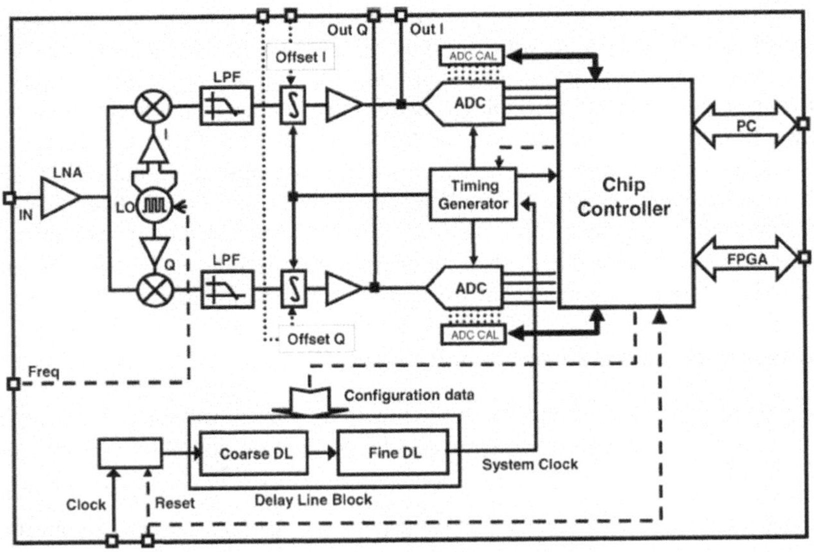

Fig. 8. Quadrature analog correlating architecture

4.3 System Measurements

The receiver of Fig. 4 was realized in a CMOS 0.18μm technology and the chip microphotograph is shown in Fig. 9.

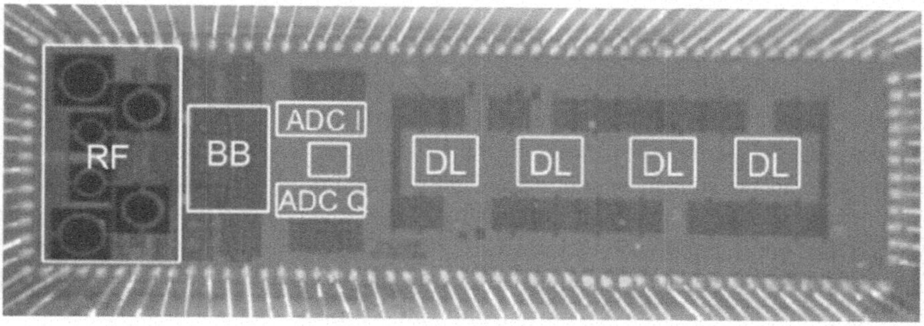

Fig. 9. Receiver chip microphotograph

All the parameter settings of the delay line and duty cycle generator, the ADC offset calibration data, variable gain amplifier (VGA) gain and bandwidth are managed by an on-chip central controller that realizes the interfacing either with a PC through a parallel link or with a field-programmable gate array (FPGA) through a high-speed bus. The analog baseband output of the integrated pulses, before ADC sampling, have been brought off-chip and a snapshot of such a measurement obtained with a high bandwidth oscilloscope is shown on Fig. 10.

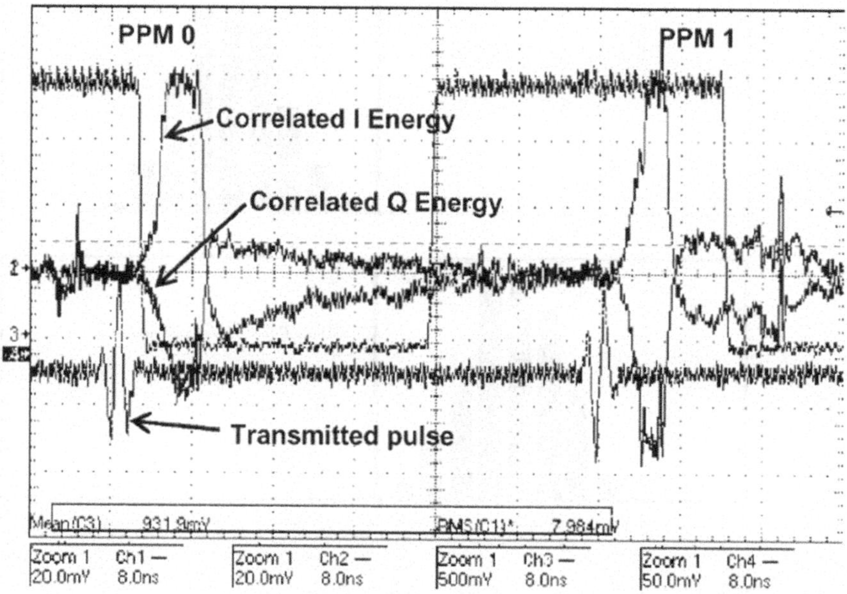

Fig. 10. Time domain measurement of the integrated pulses. Two consecutive pulses are shown each at a different PPM position.

This measurement was realized with 500 MHz pulses centered on 4 GHz. The pulse generator used in this measurement is the one presented in [11].

After ADC sampling, the data are captured either by a PC or by a FPGA. The FPGA contains a synchronization algorithm [38] that synchronizes the receiver clock to the received pulse stream. A typical distribution of the captured data around each PPM position is shown in Fig. 11. This measurement was again performed with 500 MHz pulses centered on 4 GHz.

The BER was also measured as a function of both in-band (at 250 MHz offset of the pulse center frequency) and out-of-band interferer (at 2.4 GHz) power. This is plotted in Fig. 12. In this measurement, the pulse power was kept constant at 55 dBm peak power and the system operates at 32 pulses per bit. At low interferer power, the BER converges to the error floor of about 1e-7 error rate defined by the interferer-free SNR. As the interferer power increases, the BER increases in the expected waterfall shape fashion. At higher BER than 1e-2, a higher BER is observed. This is due to the large amount of failure of the synchronization algorithm.

The power consumption of the receiver is 16mA in total from a 1.8 V supply, with 4.9 mA in the LNA-mixer, 7.5 mA in the LO generation and distribution, 1.8 mA in the analog baseband, 0.45mAin the two ADCs, and about 1mAfor the rest of the digital part. As can be observed, more than 75% of the power is consumed by the RF front-end part. This sets the need for front-end duty-cycling in order to further reduce the power consumption. Indeed, IR-UWB radios only need to operate during the pulse reception. This would provide a reduction of the power consumption by roughly 65% assuming a 20% front-end duty-cycle.

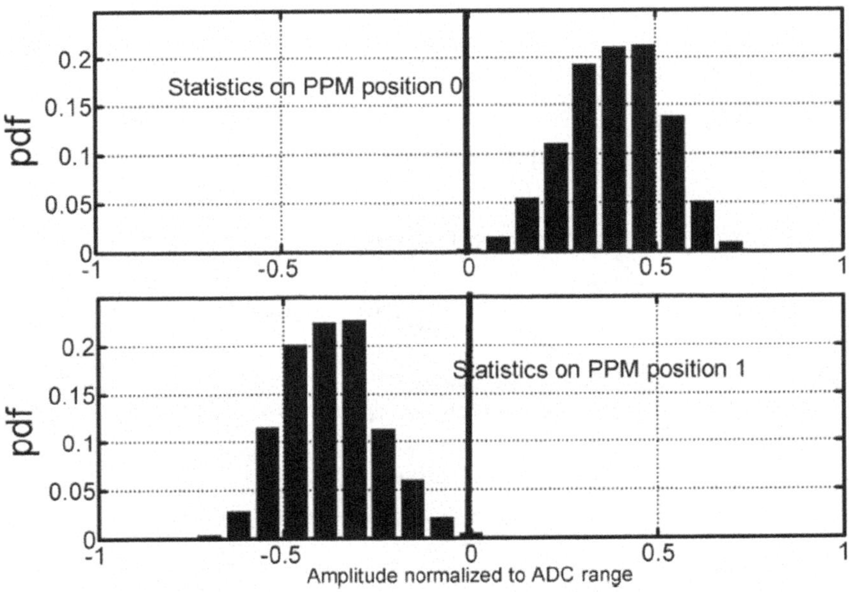

Fig. 11. Distribution of the sampled data around both PPM positions

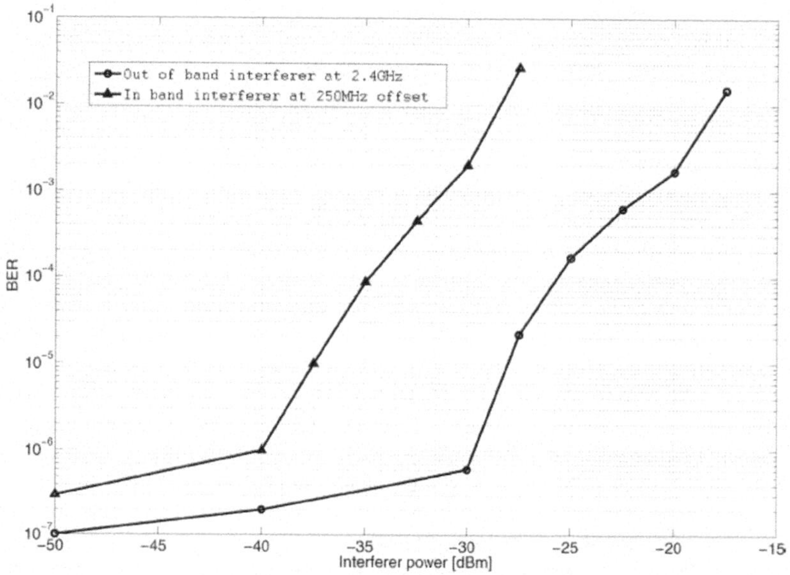

Fig. 12. BER as a function of both in-band and out-of-band interferer power. The pulse power is kept at -55 dBm and a bit is spread over 32 pulses.

5 Conclusions

In this chapter, the use of UWB technology in medicine area has been addressed. The researchers studies, as well as the companies interests, in these fields is continually growing up during the last years. This remarks the importance of the research in this field. The use of UWB signal in medicine started in 1993 with the MIR radar. Since then, several applications of UWB technology in this field have been found. This chapter focused on two main applications: the use of the UWB technology to implement a radar, and the implementation of a UWB radio for WBAN. In section 3 the radar technique has been described in detail. Section 4 describes a UWB receiver implementation in 0.18μm CMOS technology.

References

1. McEwan, T.: Body monitoring and imaging apparatus and method. United States Patent 5,766,208 (June 16, 1998)
2. McEwan, T.: Body monitoring and imaging apparatus and method. United States Patent 5,573,012 (November 12, 1996)
3. Gezici, S., Tian, Z., Giannakis, G.B., Kobayashi, H., Molisch, A.F., Vincent Poor, H., Sahinoglu, Z.: Localization via Ultra-Wideband Radios. IEEE Signal Processing Magazine, 70–84 (July 2005)

4. Kanso, M.A., Rabbat, M.G.: Efficient Detection and Localization of Assets in Emergency Situations. In: 3rd International Symposium on Medical Information and Communication Technology (ISMICT 2008), Montreal, Canada (2008)
5. Labeau, F., Tchana, A.B., Le-Ngoc, T.: Enabling Context Aware Clinical Applications through Ultra-Wideband Localization. In: 3rd International Symposium on Medical Information and Communication Technology (ISMICT 2008), Montreal, Canada (2008)
6. Kawasaki, M., Kohno, R.: A TOA based Positioning Technique of Medical Implanted Devices. In: 3rd International Symposium on Medical Information and Communication Technology (ISMICT 2008), Montreal, Canada (2008)
7. Steele, B.G., Belza, B., Cain, K., Warms, C., Coppersmith, J., Howard, J.: Bodies in motion: Monitoring daily activity and exercise with motion sensors in people with chronic pulmonary disease. Journal of Rehabilitation Research & Development 40(5) (suppl. 2), 45–58 (2003)
8. Jovanov, E., Milenkovi, A., Otto, C., De Groen, P., Johnson, B., Warren, S., Taibi, G.: A WBAN System for Ambulatory Monitoring of Physical Activity and Health Status: Applications and Challenges. In this Proceedings
9. Verhelst, M., Dehaene, W.: System design of an ultra-low-power, low data rate, pulsed UWB receiver in the 0-960 MHz band. In: Proc. IEEE Int. Conf. Communications, Seoul, Korea, May 2005, vol. 4, pp. 2812–2817 (2005)
10. O'Donnell, D., Chen, M.S.W., Wang, S.B.T., Brodersen, R.W.: An integrated, low-power, ultra-wideband transceiver architecture for low-rate indoor wireless system. Presented at IEEE CAS Workshop Wireless Commun. Networking, Pasadena, CA (September 2002)
11. Ryckaert, J., Desset, C., Fort, A., Badaroglu, M., De Heyn, V., Wambacq, P., Van der Plas, G., Donnay, S., Van Poucke, B., Gyselinckx, B.: Ultra-wide band transmitter for low-power wireless body area networks: Design and evaluation. IEEE Trans. Circuits Syst. I, Reg. Papers 52(12), 2515–2525 (2005)
12. Terada, T., Yoshizumi, S., Muqsith, M., Sanada, Y., Kuroda, T.: A CMOS ultra-wideband impulse radio transceiver for 1-Mb/s data communications and _2.5-cm range finding. IEEE J. Solid-State Circuits 41(4), 891–897 (2006)
13. Tamtrakarn, A., Ishikuro, H., Ishida, K., Takamiya, M., Sakurai, T.: A 1-V 299_W flashing UWB transceiver based on double thresholding scheme. In: 2006 Symp. VLSI Circuits Dig. Tech. Papers., Honolulu, HI, June 2006, pp. 202–203 (2006)
14. O'Donnell, I.D., Brodersen, R.: A 2.3mW baseband impulse-UWB transceiver front-end in CMOS. In: 2006 Symp. VLSI Circuits Dig. Tech. Papers, Honolulu, HI (June 2006)
15. IEEE 802.15 WPAN Low Rate Alternative PHY Task Group 4a (TG4a), IEEE 802.15.4 Std.,
`http://www.ieee802.org/15/pub/TG4a.html`,
`http://standards.ieee.org/getieee802/download/802.15.4-2003.pdf`
16. Ryckaert, J., Badaroglu, M., De Heyn, V., Van der Plas, G., Nuzzo, P., Baschirotto, D'Amico, S., Desset, C., Suys, H., Libois, M., Van Poucke, B., Wambacq, P., Gyselinckx, B.: A 16mA UWB 3-to-5GHz 20Mpulses/s quadrature analog correlation receiver in 0.18um CMOS. In: IEEE ISSCC Dig. Tech. Papers, San Francisco, CA, February 2006, pp. 368–377 (2006)
17. Win, M.Z., Scholtz, R.A.: Impulse radio: How it works. IEEE Commun. Lett. 2(2), 36–38 (1998)

18. Ammer, J.: Low power synchronization for wireless communication. Ph.D. dissertation, Univ. California, Berkeley, CA (2004); [11] Lee, F., Chandrakasan, A.: A BiCMOS ultra-wideband 3.1–10.6-GHz front-end. IEEE J. Solid-State Circuits 41(8), 1784–1791 (2006)
19. Yang, C., Chen, K., Chiueh, T.: A 1.2V 6.7mW impulse-radio UWB baseband transceiver. In: IEEE ISSCC 2005 Dig. Tech. Papers, San Francisco, CA, vol. 1, pp. 442–608 (2005)
20. Verhelst, M., Dehaene, W.: Complex analog pulsed UWB-receiver in realistic 0–1 GHz channels. Presented at the 1st IEEE Australian Conf. Wireless Broadband and Ultra Wideband Communications (AusWireless 2006), Sydney, Australia (March 2006)
21. Hoctor, R., Tomlinson, H.: Delay-hopped transmitted-reference RF communications. In: Proc. 2002 IEEE Conf. Ultra Wideband Systems and Technologies, Baltimore, MD, May 2002, pp. 265–269 (2002)
22. Bagga, S., Zhang, L., Serdijn, W., Long, J., Busking, E.: A quantized analog delay for an IR-UWB quadrature downconversion autocorrelation receiver. In: Proc. 2005 IEEE Int. Conf. Ultra-Wideband, Zurich, Switzerland, September 2005, pp. 328–332 (2005)
23. Cusmai, G., Brandolini, M., Rossi, P., Svelto, F.: An interference robust 0.18 um CMOS 3.1–8 GHz receiver front-end for UWB radio. In: Proc. IEEE Custom Integrated Circuits Conf. (CICC 2005), September 2005, pp. 157–160 (2005)
24. O'Donnell, I., Brodersen, R.: An ultra-wideband transceiver architecture for low power, low rate, wireless systems. IEEE Trans. Veh. Technol. 54(5), 1623–1631 (2005)
25. Pellerano, S., Levantino, S., Samori, C., Lacaita, A.L.: A 13.5-mW 5-GHz frequency synthesizer with dynamic-logic frequency divider. IEEE J. Solid-State Circuits 39(2), 378–383 (2004)
26. Channel modeling sub-committee final report IEEE 802.15.sg3a, IEEE 802.15-02/490rl-SG3a (February 2003)
27. Razavi, B., et al.: A UWB CMOS transceiver. IEEE J. Solid-State Circuits 40(12), 2555–2562 (2005)
28. Raha, P.: A 0.6-4.2V low-power configurable PLL architecture for 6 GHz-300 MHz applications in a 90 nm CMOS process. In: 2004 Symp. VLSI Circuits Dig. Tech. Papers, Honolulu, HI, June 2004, pp. 232–235 (2004)
29. Ginsburg, B., Chandrakasan, A.: Dual scalable 500mS/s, 5b timeinterleaved SAR ADCs for UWB applications. In: Proc. IEEE 2005 Custom Integrated Circuits Conf., San Jose, CA, pp. 403–406 (2005)
30. Blazquez, R., Newaskar, P., Lee, F., Chandrakasan, A.: A baseband processor for impulse ultra-wideband communications. IEEE J. Solid-State Circuits 40(9), 1821–1828 (2005)
31. Roovers, R., Leenaerts, D.M.W., Bergervoet, J., Harish, K.S., van de Beek, R.C.H., van der Weide, G., Waite, H., Yifeng, Z., Aggarwal, S., Razzel: An interference-robust receiver for ultra-wideband radio in SiGe BiCMOS technology. IEEE J. Solid-State Circuits 40(12), 2563–2572 (2005)
32. Sjoland, H., Karim-Sanjaani, A., Abidi, A.: A merged CMOS LNA and MIXER for aWCDMAreceiver. IEEE J. Solid-State Circuits 38(6), 1045–1050 (2003)
33. D'Amico, S., Ryckaert, J., Baschirotto, A.: A up-to-1 GHz lowpower baseband chain for UWB receivers. In: Proc. Eur. Solid-State Circuits Conf. (ESSCIRC 2006), September 2006, pp. 263–266 (2006)
34. Nuzzo, P., Van der Plas, G., De Bernardinis, F., Van der Perre, L., Gyselinckx, B., Terreni, P.: A 106mW/0.8pJ power-scalable 1 GS/s 4b ADC in 0.18_m CMOS with 5.8GHz ERBW. In: Proc. 43rd ACM/IEEE Design Automation Conf., San Francisco, CA, pp. 873–878 (2006)

35. Kobayashi, T., Nogami, K., Shirotori, T., Fujimoto, Y.: A currentcontrolled latch sense amplifier and a static power-saving input buffer for low-power architecture. IEEE J. Solid-State Circuits 28(4), 523–527 (1993)
36. Lee, M.E., Dally, W., Chiang, P.: A 90 mW 4 Gb/s equalized I/O circuit with input offset cancellation. In: IEEE ISSCC 2000 Dig. Tech. Papers, 2000, pp. 252–253, 463 (2000)
37. Desset, C., Badaroglu, M., Ryckaert, J., Van Poucke, B.: UWB search strategies for minimal-length preamble and a low-complexity analog receiver. In: Proc. IEEE 7th Workshop Signal Processing Advances in Wireless Communications (SPAWC 2006), Cannes, France, July 2006, pp. 1–5 (2006)

A Wearable Force Plate System Designed Using Small Triaxial Force Sensors and Inertial Sensors

Tao Liu, Yoshio Inoue, and Kyoko Shibata

Kochi University of Technology, Japan

Abstract. The measurement of ground reaction force (GRF) and human motion in a gait measurement laboratory is accurate and easier, but expensive and constraint. The ambulatory measurement of GRF and human motion under free-living condition is inexpensive, and really desired. A wearable force plate system was developed by integrating small triaxial force sensors and 3D inertial sensors for estimating triaxial GRF in biomechanical applications. As for the measurement accuracy, we compared the developed system's measurements of the triaxial GRF and the center of pressure (CoP) with the reference measurements of a stationary force plate and an optical motion analysis system. The RMS difference of the two transverse components (x- and y- axes) and the vertical component (z-axis) of the GRF was 4.3±0.9N, 6.0±1.3N, and 12.1±1.1N respectively, corresponding to 5.1±1.1% and 6.5±1% of the maximum of each transverse component, and to 1.3±0.2% of the maximum vertical component of GRF. The RMS distance between the two systems' CoP traces was 3.2±0.8mm, corresponding to 1.2±0.3% of the length of the shoe. Moreover, based on the assessment results of the influence of the system on the natural gait, we found that gait was almost never affected. Therefore, the wearable system as an alternative device can be used to measure CoP and triaxial GRF in non-laboratory environments.

Keywords: wearable force plate, ground reaction force, triaxial force sensor, inertial sensor.

1 Introduction

In a traditional gait analysis laboratory, the stationary force plate can not measure more than one stride; moreover, in the measurements of stair ascent and descent, complex multiple systems composed of many force plates, the motion capture system based to some high-speed cameras, and a data fusion method have been constructed [1]-[2]. Therefore, these stationary measurement systems probably impose some constraints on our ability to measure ground reaction force (GRF) and body orientations, and are not feasible for measurements in everyday situations. As an alternative to conventional techniques, an easy-to-use and inexpensive measurement system which can accurately estimate the triaxial GRF and three-dimensional (3D) body orientations, and has less influence on natural gait, is really desired for biomechanical applications.

In the past study on GRF sensors, many researchers have developed some wearable sensors attached to insoles. Pressure sensors have been widely used to measure the distributed vertical component of GRF and to analyze the loading pattern on the plantar soft tissue during the stance phase of gait [3]-[4], but the transverse component of GRF

S.C. Mukhopadhyay and A. Lay-Ekuakille (Eds.): Adv. in Biomedical Sensing, LNEE 55, pp. 61–73.
springerlink.com

(friction force) could not be estimated using these sensor systems. Fong et al. proposed a method to estimate the triaxial GRF from pressure insoles, but the acceptable high accuracy of the sensor system is limited to the measurements on the same group of subjects with the same type of shoes [5]. By mounting multi-axial force sensors beneath a special shoe, some researchers have developed instrumented shoes for ambulatory measurements of triaxial GRF in a variety of non-laboratory environments, for example, Chateau et al. applied an instrumented shoe fixed under the horse's hoof to GRF measurement on any track [8]. However, these sensor systems in which the adopted commercial sensor except the mounting frames has a height of 15.7mm [6], increase the height and weight of the shoe, and probably affect natural gait. Significant differences between instrumented and normal shoes were found in the maximum GRF, and the maximum GRF averaged over all subjects differed by 56 N in a sensor system test study [7]. Moreover, these measurement systems were limited to the specific shoes, so they could not be easily reconstructed for a variety of subjects with different foot lengths. Therefore, the first problem to be resolved in our research is to make a light and thin force plate which can implement ambulatory triaxial GRF measurements on a variety of subjects, and has a lower influence on their normal gaits.

Recently some inexpensive in-chip inertial sensors including gyroscopes and accelerometers are gradually coming to practical applications on human motion analysis. To expand application scope of our wearable force plate system, an ambulatory 3D inertial sensor system can be integrated with the force plate. Schepers et al. proposed a combination sensor system including six degrees of freedom force and moment sensors and miniature inertial sensors (Xsens Motion Technologies) to estimate joint moments and powers of the ankle [9]. If 3D orientations of the foot can be obtained when we measure triaxial GRF during gait, the inverse dynamic method can be used to accurately analyze joint dynamics of the lower limb [10]. Therefore, in our research, 3D inertial sensor modules which were designed using the lower cost inertial sensor chips including triaxial accelerometer and gyroscope were mounted on each wearable force plate to construct a wearable system.

2 Prototype of a Wearable Force Plate

As shown in Fig. 1 (a), a wearable force plate (weight: 86g and size: $80\times80\times15\text{mm}^3$) was constructed using three small triaxial force sensors provided by Tec Gihan Co., Japan, in which two strengthening fiberboards as top and bottom plates were used to accurately fix the three sensors. The specifications of the applied small sensors are given in Table 1. Each small sensor calibrated using the data provided by the manufacturer can measure triaxial forces relative to the slave coordinate system ($\sum s_i$) defined on the center of the sensor, and the subscript i represents the number of the small sensor in every force plate (i=1, 2, and 3). The GRF and center of pressure (CoP) measured using the developed force plate could be expressed in a force plate coordinate system ($\sum_f$) which is located on the interface between the force plate and the ground, and the origin of the force plate coordinate system was the center of force plate (see Fig. 1 (b)). The y-axis of the force plate coordinate system was chosen to represent the anterior-posterior direction of human movement on the bottom plate, and the z-axis was made vertical, while the x-axis was chosen such that the resulting force plate coordinate system would be right-handed. We aligned the y-axis of each

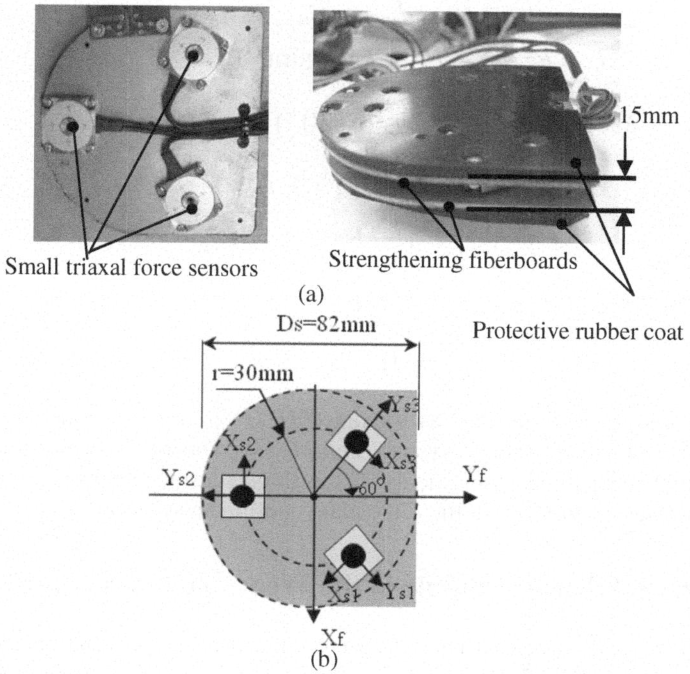

Fig. 1. Prototype of a wearable force plate (a), Coordinate systems of the force plate (b)

Table 1. Main specifications of the small triaxial force sensor used for the wearable force plate system

Type		USL06-H5-500N-C
Rated Capacity (N)	X- and Y- axis	250
	Z-axis	500
Rated Capacity (με)	X- and Y- axis	900
	Z-axis	1700
Nonlinearity (After calibration of cross effect)		Within 1.0%
Hysteresis (After calibration of cross effect)		Within 1.0%
Size (mm)		20×20×5
Weight (g)		15

sensor's slave coordinate to the origin of the force plate coordinate system, while the three origins of the salve coordinates were evenly distributed on the same circle (radius: r=30mm), and were fixed be apart of 120° with each other. Fx_i, Fy_i and Fz_i were defined as triaxial forces measured using the three triaxial sensors. Triaxial GRF and coordinates of CoP could be calculated by the following equations:

$$Fx = (Fx_1 + Fx_3)\cdot\cos(60^\circ) - Fx_2 - (Fy_3 - Fy_1)\cdot\cos(30^\circ) \qquad (1)$$

$$Fy = (Fy_1 + Fy_3)\cdot\cos(60^\circ) - Fy_2 - (Fx_1 - Fx_3)\cdot\cos(30^\circ) \qquad (2)$$

$$Fz = Fz_1 + Fz_2 + Fz_3 \tag{3}$$

$$Mx = Fz_2 \cdot r - (Fz_1 + Fz_3) \cdot \sin(30^\circ) \cdot r \tag{4}$$

$$My = (Fz_1 - Fz_3) \cdot \cos(30^\circ) \cdot r \tag{5}$$

$$Mz = (Fx_1 + Fx_2 + Fx_3) \cdot r \tag{6}$$

$$x_{COP} = My / Fz \tag{7}$$

$$y_{COP} = Mx / Fz \tag{8}$$

$$z_{COP} = 0 \tag{9}$$

where Fx, Fy and Fz were defined as triaxial GRF (F_{GRF}) measured using the force plate in the force plate coordinate system, and Mx, My and Mz indicate triaxial moments estimated from the measurements of the three sensors, while x_{COP}, y_{COP} and z_{COP} are coordinates of the CoP in the force plate coordinate system.

3 A 3D Inertial Sensor Module in the Wearable Force Plate System

As shown in Fig. 2 (a), we constructed a 3D motion sensor module composed of a triaxial accelerometer (MMA7260Q, supplied by Sunhayato Co.) on the bottom of the

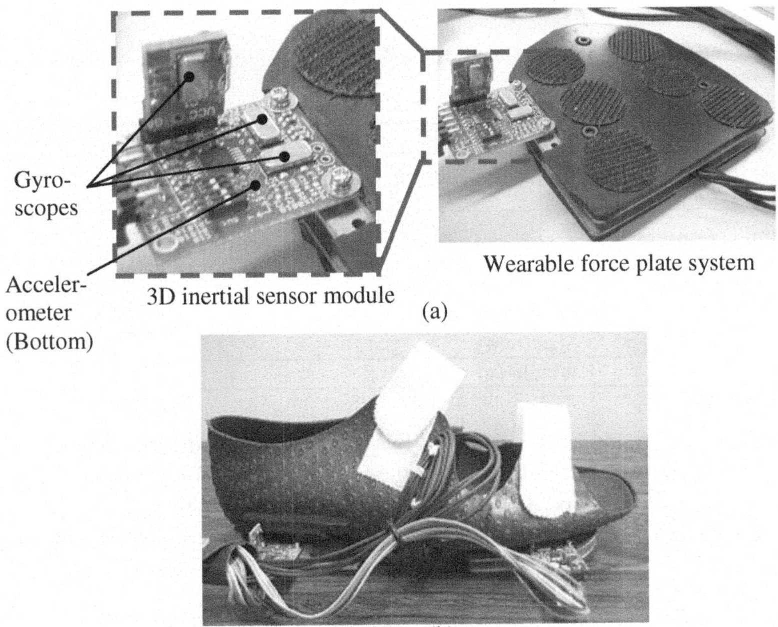

Fig. 2. 3D inertial sensor module mounted on the force plate (a), A wearable force plate system mounted beneath a shoe (b)

PCB board and three uniaxial gyroscopes (ENC-03R, supplied by Murata Co.) on the top. The module was mounted on a developed force plate to measure triaxial accelerations and angular velocities which could be used to estimate 3D orientation transformation matrix. When the force plate system is fixed under a shoe (see Fig. 2 (b)), we can implement an ambulatory GRF and CoP measurement during gait.

4 Transformation of Triaxial GRF Measured by Wearable Force Plates

Considering the bending of a shoe sole during human walking, we adopted a mechanism similar to the structure proposed by Veltink et al. [6] who mounted two small force plates beneath each shoe to measure triaxial GRF. In this paper, two force plate coordinate systems fixed to the two force plates under the heel and the forefoot were defined as $\sum_{f\_heel}$ and $\sum_{f\_toe}$ respectively. The relative position of the two force plates was aligned using a simple alignment mechanism composed of three linear guides and a ruler to let the origins of $\sum_{f\_toe}$ be on the y-axis of $\sum_{f\_heel}$, and let the y-axes of the two force plate coordinate system be collinear, before we mounted their to a shoe. The alignment mechanism and an alignment process are shown in Fig. 3. First, we can align two force plates' centers and regulate the distance between two force plates according to different foot lengths. Second, the barefoot subject is asked to stand on the

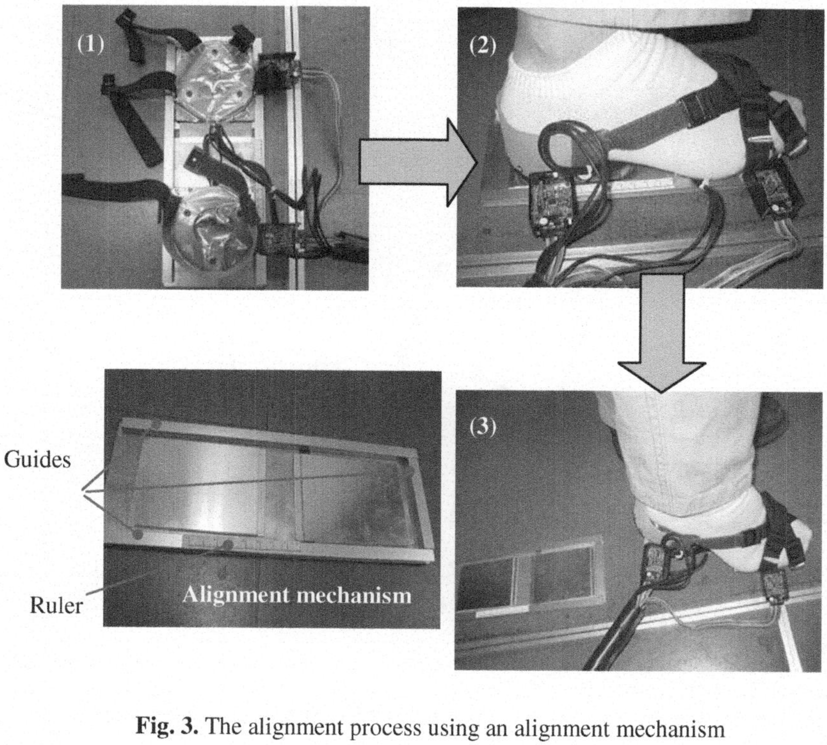

Fig. 3. The alignment process using an alignment mechanism

force plates which are fixed on the alignment mechanism, and then we band two force plates to the foot. Last, we release the two force plates from the alignment mechanism, and let the two force plates be worn on the foot during gait. For calculation purposes, such as estimating joint moments and reaction forces of the ankle during loading response and terminal stance phases [11], all vectors including joint displacement vector, GRF vector and gravity vector have to be expressed in the same coordinate system, being the global coordinate system ($\sum_g$). Moreover, the origin and orientation of this global coordinate system are renewed for each foot placement to coincide with the heel force plate coordinate system ($\sum_{f\_heel}$), when the heel is flat on the ground (see Fig. 4).

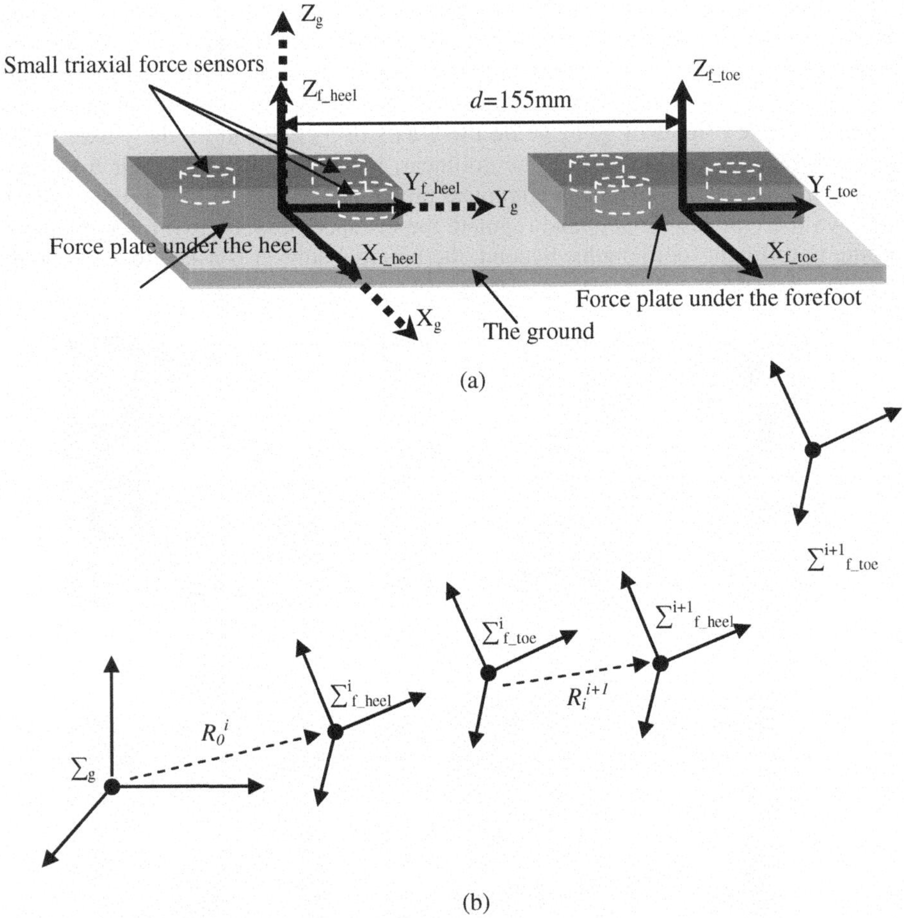

Fig. 4. Coordinate systems of the wearable force plate system (a), Coordinate transformation during the movements of the force plates (b)

The integration of the measured angular velocity vector ($\omega=[\omega x, \omega y, \omega z]$) in each force plate coordinate system was defined as $C=[Cx, Cy, Cz]$, which could be used to calculate the 3D orientation transformation matrix (R) between the global coordinate system and a force plate coordinate system by solving the following equations proposed by Bortz [12]:

$$C_i^{i+1} = [\omega x(i) + \omega x(i+1), \omega y(i) + \omega y(i+1), \omega z(i) + \omega z(i+1)] \cdot \Delta t \tag{10}$$

$$\left|C_i^{i+1}\right| = \sqrt{(Cx_i^{i+1})^2 + (Cy_i^{i+1})^2 + (Cz_i^{i+1})^2} \tag{11}$$

$$R_i^{i+1} = \frac{C_i^{i+1} \cdot C_i^{i+1T}}{\left|C_i^{i+1}\right|^2}\left(1-\cos\left|C_i^{i+1}\right|\right) + \begin{bmatrix} \cos\left|C_i^{i+1}\right| & 0 & 0 \\ 0 & \cos\left|C_i^{i+1}\right| & 0 \\ 0 & 0 & \cos\left|C_i^{i+1}\right| \end{bmatrix} + \frac{\sin\left|C_i^{i+1}\right|}{\left|C_i^{i+1}\right|}\begin{bmatrix} 0 & -Cz_i^{i+1} & Cy_i^{i+1} \\ Cz_i^{i+1} & 0 & -Cx_i^{i+1} \\ -Cy_i^{i+1} & Cx_i^{i+1} & 0 \end{bmatrix} \tag{12}$$

$$R = R_0 \cdot R_0^1 \cdot R_1^2 \cdot \cdot R_i^{i+1} \cdots \tag{13}$$

where $[\omega x(i), \omega y(i), \omega z(i)]$ is a sample vector of the triaxial angular velocities of the force plate during a sampling interval Δt, and C_i^{i+1} is a angular displacement vector in a sampling interval, and R_0 is an initial transformation matrix initialized as a unit matrix ($|R_0|=1$). If the force plate is flat on a level ground we can update R by $R= R_0$.

The triaxial GRF measured by the two force plates could be transformed to the global coordinates, and then be combined to calculate the total GRF (F_{FRG}^g) and the global coordinate vectors of CoP ($[x,y,z]_{COP}^{g\_heel}$ and $[x,y,z]_{COP}^{g\_toe}$) using the following equations:

$$F_{FRG}^g = R_g^{f\_heel} \cdot F_{FRG}^{heel} + R_g^{f\_toe} \cdot F_{FRG}^{toe} \tag{14}$$

$$[x,y,z]_{COP}^{g\_heel} = R_g^{f\_heel} \cdot [x,y,z]_{COP}^{heel} \tag{15}$$

$$[x,y,z]_{COP}^{g\_toe} = R_g^{f\_toe} \cdot [x,y,z]_{COP}^{toe} \tag{16}$$

where F_{FRG}^{heel} and F_{FRG}^{toe} are triaxial GRF measured by the two force plate under the heel and the forefoot with their coordinate systems respectively; $[x,y,z]_{COP}^{heel}$ and $[x,y,z]_{COP}^{toe}$ are coordinate vectors of CoP measured using two force plates; $R_g^{f\_heel}$ and $R_g^{f\_toe}$ are the orientation transformation matrices of the two force plate system for transforming the triaxial GRF measured by the two force plates in their attached coordinate systems into the measurement results relative to the global coordinate system.

5 Experiment Methods

A combination system composed of a force plate EFP-S-2KNSA12 (KYOWA, Japan) and an optical motion analysis system Hi-DCam (NAC Image Tech., Japan) was used as a reference measurement system to verify the measurement results of the developed system (see Fig. 5 (a)). As shown in Fig. 5 (b), A young volunteer (age=29 years, height=170 cm, weight=66 kg) was required to wear the force plate system to walk on the stationary force plate in the capture region of Hi-DCam, and the signals from the two measurement systems were simultaneously sampled at a rate of 100 samples/s.

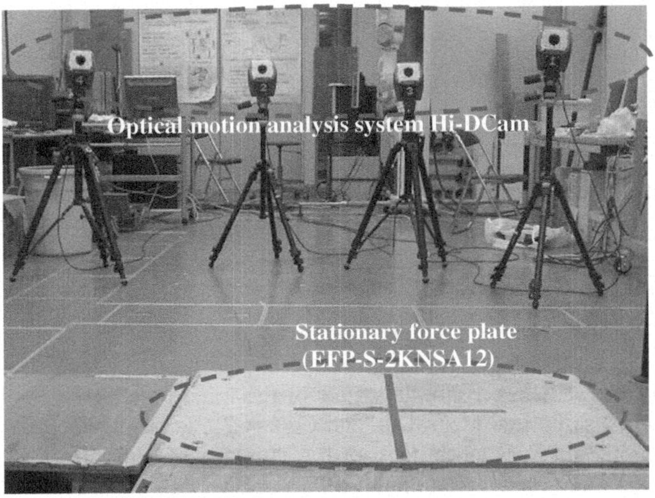

(a)

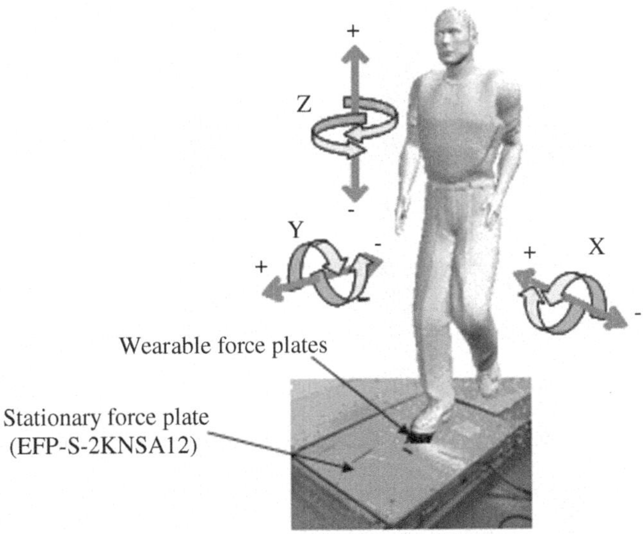

(b)

Fig. 5. Reference measurement system (a), GRF measurement verification experiment (b)

Besides verifying the accuracy of the wearable force plate system, we also used the parameters including stride length (SL), stride width (SW), maximum lateral foot excursion (ME), single stance time (SST), double stance time (DST), stride time (ST), maximum GRF (MaxF) and minimum GRF (MinF) proposed by Liedtke et al. [7] to assess the influence of the wearable system on natural gait. When the subject walked across the stationary force plate with normal shoes and walked with the wearable force plate system respectively, we repeatedly measured foot motion and triaxial GRF using the reference sensor system using the reference measurement system for ten times. Each parameter was determined for a stride and averaged over the repeated ten walking trials by a normal shoe, and then was compared with the results by the wearable system in place under the same conditions.

6 Results

A group of representative experimental results of GRF and CoP measurement during gait is plotted in Fig. 6. The triaxial forces' measurements of the small triaxial force sensors (see Fig. 6 (a)) in the two force plates (one under the forefoot and another under the heel) were used to calculate triaxial GRF and CoP according to equations (1)-(3) and equations (7)-(9) in each force plate. Equations (14)-(16) was adopted for calculating total triaxial forces and CoP in the global coordinate system based on the force plate orientation measurements (see Fig. 6 (b) and (c)). As shown in Fig. 7, comparisons of the three components of GRF and CoP trajectory measured by the wearable system and the reference measurement systems were demonstrated in the

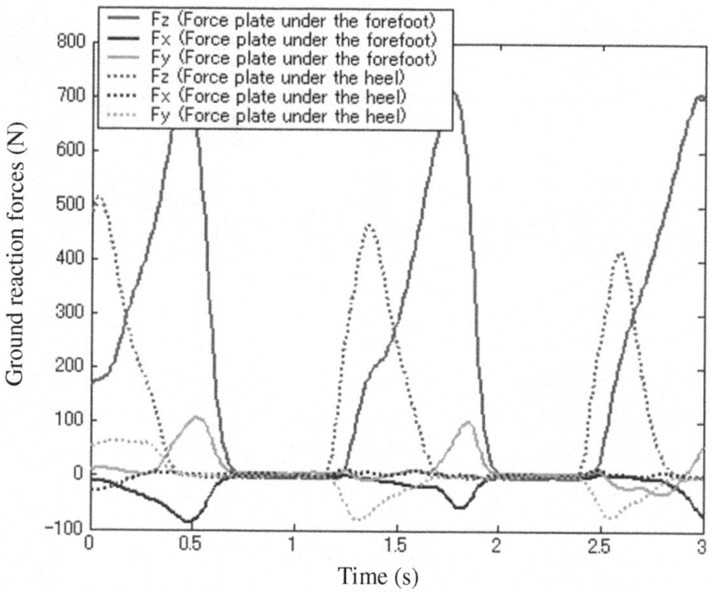

(a)

Fig. 6. GRF measurements of the two force plates (a), Orientations of the two force plates (b), Origins of the two force plates' coordinate system (c)

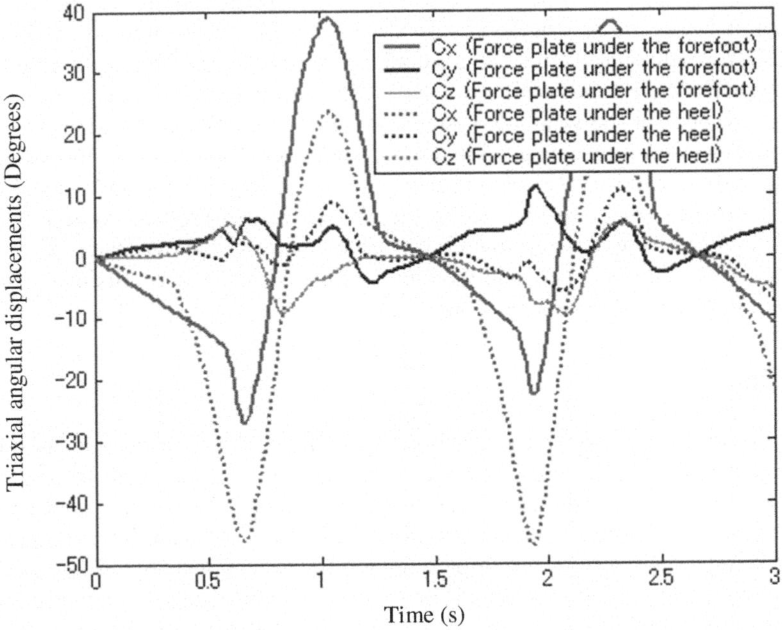

(b)

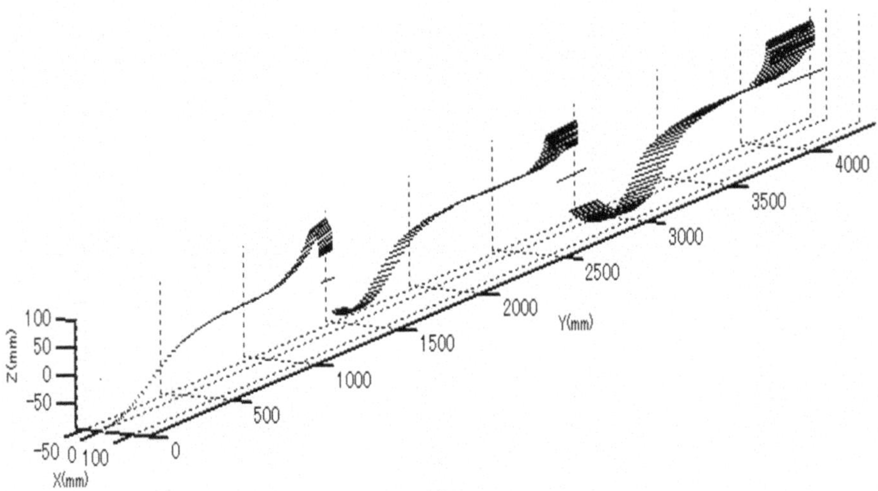

(c)

Fig. 6. (*continued*)

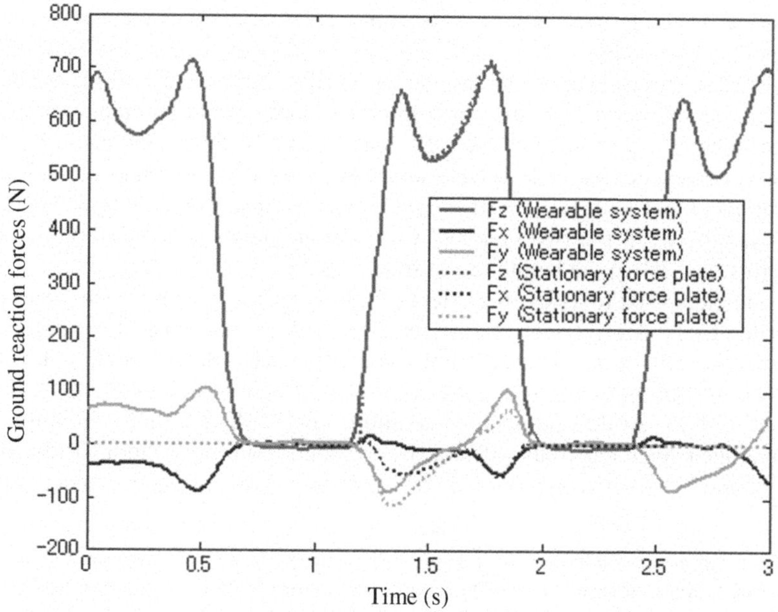

(a)

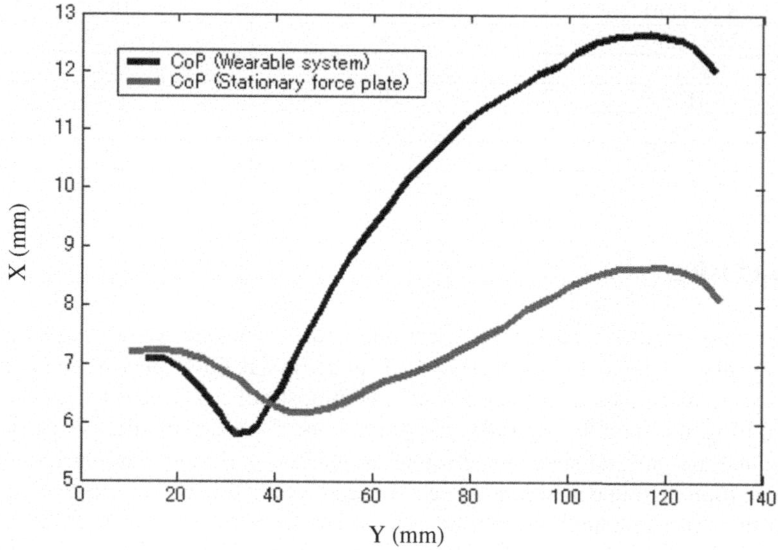

(b)

Fig. 7. Comparison results of the triaxial GRF measurement (a), CoP measurements of the two systems (b) representative walking trial. The results show good correspondence between the measurements of wearable system and the reference devices, which was examined by RMS (Root Mean Square) and standard deviation (SD) for the ten walking trials. The RMS difference (RMS±SD) of the two transverse components (x- and y- axes) and the vertical component (z-axis) of the GRF was 4.3±0.9N, 6.0±1.3N, and 12.1±1.1N respectively, corresponding to 5.1±1.1% and 6.5±1% of the maximum of each transverse component and to 1.3±0.2% of the maximum vertical component of GRF. The RMS distance between the two systems' CoP measurements was 3.2±0.8mm, corresponding to 1.2±0.3% of the length of the shoe.

The parameters used to assess the effect of the wearable system on gait were averaged over ten trials. An overview is presented in Table 2, and none of the parameters for the two shoe conditions showed significant differences. Stride length (SL), stride width (SW), maximum lateral foot excursion (ME), single stance time (SST), double stance time (DST), stride time (ST), maximum GRF (MaxF), and minimum GRF (MinF) averaged over ten trials differed by 12.2mm, 6.9mm, 2.8mm, 0.15s, 0.02s, 0.04s, 1.7N and 5.3N between a normal shoe and the wearable system respectively.

Table 2. Gait parameters including stride length (SL), stride width (SW), maximum lateral foot excursion (ME), single stance time (SST), double stance time (DST), stride time (ST), maximum GRF (MaxF) and minimum GRF (MinF) averaged over ten walking trials

	Normal shoe	The wearable system
SL (mm)	1441.3	1429.1
SW (mm)	83.1	100.0
ME (mm)	27.4	30.2
SST (s)	0.83	0.98
DST (s)	0.20	0.22
ST (s)	1.27	1.31
MaxF (N)	700.9	699.2
MinF (N)	554.3	549.0

7 Conclusion

By integrating small triaxial force sensors and inertial sensors, we developed a wearable force plate system for measuring CoP and triaxial GRF in a number of non-laboratory environments, and natural gait was almost never affected by the wearable system during the ambulatory GRF measurements. Significant differences between instrumented and normal shoes were found in maximum ground reaction force [7], in which maximum ground reaction force averaged over all subjects differed by 56 N between their instrumented and normal shoes. However, as shown in Table 2, there were not significant differences between our wearable system and a normal shoe, because we adopted the small triaxial force sensors in development of the wearable force plate (size: 80×80×15mm^3) which allows natural or near-natural gait.

The force measurements by the wearable system in x- and y- axes demonstrated both amplitude and phase shifts from the reference measurements (Fig. 7), and the discrepancy in CoP trajectory is slightly larger than the results reported by Veltink [6]. The output signals of the force plate and motion capture data in the reference sensor system were filtered by a low-pass filter with cut-off frequency of 10Hz, which must lead to phase shifts in the measurements between the reference system and the wearable system. The most likely source of amplitude error in the triaxial GRF measurement was in the orientation estimate of the moving force plate system using triaxial inertial sensors which could only implement x- and y- axial angular displacements' re-calibration. In the future we will integrate a triaxial magnetic sensor [13] for estimating the heading angle (z-axial angular displacement) during gait, because the z-axial (vertical axis) cumulative error induced by the drift effect of the gyroscope sensor could be re-calibrated using the accelerometer in the wearable system. On the other hand, only six triaxial force sensors were used to design the prototype of the wearable system to measure CoP and triaxial GRF, and if we fix more triaxial force sensors beneath the shoe, the precision of the wearable system will be improved.

References

1. Stacoff, A., Quervain, I., Luder, G., List, R., Stussi, E.: Ground reaction forces on stairs. Part II: Knee implant patients versus normal. Gait and Posture 26(1), 48–58 (2007)
2. Stacoff, A., Diezi, C., Luder, G., Stussi, E., Kramers-De Quervain, I.: Ground reaction forces on stairs: Effects of stair inclination and age. Gait and Posture 21(1), 24–38 (2005)
3. Faivre, A., Dahan, M., Parratte, B., Monnier, G.: Instrumented shoes for pathological gait assessment. Mechanics Research Communications 31(5), 627–632 (2004)
4. Zhang, K., Sun, M., Lester, K., Pi-Sunyer, X., Boozer, N., Longman, W.: Assessment of human locomotion by using an insole measurement system and artificial neural networks. Journal of Biomechanics 38(11), 2276–2287 (2005)
5. Fong, P., Chan, Y., Hong, Y., Yung, H., Fung, Y., Chan, M.: Estimating the complete ground reaction forces with pressure insoles in walking. Journal of Biomechanics 41(11), 2597–2601 (2008)
6. Veltink, H., Liedtke, C., Droog, E., Kooij, H.: Ambulatory measurement of ground reaction forces. IEEE Trans. Neural Syst. Rehabil. Eng. 13(3), 423–527 (2005)
7. Liedtke, C., Fokkenrood, W., Menger, T., van der Kooij, H., Veltink, H.: Evaluation of instrumented shoes for ambulatory assessment of ground reaction forces. Gait and Posture 26(1), 39–47 (2007)
8. Chateau, H., Robin, D., Simonelli, T., Pacquet, L., Pourcelot, P., Falala, S., Denoix, M., Crevier-Denoix, N.: Design and validation of a dynamometric horseshoe for the measurement of three-dimensional ground reaction force on a moving horse. Journal of Biomechanics 42(3), 336–340 (2009)
9. Schepers, M., Koopman, M., Veltink, H.: Ambulatory assessment of ankle and foot dynamics. IEEE Trans. Biomed. Eng. 54(5), 895–900 (2007)
10. Winter, A.: Biomechanics of normal and pathological gait: implications for understanding human locomotor control. Journal of Motor Behavior 21(4), 337–355 (1989)
11. Parry, J.: Gait analysis normal and pathological function. Slack Incorporated, 149–158 (1992)
12. Bortz, E.: A new mathematical formulation for strapdown inertial navigation. IEEE Trans. Aero. Ele. 7(1), 61–66 (1970)
13. Zhu, R., Zhou, Z.: A Small Low-Cost Hybrid Orientation System and Its Error Analysis. IEEE Sensors Journal 9(3), 223–230 (2009)

Optical Ranging in Endoscopy: Towards Quantitative Imaging

Agnese Lucesoli[1,2], Luigino Criante[2,3], Andrea Di Donato[1], Francesco Vita[2,3], Francesco Simoni[2,3], and Tullio Rozzi[1]

[1] Università Politecnica delle Marche, Department of Biomedical, Electronic and Telecommunication Engineering, Ancona, Italy
[2] HEOS s.r.l. Ancona, Italy
www.heosphotonics.com
[3] Università Politecnica delle Marche, Department of Physics and Materials Engineering and CNISM-MATEC, Ancona, Italy

Abstract. Nowadays endoscopic analysis is limited to a direct and qualitative view of internal anatomy. On the other hand, the measurement of the actual size of anatomical objects could be a powerful instrument both in research and in clinical survey. For instance, an important application could be monitoring lesion size, both during diagnosis and in follow-up. The foremost obstacle to quantitative imaging is the incapability of measuring the distance between the endoscopic probe and the anatomical object under examination, since the dimension of the object in the image depends on that distance. This problem has not been solved yet in a satisfactory way.

In this Chapter we describe our work to address this problem by means of an optical measurement of the distance between the endoscope distal tip and the anatomical wall. We make use of Fiber Optics Low Coherence Interferometry to realize an absolute distance sensor compatible with endoscope technology. The result is a system integrating a clinical endoscope and an optical distance sensor, equipped with a software that allows an user to acquire an endoscopic image, select a region of interest, and obtain its quantitative measure.

Keywords: quantitative endoscopy, Fiber Optics Low Coherence Interferometry, optical sensors, medical imaging.

1 Introduction

In medicine, endoscopy is the gold standard technique to investigate hollow organs, such as the upper part of the gastrointestinal tract (gastroscopy), the colon (colonscopy) and the airways (bronchoscopy). Other cavities frequently inspected by endoscopy are the urinary tract and the female reproductive system. Through a small incision, endoscopes allow to observe normally closed body cavities as well, such as the abdominal or pelvic cavity, the interior of a joint or blood vessels for cardiologic applications [1],[2].

The present technology provides substantially two kinds of endoscope. In the traditional one the imaging system is a bundle of optical fibers accurately aligned so that the geometrical order of the fibers in one end is identical to the order in the other end

S.C. Mukhopadhyay and A. Lay-Ekuakille (Eds.): Adv. in Biomedical Sensing, LNEE 55, pp. 74–92.
springerlink.com

(coherent fiber bundle or fiberscope). Then, each fiber represents a "pixel" of the whole image. Superior image quality can be obtained by using solid-state charge-coupled-device (CCD) cameras at the endoscope tip, but such endoscopes are thicker than current fiberscopes and cannot be used for microendoscopy purposes. Both the imaging systems use a wide angle lens at the distal tip to enlarge the field of view, and endoscopic examination consists in a direct and qualitative observation of anatomical objects by means of an eyepiece or by a monitor of a workstation: no post-processing is usually performed. Clinical endoscopes for diagnostics are generally provided with one or more operative channels, i.e. hollow channels parallel to the imaging system used to insert forceps for biopsies, surgical laser, gases, etc. A cross section of an endoscope is shown in Fig. 1.

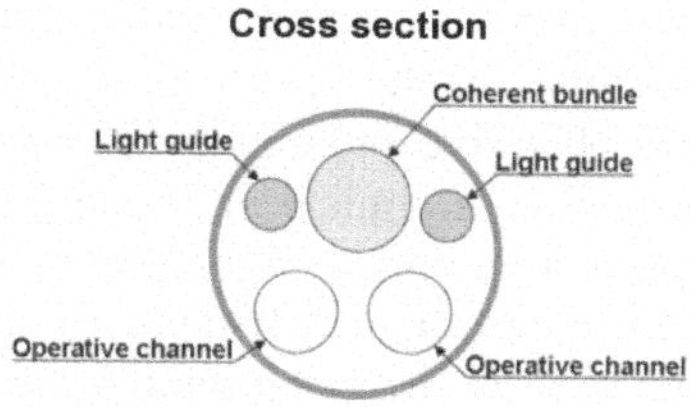

Fig. 1. Cross section of a typical endoscope, that includes the imaging channel (either fiber bundle or CCD device), two channels providing white light illumination of the field of view, and two operative channels

On the other hand the possibility of performing quantitative measurements on endoscopic images could be a powerful instrument both in research and in clinical investigation. For instance, evaluating the actual size of lesions could help clinicians during both diagnosis and follow-up. A further application could be to bronchoscopy, where the capacity for accurate measurement of the airway lumen is an important, but yet to be realized, goal. In fact, as the role of flexible endoscopes continues to expand in clinical and research work, getting accurate real time measurements of airway cross-sectional areas and lesion size would be a major improvement in the diagnosis and treatment of diseases of the respiratory system [3],[4].

A range of techniques have been developed to measure organ and lesion dimensions, such as X-Ray Computed Tomography, Magnetic Resonance Imaging or ultrasounds. However, endoscopy is still the technique of choice for hollow organ investigation because of its capability to provide direct images of lesions and, at the same time, dynamic observation of organ functionality. Nevertheless, until now no system has been successfully employed for the endoscopic quantitative measurement of anatomical dimensions. In fact, by means of current flexible endoscopes it is impossible to measure the actual size of the anatomical objects imaged, just owing to the presence of the above mentioned wide-angle lens that magnifies details depending on their (unknown and variable during the examination) distance from the endoscope tip. The major problem is the lack of calibration data and of reference points to make objective comparisons. A further uncertainty factor, that makes more difficult the measurement of areas, is the peripheral distortion (barrel distortion) induced in the images by the lens. Several methods, based on mathematical models, have been proposed for the distortion correction but they do not solve the calibration problem.

In clinical practice empirical methods are used to obtain an approximate measure of lesion size, but unfortunately they are inaccurate, hardly repeatable and often invasive. During the last years many approaches to this problem have been proposed. If possible, an object of known size is put near the organ of interest for image calibration, but such solution is not always applicable in clinical practice [3],[5]. Other authors propose the projection of a pattern on the field of view, so that the dimension of anatomical objects can be computed by processing the pattern image [4],[6].

A different approach is based on the measurement of the distance between endoscope tip and anatomical walls. At present, this measurement may be performed either by invasive methods, such as the insertion of a marker in the endoscope tip [7], or by indirect ways, i.e. by manually measuring the relative displacement between the endoscope tip and the anatomical wall [8]. In conclusion, with current techniques the evaluation of the dimensions is given by qualitative and subjective approximations.

The approach described in this Chapter is aimed at measuring the unknown distance tip-anatomical wall in an objective and repeatable way, by means of an absolute distance optical sensor integrated in a clinical endoscope. The basic principle of an optical ranging device is to project an optical signal onto an object and to process the reflected or scattered signal to determine the distance. In endoscopic application the probe signal should be projected from the distal endoscope tip, then a fiber optic implementation of the sensor is needed. An easy and effective solution could be the insertion of the fiber probe into an operative channel of the endoscope. Furthermore, we need to select, among several techniques of optical ranging, the most appropriate for the working distances considered (1-50 mm) and for the unstable conditions in the human body cavities, that may produce noise in the optical signal. Optical distance measurement methods can be technically classified into three categories: triangulation, time-of-flight (TOF) and interferometry [9]. Optical triangulation technique exploits the reflection of coherent light incident at a well-defined angle, and it is based on the geometric analysis of the incident and reflected beams. However, this method does not fit endoscopic probes, where the room for an optical sensor is very small. In a different way, TOF-based systems measure the time delay between a light pulse emission and the return of its echo resulting from the reflectance off an object, but they are inadequate in applications involving distances shorter than 1 m. Then interferometric techniques could be the right solution, since they are suitable both for the range of distances involved and for fiber optic implementation. However, the sensitivity to noise that characterizes high coherence interferometry is a strong drawback for the weak signals reflected by biological tissues. The technique that seems to be the most promising for our needs is Low Coherence Interferometry (LCI). Its use in optical ranging is well-known [10] and it is already applied in medical field to Optical Coherence Tomography (OCT) [11].

In the following we describe the design and implementation of a Fiber Optic LCI sensor for absolute distance measurement and its integration with a clinical endoscope. Calibration procedure and tests on ex-vivo biological samples are also reported.

2 Low Coherence Interferometry

Low Coherence Interferometry (LCI), and its fiber optic implementation (Fiber Optic Low Coherence Interferometry - FOLCI) attracted much interest in recent years and became an important technique for the absolute remote measurement of quasi-static parameters, such as distance, displacement, temperature, pressure, strain, refractive index and dispersion [12],[13]. One of the most distinguishing features of FOLCI, when compared to the conventional intensity-based fiber optic sensors, is that measurement accuracy is virtually insensitive to optical power fluctuations.

LCI is a phenomenon well described by the theory of optical coherence [14]. Light is deterministic or "coherent" if the dependence of the wavefunction on time and position is perfectly periodic and predictable (e.g. laser source). If, on the contrary, the source does not emit deterministic light, or such determinism is lost during propagation, the light is called random or incoherent, and the dependence of the wavefunction on time and position is not totally predictable, and can be described by means of statistical methods.

An arbitrary optical wave is described by a wavefunction $u(\mathbf{r},t) = \mathrm{Re}\{U(\mathbf{r},t)\}$, where $U(\mathbf{r},t)$ is the complex wavefunction. For incoherent light, both functions $u(\mathbf{r},t)$ and $U(\mathbf{r},t)$ are random and are characterized by a number of statistical averages. The average intensity is then defined as

$$I(\mathbf{r},t) = \left\langle |U(\mathbf{r},t)|^2 \right\rangle \tag{1}$$

where the average is calculated over several trials of the wave produced under the same conditions, and the quantity $|U(\mathbf{r},t)|^2$ is called random or instantaneous intensity. Let us consider the fluctuations of stationary light (i.e. $I(\mathbf{r},t) = I(\mathbf{r})$) at a fixed position $\mathbf{r}$ as a function of time. For the sake of brevity, we drop the $\mathbf{r}$ dependence, so that $U(\mathbf{r},t) = U(t)$ and $I(\mathbf{r}) = I$. The random fluctuations of $U(t)$ are characterized by a time scale representing the "memory" of the random function. Fluctuations at points separated by a time interval longer than the memory time are independent, so that the process "forgets" itself. A statistical quantity describing the random fluctuations of $U(t)$ is the autocorrelation function. In case of optical wavefunctions, the autocorrelation is also called *temporal coherence function*, and is defined as a function of the time delay τ:

$$G(\tau) = \langle U^*(t)U(t+\tau) \rangle \tag{2}$$

It can be easily shown that the intensity becomes $I = G(0)$. The temporal coherence function carries information about both the intensity I and the degree of correlation, or *coherence*, of the stationary light. The normalized autocorrelation function, or *complex degree of temporal coherence* provides a measure of coherence independently of the intensity:

$$g(\tau) = \frac{G(\tau)}{G(0)} \tag{3}$$

Its absolute value $|g(\tau)|$ is a measure of the degree of correlation between $U(t)$ and $U(t + \tau)$, and it cannot exceed unity:

$$0 \le |g(\tau)| \le 1 \tag{4}$$

Usually, $|g(\tau)|$ drops from its largest value $|g(0)| = 1$ as τ increases and the fluctuations become uncorrelated. A measure of the memory time of the fluctuations is given by the *coherence time* τ_c. In general, τ_c is the width of $|g(\tau)|$. The most common definition of τ_c is the power-equivalent width

$$\tau_c = \int_{-\infty}^{+\infty} |g(\tau)|^2 \, d\tau \tag{5}$$

Light for which the coherence time τ_c is much longer than the differences of the propagation time encountered in the optical system of interest is effectively coherent. Thus light is effectively coherent if the distance $c\tau_c$ is much greater than all Optical Path Differences (OPD) encountered. On the contrary, if the OPDs of the system are greater than $c\tau_c$ the light behaves incoherently. The distance $l_c = c\tau_c$ is called *coherence length*.

The *intensity spectral density* of an incoherent light is defined as a function of the frequency ν:

$$S(\nu) = \lim_{T\to\infty} \frac{1}{T} \left\langle |V_T(\nu)|^2 \right\rangle \tag{6}$$

where

$$V_T(\nu) = \int_{-T/2}^{T/2} U(t) \exp(-j2\pi\nu t) dt \tag{7}$$

is the Fourier transform of $U(t)$ observed over a time window T. The autocorrelation function $G(\tau)$, defined by Eq. (2), and the spectral density can be shown to form a Fourier pair by the *Wiener-Kintchin theorem*

$$S(\nu) = \int_{-\infty}^{\infty} G(\tau) \exp(-j2\pi\nu\tau) d\tau \tag{8}$$

The Wiener-Khinchin theorem shows that the coherence of a source strictly depends on its spectrum: the broader the spectral width $\Delta\nu$ of the source, the narrower its degree of coherence. The spectral width $\Delta\nu$ is usually defined as the FWHM of $S(\nu)$. An alternative but useful definition is

$$\Delta\nu_c = \frac{\left(\int_0^{\infty} S(\nu) d\nu \right)^2}{\int_0^{\infty} S^2(\nu) d\nu} \tag{9}$$

that allows relating the spectral width and the coherence time as $\Delta\nu_c = 1/\tau_c$. The coherence length can be related to the spectral width as well:

$$l_c = c / \Delta\nu_c = c / \left(\frac{\nu_0}{\lambda_0} \Delta\lambda \right) = \frac{\lambda_0^2}{\Delta\lambda} \tag{10}$$

where ν_0 and λ_0 are the central frequency and the central wavelength of the light source. This important relation directly connects the coherence length of a source to its spectral characteristics.

Let us now consider a partially coherent wave $U(t)$ with intensity I_0 and complex degree of temporal coherence $g(\tau) = \langle U^*(t) U(t+\tau) \rangle / I_0$. If $U(t)$ is added to a replica of itself delayed by τ, $U(t+\tau)$, it can be easily shown that the intensity of the superposition is given by

$$I = \left\langle |U(t) + U(t+\tau)|^2 \right\rangle = 2I_0 \left(1 + |g(\tau)| \cos\varphi \right) \tag{11}$$

where $\varphi = \arg\{g(\tau)\}$. There are two relevant borderline cases:

- for a completely coherent wave $g(\tau) = \exp(j\varphi)$ and $|g(\tau)| = 1$, thus $I = 2I_0(1 + \cos\varphi)$ is a cosinusoidal function with constant amplitude for every value of φ;
- for a completely incoherent wave $g(\tau) = 0$, $I = 2I_0$ and there is no interference.

In the general case, the intensity I versus the phase φ assumes the form of a sinusoidal pattern modulated by the shape of $|g(\tau)|$. The strength of the interference is measured by the *visibility* K:

$$K = \frac{I_{\max} - I_{\min}}{I_{\max} + I_{\min}} \tag{12}$$

that in our case becomes $K = |g(\tau)|$.

In a practical implementation, a wave may be added up to a time-delayed replica of itself by using a beamsplitter (or a directional coupler in case of fiber optic implementation) to generate two identical waves, one of which is made to travel a longer optical path before the two waves are recombined using another - or the same - beamsplitter/coupler. These possibilities correspond to the use of a Mach-Zender or a Michelson interferometer, respectively.

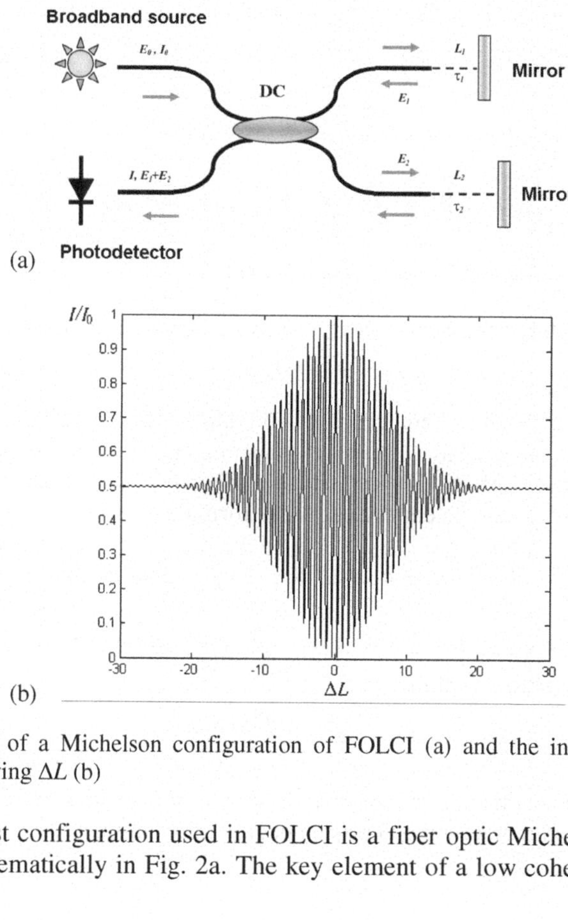

Fig. 2. Scheme of a Michelson configuration of FOLCI (a) and the interferometric signal obtained by varying ΔL (b)

The simplest configuration used in FOLCI is a fiber optic Michelson interferometer, shown schematically in Fig. 2a. The key element of a low coherence system is a broadband light source (i.e. a LED or a Superluminescent Light Emitting Diode or SLED), whose coherence length is much less than the one of a typical laser. A stationary optical signal with amplitude U_0 and intensity I_0 is launched from the source into one arm of a bidirectional fiber coupler and is split in two signals of amplitude U_1 and U_2, travelling in two identical optical fiber arms. The signals are reflected by two reflectors positioned at distances L_1 and L_2 from the arm end (their reflectance is α and β respectively), and they are then recombined by the coupler. At the time t, the two amplitudes U_1 and U_2 at the photodetector are:

$$\begin{cases} U_1(t) = \alpha k_c k_t U_0(t+\tau_1)\exp\left[j2\pi\nu(t+\tau_1)\right] \\ U_2(t) = \beta k_c k_t U_0(t+\tau_2)\exp\left[j2\pi\nu(t+\tau_2)\right] \end{cases} \tag{13}$$

where τ_1 and τ_2 are the time delays suffered by the two beams during propagation through the interferometer, while k_t and k_c are the amplitude transmission and coupling coefficients of the directional coupler. For a 3dB directional coupler, that splits I_0 in two signals of equal intensity $I_0/2$, we have

$$\begin{cases} k_t = 1/\sqrt{2} \\ k_c = j/\sqrt{2} \end{cases} \tag{14}$$

In this case the intensity I recorded by the photodetector is given by

$$\begin{aligned} I &= \left\langle \left| U_1(t) + U_2(t) \right|^2 \right\rangle \\ &= \left\langle \left| \frac{j}{2}\left[\alpha U_0(t+\tau_1) e^{j2\pi\nu(t+\tau_1)} + \beta U_0(t+\tau_2) e^{j2\pi\nu(t+\tau_2)} \right] \right|^2 \right\rangle \\ &= \frac{1}{4}\left[\alpha^2 I_0 + \beta^2 I_0 + \alpha\beta \langle U_0(t) * U_0(t+\tau) \rangle \cos(2\pi\nu\tau) \right] \\ &= \frac{I_0}{4}\left[\alpha^2 + \beta^2 + \alpha\beta |g(\tau)| \cos(2\pi\nu\tau) \right] \end{aligned} \tag{15}$$

where τ is the time delay between the two beams as shown in the following relation

$$\tau = \tau_2 - \tau_1 = 2(L_2 - L_1)/c = 2\Delta L/c \tag{16}$$

and $|g(\tau)|$ is the degree of coherence of the broadband source. The sources usually employed in interferometry have Gaussian spectrum, characterized by a central wavelength λ_0 and a bandwidth $\Delta\lambda$. In this case the degree of coherence is Gaussian as well, thus the intensity may be expressed by the formula

$$I = \frac{I_0}{4}\left\{ \alpha^2 + \beta^2 + \alpha\beta \exp\left[-\left(\frac{4\Delta L}{l_c} \right)^2 \right] \cos(2k_0\Delta L) \right\} \tag{17}$$

where $l_c = \lambda_0^2/\Delta\lambda$ (see Eq. 10) and k_0 is the light beam propagation constant in air of the light beam. The fringe visibility is:

$$K = \frac{2\alpha\beta |g(\Delta L)|}{\alpha^2 + \beta^2} \tag{18}$$

A plot of the intensity I as a function of ΔL is shown in Fig. 2(b). We can observe the sinusoidal wave modulated by the Gaussian degree of coherence.

The principle described above is at the basis of all the LCI techniques. While scanning one of the distances L_1 or L_2, interference occurs at the photodetector as long as the Optical Path Difference (OPD) between the two arms (in our case OPD = $2\Delta L$) is shorter than the coherence length l_c. The identification of the central fringe position, corresponding to zero OPD, is normally used to determine the unknown optical path of one arm, called *probe arm*, by knowing the optical path of the other one, called *reference arm*.

3 Absolute Distance Measurement

The principle of FOLCI can be applied to build an optical sensor for absolute distance measurement to be combined with an endoscope. In its endoscopic implementation, the probe arm will be inserted in one of the operative channels, so that the endoscope tip and the probe arm tip coincide, while the reference arm will be moved to scan a range of typical working distances for an endoscopic inspections. Thanks to the low coherence of the source, interference will occur in the detected signal only for small difference of the the optical paths of the interferometer arms allowing high resolution in the distance measurement, and the central fringe position will correspond to the real distance between the endoscope tip and the anatomical wall.

We have applied the principle of FOLCI to build an optical sensor for absolute distance measurement integrated in a commercial endoscope. In its endoscopic implementation, the probe arm is inserted in one of the operative channels, so that the endoscope tip and the probe arm tip coincide, while the reference arm is moved to scan a range of typical working distances for an endoscopic examination.

In this application of FOLCI it is worth noting that biological tissues do not have as much reflectance as a mirror. On the contrary, most of the light is partly absorbed and partly scattered, while back-reflection is just a small percent of the incident light [15]. Scattering depends on the ultrastructure of tissues (membranes, collagen fibers, nuclei, cells, etc.), as photons are strongly scattered by those structures whose size matches the photon wavelength. The analysis of the optical properties of biological tissues is beyond the purposes of this Chapter, however we must take into account that the signal collected by the probe arm is very low, reducing the fringe visibility of the interferogram.

As explained in the previous section, owing to the limited coherence length of the source, optical interference is observed only when the Optical Path Difference between the probe arm and the reference arm is smaller than the coherence length (OPD $< l_c$). The interferogram is modulated by the envelope represented by the degree of coherence $|g|$ of the source, and the position of the central fringe, obtained for $\Delta L = 0$, indicates the real distance between the endoscope tip and the anatomical wall. For a source with a Gaussian spectrum centred at a wavelength λ_0 and with a spectral width $\Delta\lambda$ the FWHM of the envelope function is [16]:

$$\frac{2\ln(2)\lambda_0^2}{\pi\Delta\lambda} \approx 0.44 l_c \tag{19}$$

According to theory, the position of the probe reflector can be determined very precisely by locating the central fringe of the interferogram, providing submicron resolution [17]. On the other hand, the correct identification of the central fringe in the interference signal is a quite critical task. This occurs because the most suitable light sources currently used, such as LEDs, superluminescent diodes (SLEDs) and multimode laser diodes (MLDs), have a relatively large coherence length. Therefore, the intensity difference between the central fringe and its adjacent fringes may be very small and buried by noise. A high signal-to-noise ratio (SNR> 50 dB), is required to identify the central fringe without ambiguity, but this is difficult to realize in the practice. The minimum SNR to identify the central fringe can be easily calculated to be:

$$SNR_{\min} = -20\log(\Delta I) \tag{20}$$

where ΔI is the normalized intensity difference between the central fringe and the adjacent ones. Therefore, in order to lower $\mathrm{SNR}_{\min}$ it is necessary to find a way to enhance ΔI. To this aim, a wide range of new signal processing schemes have been introduced, that have considerably enhanced the performance of the sensors based upon FOLCI.

A class of solutions to the problem of the central fringe identification is based on the synthesis of new sources with lower coherence length. This is achieved by combination of two or more broadband sources [18]. This methods can decrease the requirements of $\mathrm{SNR}_{\min}$ of about 32-38 dB. On the other hand, methods requiring two or more sources complicate the setup and the signal acquisition and increase the cost of the sensor.

Different techniques using a single source are aimed at increasing the relative value of the central fringe and suppressing the peak value of the other fringes, lowering in this way $\mathrm{SNR}_{\min}$.

Multi-squaring is an example of this kind of methods [19]. It simply consists in a series of power raising of the interference signal. According to Eq. 17 (for the sake of simplicity we consider $\alpha = \beta = 1$), the n-th power of the normalized intensity is

$$I^n(\Delta L) = \frac{1}{2^n}\left[1 + e^{-\left(\frac{4\Delta L}{l_c}\right)^2}\cos(2\pi\Delta L)\right]^n \tag{21}$$

The zero-order fringe is obtained by imposing $\Delta L = 0$ and its intensity is $I(0) = I_0 = I_0^n = 1$, the subscript 0 indicating the fringe order. The nearest fringe to the central one has its maximum at $\Delta L = \lambda/2$, and its intensity raised to the n-th power is

$$I_1^n = \frac{1}{2^n}\left[1 + e^{-(2\lambda/l_c)^2}\right]^n \tag{22}$$

By defining $\Delta I_{01}^{(n)} = I_0^n - I_1^n = 1 - I_1^n$, it can be demonstrated that $\Delta I_{01}^{(n)} > \Delta I_{01}^{(n-1)}$. Therefore, the difference between the zero-order and the first-order fringe is enhanced and $\mathrm{SNR}_{\min}$ is reduced as n increases.

An analogous method to identify and enhance the central fringe is based on a *fringe transform* of the interference pattern [20]. It considers the *ac* component of the normalized signal

$$I_{ac}(\Delta L) = \exp\left[-(4\Delta L / l_c)^2\right]\cos(4\pi\Delta L / \lambda) \tag{23}$$

Its value at the position $\Delta L = \lambda/2$, corresponding to the first-order fringe, is

$$I_{ac}(\lambda / 2) = \exp\left[-(2\lambda / l_c)^2\right] \tag{24}$$

It can be easily shown that the square of $I_{ac}(\Delta L)$ is

$$I_{ac}^2(\Delta L) = 1/2\exp\left[-2(4\Delta L/l_c)^2\right]\cos(1+8\pi\Delta L/\lambda) \tag{25}$$

whose cosinusoidal component has a doubled frequency with respect to $I_{ac}(\Delta L)$. It means that the first order fringe of I_{ac}^2 corresponds to the first negative peak of I_{ac}. Therefore, if we add I_{ac} and I_{ac}^2 we can suppress the negative peaks of I_{ac}, while at the position $\Delta L = \lambda/2$ the sum is

$$\begin{aligned}\frac{1}{2}\left[I_{ac}(\lambda/2)+I_{ac}^2(\lambda/2)\right] &= \frac{1}{2}\left\{\exp\left[-(2\lambda/l_c)^2\right]+\exp\left[-2(2\lambda/l_c)^2\right]\right\}\\ &< \exp\left[-(2\lambda/l_c)^2\right] = I_{ac}(\lambda/2)\end{aligned} \tag{26}$$

resulting in the reduction of the first order fringe with respect to the central one. Based on this principle, the *fringe transform* of the signal is defined as

$$I_n(\Delta L) = \left\{\frac{1}{2}\left[I_{ac}(\Delta L)+I_{ac}^2(\Delta L)\right]\right\}^2 \tag{27}$$

and the following relation holds

$$\left\{\frac{1}{2}\left[I_{ac}(\lambda/2)+I_{ac}^2(\lambda/2)\right]\right\}^n < \left\{\frac{1}{2}\left[I_{ac}(\lambda/2)+I_{ac}^2(\lambda/2)\right]\right\}^{n-1} < \ldots < I_{ac}(\lambda/2) \tag{28}$$

Therefore, the amplitude of the first order fringe can be reduced by increasing the transform order n.

The multi-squaring and the fringe transform methods are based on the identification of the central fringe of the interference signal. Then the measurement of zero OPD position has a resolution equal to the width of the central fringe, that in case of Michelson configuration is $\lambda/4$. However, in presence of very low SNR, the central fringe may not to be the highest and, in the worst cases, the signal may be completely hidden by noise and false maxima may be present outside the interferogram. In these cases, algorithms based on the central fringe detection are likely to fail.

An opposite approach is based on the computation of the signal envelope. Because this approach exploits the whole interferogram it is intrinsically less affected by noise, even if characterized by a worse resolution. However this latter limitation is not relevant for endoscopic applications, where a submicrometric resolution is not required.

One of the most common methods to compute the signal envelope is based on the Hilbert transform [21]. The Hilbert transform of a real-valued function $f(t)$ is defined by

$$h(t) = \frac{1}{\pi}\int_{-\infty}^{\infty}\frac{f(t')dt'}{t'-t} = -\frac{1}{\pi t} * f(t) \tag{29}$$

where the symbol '*' denotes the convolution operator. We can associate with $f(t)$ a complex function

$$f(t) - jh(t) \tag{30}$$

whose real part is $f(t)$ itself and whose imaginary part is its Hilbert transform. Such complex function is called analytic signal. The envelope of a function $f(t)$, that we denote as $E_f(t)$, is defined as the amplitude of its analytic signal

$$E_f(t) = |f(t) - jh(t)| = \sqrt{f^2(t) + h^2(t)} \tag{31}$$

Fig. 3 shows an interference signal and its envelope computed by Hilbert transform, corresponding to the degree of coherence $|g(\Delta L)|$.

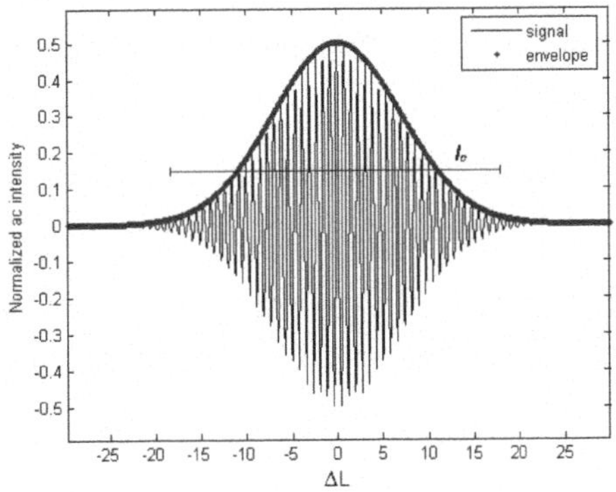

Fig. 3. Normalized *ac* component of an interferogram and its envelope computed by Hilbert transform

4 Experimental Tests

A sketch of the whole system, including both the distance sensor and the endoscopic imaging system, is shown in Fig. 4.

Let us first describe the optical distance sensor [22]. Low Coherence Interferometry is performed by means of a Superluminescent Light Emitting Diode (λ_0 = 1557.18 nm, $\Delta\lambda$ = 61.3 nm, maximum power 190 μW). From Eq. 10 we calculate the coherence length of the source

$$l_c = \lambda^2 / \Delta\lambda \approx 40\mu m \tag{32}$$

An optical signal is sent from the source to a fiber optic interferometer, consisting in a 3dB broad band (60 nm) directional coupler. An optical isolator between the source and the coupler prevents the backward signal of the interferometer affecting the light emission. One arm of the interferometer, that is inserted in the operative channel of the endoscope, is the probe (or sample) arm, while the other arm, used as reference, is placed externally. A focusing GRIN lens (focal length ≈ 10mm, focal spot diameter ≈ 50μm) was placed at the tip of the sample arm to reduce the focal spot. In this way scattering from the sample are minimized and the light collected by the probe arm is maximized. In the reference arm, where light is reflected by a mirror (reflection coefficient R ≈ 100%), a collimating GRIN lens is used. The reference arm optical path is scanned moving a reference mirror by an automatic stepping motor (travel range 25 mm; full-step resolution 5 μm; maximum speed 20 mm/s; accuracy ±100 μm over 50 mm). Moreover, in all the measurements we set the minimum microstep resolution to 20 nm.

The interference signal is detected by a photodetector connected to a Data Acquisition (DAQ) PC-Card. In this way, we measure an interference intensity *vs* the reference mirror position. As shown in the previous Section, if the fiber branches used in the two arms have the same length, the real distance L_1 of the sample is equal to the mirror position L_2 in correspondence of the maximum of the interference pattern. The moving and acquiring process is completely automated using a PC system and a Labview software is used to acquire and process the experimental data.

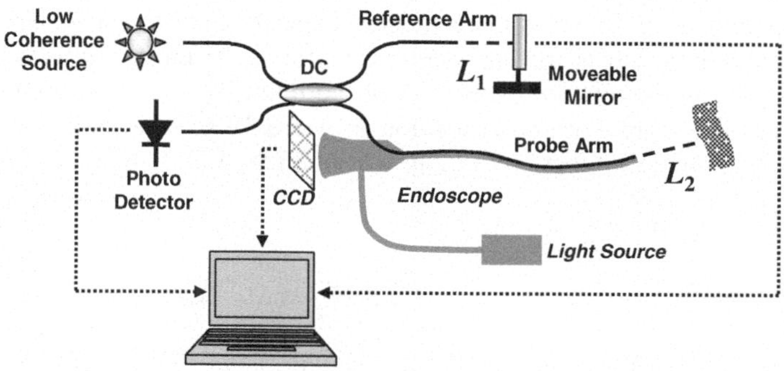

Fig. 4. Scheme of the system combining the optical distance sensor and a clinical endoscope

First tests were performed with a mirror at the probe arm, in order to calibrate the system with the maximum reflected power. The output of the software, shown in Fig. 5, is a plot of the intensity recorded by the photodetector as a function of the absolute distance of the reference mirror L_2. We can recognize the interference pattern shown in Fig. 2. The small asymmetry in the measured interferogram is only due to a small mismatch of the lengths of the optical fibers used in the two arms. In fact, few millimetres of fiber path difference correspond to a strong difference of chromatic dispersion in the two arms, that is enough to induce the pattern interference asymmetry [12].

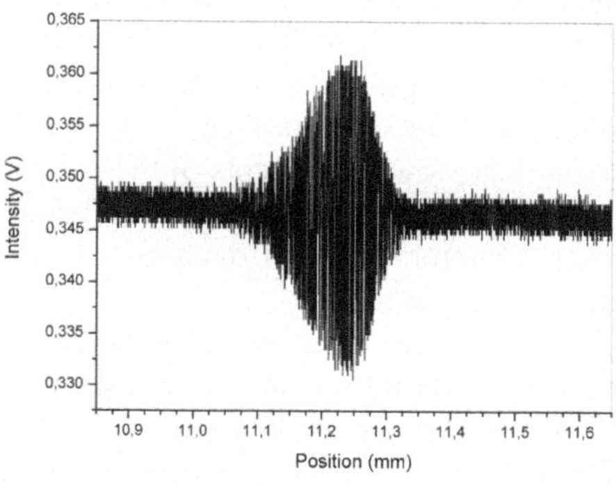

Fig. 5. A typical interferogram obtained with our system (reproduced from [22])

To obtain an absolute distance measurement, it is necessary to know exactly the position L_2 during the acquisition. To this aim, we move the stepping motor at the nearest position in front to the reference arm tip. Such distance, that we call L_0, is measured by a caliper with a sensitivity of 0.05 mm. During the scanning process the reference position is simply computed from the motor speed v_m and the time elapsed t: $L_2 = L_0 + n_f d_f + v_m t$ where n_f is the effective refractive index of the mode propagating into the fiber and d_f is the length difference between the fibers of the two arms. $L_0 + n_f d_f$ is the minimum probe distance that can be measured by the interferometer, while the maximum distance, limited by the motor travel range, is $L_0 + + n_f d_f + 25$ mm. An important parameter to be set is the acquisition speed of the motor v_m. As mentioned above, the maximum speed of our stepping motor is 20 mm/s. However, since the value of v_m affects the temporal spectrum of the intensity signal, it represents a constraint for the sample rate of the acquisition software. If we express the *ac* component of the interferogram as a function of time, we obtain the following relations

$$\Delta L(t) = L_2(t) - L_1 = L_0 + n_f d_f + v_m t - L_1 = \Delta L_0 + v_m t \tag{33}$$

$$I_{ac}(t) = \exp\left\{-4\left[\left(\Delta L_0 + v_m t\right)/l_c\right]^2\right\}\cos\left[4\pi\left(\Delta L_0 + v_m t\right)/\lambda_0\right] \tag{34}$$

By applying the Fourier transform properties, the frequency spectrum of $I_{ac}(t)$ is

$$F_{ac}(f) = \frac{\sqrt{\pi} l_c}{4 v_m} \exp(2\pi j \Delta L_0 f / v_m)\left\{\exp\left[-\frac{\pi^2 l_c^2}{4 v_m^2}\left(f - \frac{2 v_m}{\lambda_0}\right)^2\right] + \exp\left[-\frac{\pi^2 l_c^2}{4 v_m^2}\left(f + \frac{2 v_m}{\lambda_0}\right)^2\right]\right\} \tag{35}$$

Considering only the positive part of the spectrum, $F_{ac}(f)$ is represented by a Gaussian function centered at $2v_m/\lambda_0$ and whose FWHM is $\approx 0.9 v_m/l_c$. Then the maximum frequency of the signal can be approximated as $f_{max} \approx 2v_m/\lambda_0 + v_m/l_c$, that in our case becomes $f_{max} \approx 1.25\times10^6 v_m$. According to the Nyquist-Shannon sampling theorem [23], the sample rate f_s must be suitably selected so that

$$f_s \geq 2 f_{max} \approx 2.50\times10^6 v_m \tag{36}$$

The acquisition board used in these tests has a sample rate of 1.25 MHz, allowing a maximum speed $v_m \approx 500$ mm/s, much higher than the maximum speed allowed by our motor. On the other hand, during an endoscopic examination it would be useful a further increase of the scan speed (by changing the stepping motor), in order to reduce the scanning time and avoid that slow movements of the patient or the clinician affect the measurement. Reducing the acquisition time could also allow a real-time measurement of the endoscope-to-anatomical wall distance. However, attention must be paid to the correct signal sampling problem.

The signal acquired by the software described above is processed to obtain the distance measurement. In the first tests we compared three post processing algorithms in order to choose the most suitable for our application: 1. envelope computation by Hilbert transform; 2. fringe transform of order 32; 3. multi-squaring of order 32. Both the fringe transform and the multisquaring were stopped at the 32nd order, that is enough to obtain a good visibility of the central fringe. The graphical output of the signal processing software is shown in Fig. 6.

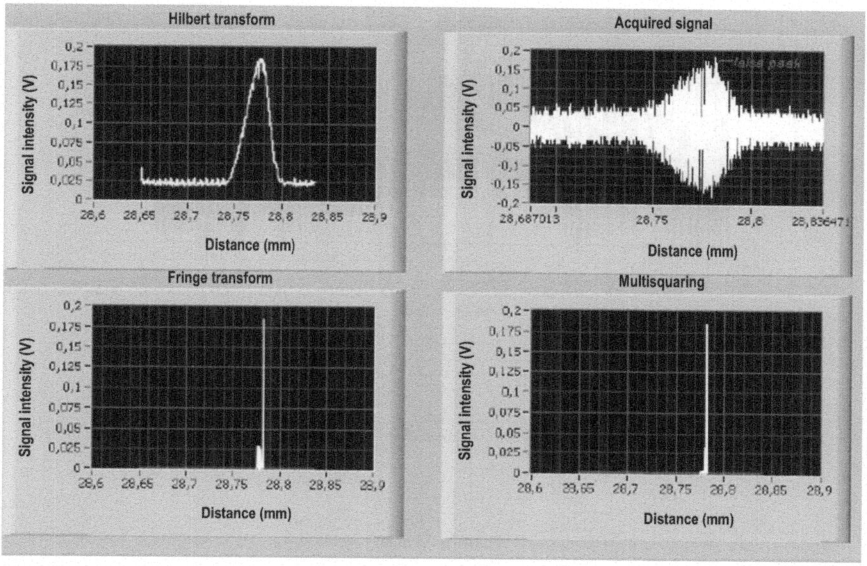

Fig. 6. Graphical interface of the processing software. The four plots represent the processed signal. Left-top: envelope by Hilbert transform. Right-top: the acquired signal. Left-bottom: fringe transform. Right-bottom: multisquaring.

First tests were performed with a mirror at the probe arm. The intensity reflected by each mirror and recoupled into the detection arm was about 40 μW, giving an interference pattern with good fringe visibility. We considered many distances of the probe mirror L_1 ranging from 5 mm to 50 mm, all measured by a calliper with sensitivity of 0.05mm. We repeated each measurement 10 times, with different motor speed, always obtaining a standard deviation of about 20 μm and an average value in agreement with the calliper measurement.

Once the distance sensor was calibrated as described previously, we tested it on biological samples. We used *ex-vivo* pig trachea, dipped in warm water (37°C) in order to simulate the real environmental conditions of humidity and temperature. As we have seen in the previous sections, a minimum SNR is necessary to identify the maximum of the interference pattern. Unfortunately, biological samples are characterized by a very low reflectivity of the anatomical walls. Using the lens above mentioned (focal length: 1 cm and a focal spot diameter < 100 μm), we reduced the scattering from the tissue, obtaining a good reflected signal (in a range from few nW to 0.2 μW in the best cases). Nevertheless, in the real conditions of an endoscopic examination the SNR of the interferogram may be very low, making it difficult for the acquisition software to detect the correct distance value. In these cases the peak detection algorithm based on the Hilbert transform is the most suitable for the correct detection, even in case of very weak probe signals, down to 2-3 nW. An example of measurement on biological sample, where the SNR is particularly low, is shown in Fig. 7. It is clearly visible how the Hilbert transform allows to get rid of both the high frequency noise and the slow average intensity fluctuation of the signal, providing a very fine reconstruction of the envelope function.

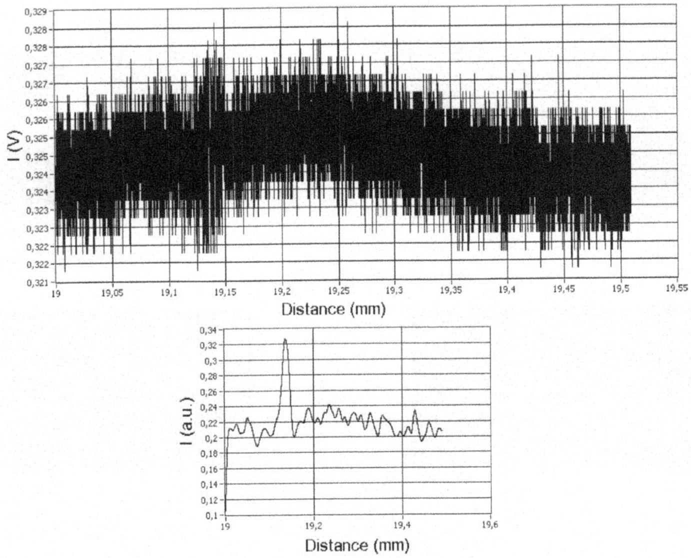

Fig. 7. Top: acquired signal as a function of the reference mirror position L_2. Bottom: signal envelope computed by Hilbert transform. The signal intensity reflected by the sample was about 3 nW in this case.

5 Measurement of Dimensions

Once the absolute distance sensor is realized and tested on biological samples, endoscopic image processing is possible. While the sensor is measuring the sample/tip distance, the image of the sample is captured simultaneously by the endoscopic vision system and saved by a dedicated software. The light source (a standard bulb lamp) of the endoscopic system is used to obtain the needed illumination of the sample. The size measurement of the anatomical object under investigation can be divided in two parts: first, the system determines the real distance L_1 from the endoscope tip to the anatomical wall and then it exploits this data to compute the actual dimension and area. By means of an interactive software, the user is able to acquire an endoscopic image, select a region of interest (ROI) and obtain its quantitative measure of area or length. The graphical output of the signal processing software is shown in Fig. 6.

Since the distance sensor can be inserted in any commercial endoscope - equipped with one or more operative channels – with any field of view and magnification system, calibration curves of the imaging system as function of the distance are necessary. By means of targets of known size we acquired the same image at different distances L_1. For each image, we selected the observed object and obtained by the software its length or/and area in pixels. From the known actual dimension of the target, we obtained a calibration curve as a function of the distance (see Fig. 8). A typical behaviour of the calibration curves is shown in Fig. 9 [24]. In this way we are able of providing the correlation between distance (here we indicate L_1 = d) and the

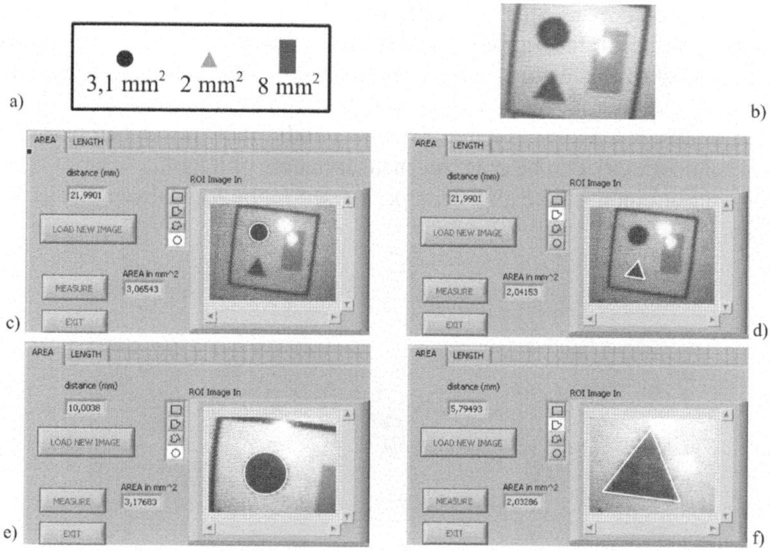

Fig. 8. Images acquired and processed with our system, a) known-size target, b) target imaged by the endoscopic imaging system, c-d) area measurement of the two different calibration target at d = 22 mm, e) area measurement of the circle target at d = 10 mm, f) area measurement of the triangle target at d = 5.8 mm (reproduced from [24])

relative dimension (y) corresponding to one pixel. Concerning the 1-dimensional measurement, the analytical model that best fits the experimental data is a linear one

$$y_{length} = A \cdot d + B \text{ [mm/pixel]} \tag{37}$$

while the analytical model best fitting the experimental data of the 2-dimensional measurement (areas) is

$$y_{area} = C \cdot d^{\beta} \text{ [mm}^2\text{/pixel]} \tag{38}$$

in which we expect β to be 2 (quadratic dependence). For our imaging system, we computed A = 0.0155 ± 0.0005 pixel^{-1}, B = - 0.0005 ± 0.0001 mm·pixel^{-1}, C = 10^{-4} mm^2·pixel^{-1} and β = 1.95 ± 0.05.

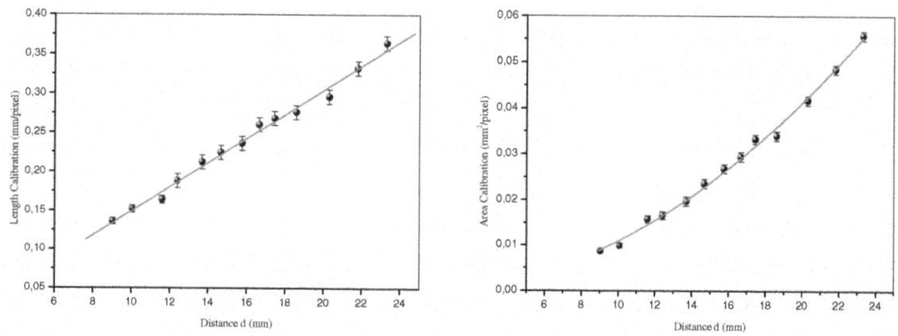

Fig. 9. Length (left) and area (right) calibration curves as a function of the distance d (reproduced from [24])

In the measurements of area, the difference between the nominal values and the corresponding value on the calibration curve does not exceed 2.5%. It is worth considering that most of this error is strictly connected to the operator who is not able to perfectly select the contour of the object under examination because of the low resolution of the endoscope viewer system (a fiber bundle that introduces pixilation artifacts). The same remark can be done for measurements of length.

Similar measurements have also been performed on biological sample, as shown in Fig. 10 [24],[25].

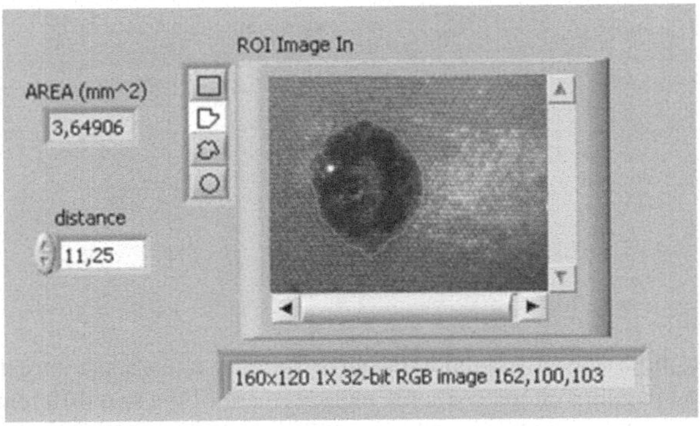

Fig. 10. Area measurement of a spot on a *ex-vivo* pig trachea (reproduced from [25])

6 Conclusions

In this Chapter we have described the realisation and test of an absolute size measurement system combining a distance sensor, based on the optical FOLCI technique, with an image processing software and easily integrable with current clinical endoscopic probes. Tests performed on biological samples have shown that the system is very sensitive, providing correct and repeatable quantitative endoscopy analysis even in case of very low signal-to-noise ratios (error of 2.5% on area measurement). Thanks to the effectiveness of FOLCI and to our signal processing technique based on the Hilbert Transform, we obtained accurate values of distance even in case of very low collected signal (down to 2 – 3 nW), also in the worst case when noise or/and false peaks make difficult to perform a correct data analysis. Further improvements can be made for enhancing the light collected by the probe arm, for instance by increasing the source power up to the safety limits of the maximum laser radiation exposure for biological tissues, or by techniques of noise stifling (for instance using a lock-in amplifier). Moreover, it would be useful to solve the problem of the above described barrel distortion due to the imaging lens.

The scanning range of the motorized translator we used is 26 mm, which is an appropriate interval to identify the correct distance value in an endoscopic examination. However, in principle, the sensor can work in any other range of distance by simply changing the optical and mechanical characteristics of the focusing GRIN lens and of the translator, respectively.

The work carried out until now has demonstrated the possibility of integrating an optical distance sensor with a clinical endoscope, in order to provide quantitative dimensional data from endoscopic images. The experimental results show that our system provides correct size measures if the object under investigation is in front of the endoscope tip. On the other hand, the endoscopic inspection of an anatomical cavity often allows only lateral view of lesions. To solve this problem, we can improve our sensor by introducing a prism that deflects the optical beam, in order to measure the distance of lateral objects. A further obstacle to our system can be the irregular 3D shape of many anatomical objects. To have a 3D view of such objects, we could use different sensing fibers, measuring the distance of different positions in the FOV at the same time.

In conclusion we described a technique that, once clinical tests are performed, could allow to obtain quantitative measurements in a endoscopic examination with a suitable accuracy of the system (error on area measurement of 2.5%), taking advantage of the knowledge of the distance between the endoscope tip and the anatomical object under investigation. Our integrated system consists in a technical solution that can be easily built in a traditional clinical endoscope - maintaining all the advantages of such imaging technique- and in a intuitive graphic tool that allows easily selecting the region of interest. Moreover, it could allow faster and easier introduction of the device in the clinical field than a completely new imaging technique, avoiding time-consuming and expensive training.

References

1. Meyers, R.A.: Encyclopedia of lasers and optical technology. Academic Press, San Diego (1991)
2. Vitale, G.C., Davis, B.R., Tran, T.C.: The advancing art and science of endoscopy. The American Journal of Surgery 190, 228–233 (2005)
3. Masters, I.B., Eastburn, M.M., Wootton, R., Ware, R.S., Francis, P.W., Zimmerman, P.V., Chang, A.B.: A new method for objective identification and measurement of airway lumen in paediatric flexible videobronchoscopy. Thorax 60, 652–658 (2005)
4. Dörffel, W.V., Fietze, I., Hentschel, D., Liebetruth, J., Rückert, Y., Rogalla, P., Wernecke, K.D., Baumann, G., Witt, C.: A new bronchoscopic method to measure airway size. European Respiratory Journal 14, 783–788 (1999)
5. McFawn, P.K., Forkert, L., Fisher, J.T.: A new method to perform quantitative measurement of brobchoscopic images. European Respiratory Journal 18, 817–826 (2001)
6. Rosen, D., Minhaj, A., Hinds, M., Kobler, J., Hillman, R.: Calibrated Sizing System for Flexible Laryngeal Endoscopy. In: 6th International Workshop: Advances in Quantitative Laryngology, pp. 1–8 (2003)
7. Santos, M.C., Strande, L., Doolin, E.J.: Airway measurement using morphometric analysis. Annals of Otology, Rhinology & Laryngology 104, 835–838 (1995)
8. Truitt, T.O., Adelman, R.A., Kelly, D.H., Willging, J.P.: Quantitative endoscopy: Initial accuracy measurements. Annals of Otology, Rhinology & Laryngology 109, 128–132 (2000)
9. Amann, M.C., Bosch, T., Lescure, M., Myllylä, R., Roux, M.: Laser ranging: a critical review of usual techniques for distance measurement. Optical Engineering 40, 10–19 (2001)

10. Danielson, B.L., Boisrobert, C.Y.: Absolute optical ranging using low coherence interferometry. Applied Optics 30, 2975–2979 (1991)
11. Schmitt, J.M.: Optical Coherence Tomography (OCT): A Review. IEEE Journal of Selected Topics in Quantum Electronics 3, 1205–1215 (1999)
12. Flavin, D.A., McBride, R., Jones, J.D.C.: Interferometric Fiber Optic Sensing Based on the Modulation of Group Delay and First Order Dispersion: Application to Strain-Temperature Measurand. Journal of Lightwave Technology 13, 1314–1323 (1995)
13. Loret, S.L., Inaudi, D.: Amplitude modulation of a low-coherence source, applications to distance and dynamic displacement sensing. Optics and Lasers in Engineering 32, 325–337 (2000)
14. Saleh, B.E.A., Teich, M.C.: Fundamental of Photonics. John Wiley & Sons, Chichester (1991)
15. Cheong, W.F., Prahl, S.A., Welch, A.J.: A review of the optical properties of biological tissues. IEEE Journal of Quantum Electronics 26, 2166–2185 (1990)
16. Bouma, B.E., Tearney, G.J.: Handbook of Optical Coherence Tomography. Marcel Dekker, New York (2002)
17. Dufour, M.L., Lamouche, G., Detalle, V., Gauthier, B., Sammut, P.: Low-coherence interferometry, an advanced technique for optical metrology in industry. Industrial Materials Institute, National Research Council, Boucherville, Quebec, Canada
18. Rao, Y.J., Jackson, D.A.: Recent progress in fiber optic low-coherence interferometry. Measurement Science And Technology 7, 981–999 (1996)
19. Song, G., Wang, X., Qian, F., Fang, Z.: The determination of the zero-order interferometric region and its central fringe in white-light interferometry. Optik 112, 26–30 (2001)
20. Wang, D.N., Shu, C.: Discrete fringe pattern to reduce the resolution limit for white light interferometry. Optics Communication 162, 187–190 (1999)
21. Bracewell, R.N.: The Fourier Transform and its Applications, 3rd edn. McGraw-Hill, Boston (2000)
22. Lucesoli, A., Criante, L., Farabollini, B., Bonifazi, F., Simoni, F., Rozzi, T.: Distance optical sensor for quantitative endoscopy. Journal of Biomedical Optics 13(1), 010504/3 (2008)
23. Oppenheim, A.V., Schafer, R.W., Buck, J.R.: Discrete-time signal processing. Prentice Hall, Upper Saddle River (1999)
24. Criante, L., Lucesoli, A., Farabollini, B., Bonifazi, F., Rozzi, T., Simoni, F.: Size measurement in endoscopic images by low coherence interferometry. Journal of Optics A, Pure and Applied Optics 11, 034007(7 p.) (2009)
25. Lucesoli, A., Criante, L., Farabollini, B., Bonifazi, F., Simoni, F., Di Donato, A., Rozzi, T.: Quantitative Endoscopy by FOLCI-Based Distance Sensor. In: IEEE Sensors 2008, Lecce, Italy, October 26-29, pp. 870–873 (2008)

Validation of Denoising Algorithms for Medical Imaging

Fabrizio Russo

D.E.E.I. – University of Trieste, Trieste, Italy

Abstract. The development of techniques for image denoising is a very challenging issue because noise must be removed without destroying image features that are important for medical diagnosis. This paper shows how validation of denoising algorithms can be performed by means of a vector measure of the image quality that takes into account noise cancellation and detail preservation by resorting to a fuzzy segmentation of the image data. Results of computer simulations show that the method overcomes the limitations of current techniques based on scalar image quality measurements. Furthermore it is conceptually simple and easy to implement.

Keywords: medical imaging, noise cancellation, image quality assessment (IQA), vector root mean squared error (VRMSE).

1 Introduction

The development of effective algorithms for noise removal is an important issue in medical imaging because noise can significantly reduce the accuracy of operations such as feature extraction and object recognition that play a key role in medical diagnosis. However, image denoising is a very difficult task since fine details embedding diagnostic information should not be destroyed during noise removal. Indeed, cancellation of noise and preservation of image features are two conflicting goals that should be carefully taken into account in the validation of any new filter. A classical validation procedure is briefly described in Fig.1a. In this procedure a test image is degraded by an assigned amount of noise and the noisy data are processed by the filter. The filter's output is compared to the original noise-free picture in order to assess the quality of the filtering. The most commonly used image quality measurements resort to the evaluation of pixelwise differences between the original and the filtered images, such as the mean squared error (MSE) and the peak signal-to-noise ratio (PSNR) [1-5]. The advantages of these methods are low computational complexity and ease of implementation, However, it is well known that the MSE and PSNR by themselves cannot characterize the behavior of a filter with respect to noise cancellation and detail preservation, so visual inspection is generally necessary to appraise these important aspects. In the general field of image quality assessment, some interesting methods have recently been proposed to overcome some limitations of MSE and PSNR.

S.C. Mukhopadhyay and A. Lay-Ekuakille (Eds.): Adv. in Biomedical Sensing, LNEE 55, pp. 93–105.
springerlink.com

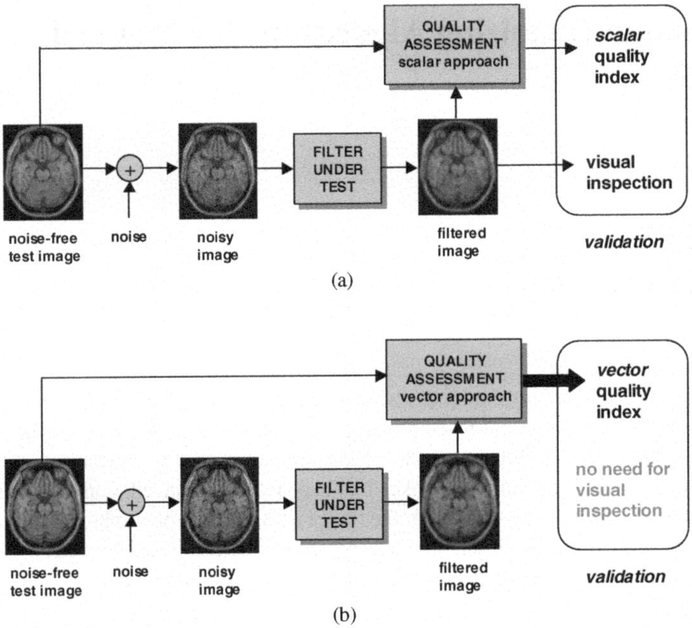

Fig. 1. Validation procedures adopting scalar (a) and vector (b) approaches for quality assessment of denoising filters

Such methods include the *universal quality index* (UQI) [6], the *structural similarity index* (SSIM) [7-8] and the M-SVD technique [9]. Some of them have been adopted for performance evaluation of denoising algorithms for medical imaging [10]. In this respect, it should be observed that all the mentioned methods yield a different quality index in the form of a scalar value. While a *scalar approach* is enough to address many different sources of image degradation, it does not seem adequate to describe the visual quality of a filtered picture for validation purposes, as shown in the next section. A *vector measure* (Fig.1b) whose components give the necessary information about noise cancellation and detail preservation represents a more appropriate solution [11-12]. This chapter will describe how the *vector root mean squared error* (VRMSE) can offer a simple and effective way to analyze the performance of a denoising technique without the need for visual inspection.

2 Limitations of Current Approaches

As mentioned in the previous section, MSE and PSNR have been commonly adopted to validate methods for image filtering. These quality indexes are formally defined as follows. Let us suppose we deal with digitized images having L gray levels. Let $x_{i,j}$ be the pixel luminance at location [i,j] in the noisy image. Let $y_{i,j}$ and $r_{i,j}$ be the pixel luminances at the same location in the filtered and reference images, respectively.

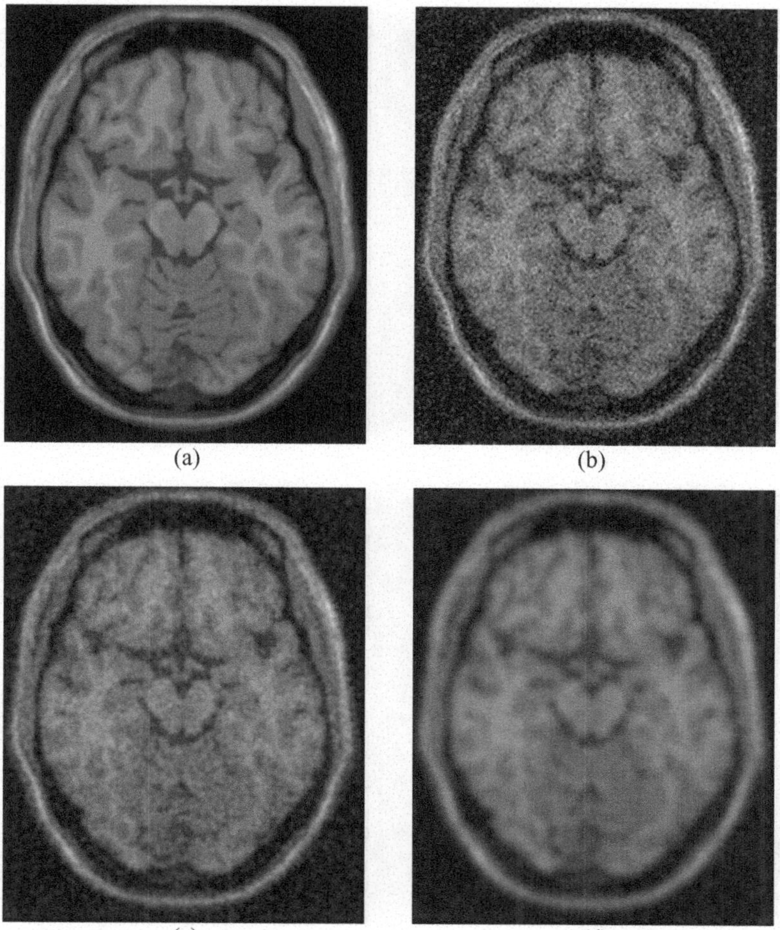

(a) (b) (c) (d)

Fig. 2. Example of filtered data having the same MSE and different image quality: (a) noise-free image, (b) image corrupted by Gaussian noise with standard deviation σ=23.3, (c) result yielded by the five-point arithmetic mean filter (MSE=122.5), (d) result yielded by the 5×5 arithmetic mean filter (MSE=122.5)

Thus, MSE and PSNR are given by the following relationships:

$$MSE = \frac{1}{N}\sum_i \sum_j (y_{i,j} - r_{i,j})^2 \tag{1}$$

$$PSNR = 10 \ \log_{10} \frac{\sum_i \sum_j (L-1)^2}{\sum_i \sum_j (y_{i,j} - r_{i,j})^2} \tag{2}$$

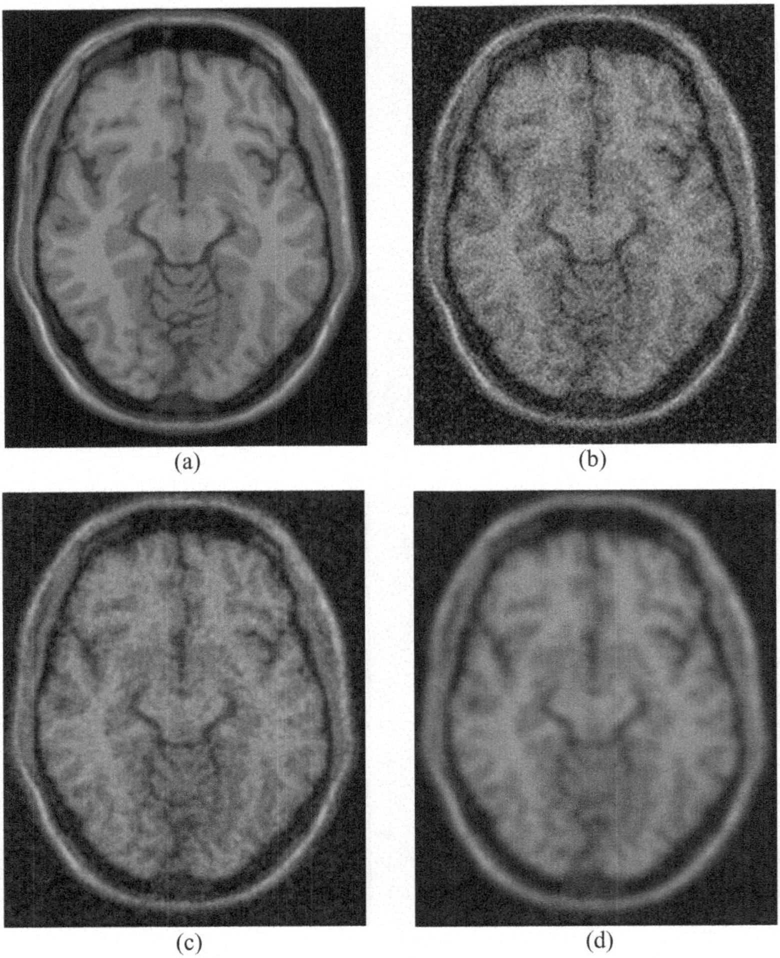

Fig. 3. Example of filtered data having the same UQI and different image quality: (a) noise-free image, (b) image corrupted by Gaussian noise with standard deviation σ=20.8, (c) result yielded by the five-point arithmetic mean filter (UQI=0.65246), (d) result yielded by the 5×5 arithmetic mean filter (UQI=0.65246)

where N denotes the number of processed pixels. The limitations of such quality indexes can be highlighted by considering a simple experiment dealing with T1-weighted simulated BrainWeb data [13-17]. In this experiment, the well-known behavior of arithmetic mean filters are exploited in order to achieve two output pictures that have the same MSE but different image quality. The noise-free data are graphically represented in Fig.2a. The image in Fig.2b has been generated by adding zero-mean Gaussian noise with standard deviation σ=23.3. The results yielded by two arithmetic mean filters with different window sizes are shown in Fig.2c (5-point cross window) and 2d (5×5 window).

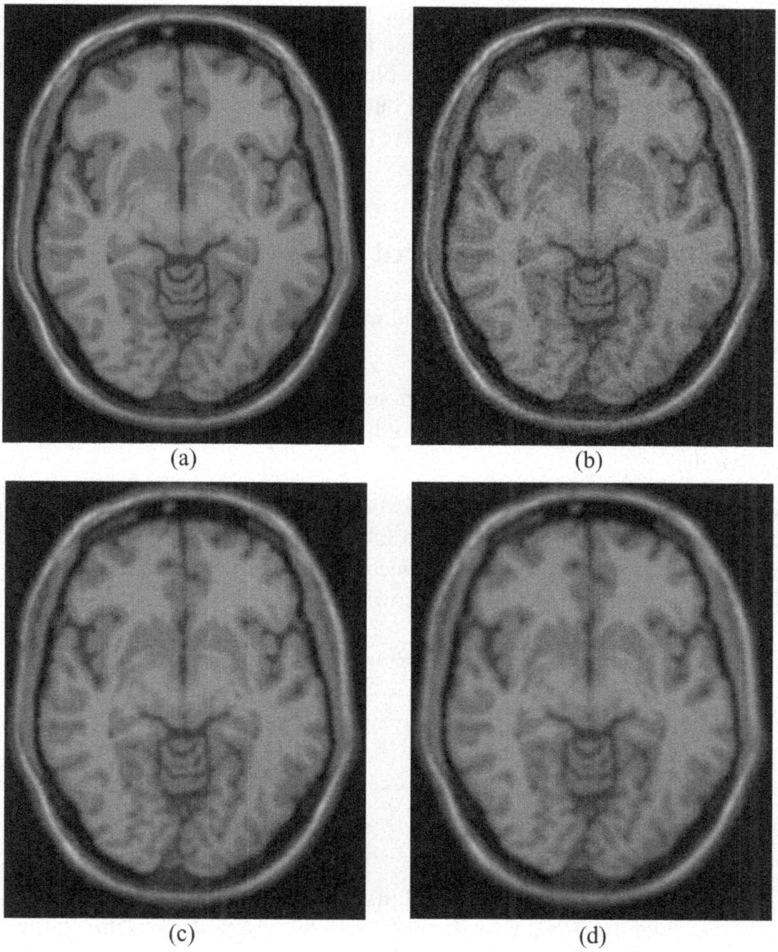

Fig. 4. Example of filtered data having the same MSSIM and different image quality: (a) noise-free image, (b) image corrupted by Gaussian noise with standard deviation σ=8.5, (c) result yielded by the five-point arithmetic mean filter (MSSIM=0.8660), (d) result yielded by the 3×3 arithmetic mean filter (MSSIM=0.8660)

Both filtered pictures have the same MSE=122.5, however their visual quality is quite different. Not surprisingly, the 5×5 window operator is more effective in removing the noise and this smoothing effect can be appraised especially in the uniform (or slightly varying) areas of the picture. The price to be paid is a very annoying blurring of the edges in the processed picture (Fig.2d). Obviously, the 5-point window filter shows a complementary behavior. Indeed, it can be observed that the image blurring is more limited and the filtered data are still noisy. The mentioned UQI and MSSIM [7-8] are excellent techniques for general-purpose image quality assessment. In the specific field of performance evaluation of denoising algorithms, however, such methods cannot replace the visual inspection of processed data. Indeed, Fig.3

shows two filtered pictures having the same UQI=0.65246 but a very different quality from the point of view of noise cancellation and detail preservation. Clearly, the noise is more perceivable in Fig.3c, while the blurring is much more apparent in Fig.3d. Similarly, a MSSIM evaluation is carried out for images processed by the 5-point and the 3×3 arithmetic mean operators (Fig.4). The filtered pictures represented in Fig4c and 4d have the same MSSIM=0.8660.

3 The Vector Root Mean Squared Error (VRMSE)

As shown in the previous examples, visual inspection of filtered pictures typically focuses on the following subsets of pixels:

- uniform (or slightly varying) regions in order to highlight the residual noise,
- non-uniform (edge) regions in order to appraise the preservation of image details.

Clearly, *uniform* or *slightly varying* areas denote fuzzy concepts [18]. Hence, fuzzy membership functions can be adopted to take into account how much a given pixel belongs to an uniform region or to an object border. The mentioned features can be computed by defining the VRMSE as follows:

$$\mathbf{VRMSE} = [\, RMSE_{UN}\,,\, RMSE_{ED}\,] \tag{3}$$

$$RMSE_{UN} = \sqrt{\frac{1}{N}\sum_i \sum_j (1-\delta_{i,j})(y_{i,j}-r_{i,j})^2} \tag{4}$$

$$RMSE_{ED} = \sqrt{\frac{1}{N}\sum_i \sum_j \delta_{i,j}\,(y_{i,j}-r_{i,j})^2} \tag{5}$$

where N is the total number of processed pixels and $\delta_{i,j}$ is a membership function that evaluates how much the pixel $r_{i,j}$ belongs to an edge. According to this definition, the degree of membership of $r_{i,j}$ to a uniform region is given by $1-\delta_{i,j}$. A map of edge gradients of the reference image can be used to extract this information [11-12]. In this paper $\delta_{i,j}$ is evaluated by means of the fuzzy edge detector defined by the following relationships:

$$\delta_{i,j} = MAX\left\{ \delta_{i,j}^{(A)}, \delta_{i,j}^{(B)} \right\} \tag{6}$$

$$\delta_{i,j}^{(A)} = MIN\left\{ 1, k_1(1-a_{i,j}) \right\} \tag{7}$$

$$\delta_{i,j}^{(B)} = MIN\left\{ 1, k_1(1-b_{i,j}) \right\} \tag{8}$$

$$a_{i,j} = \frac{\mu_{sim}(x_{i,j},x_{i,j-1}) + \mu_{sim}(x_{i,j},x_{i,j+1}) + \mu_{sim}(x_{i,j},x_{i-1,j}) + \mu_{sim}(x_{i,j},x_{i+1,j})}{4} \tag{9}$$

$$b_{i,j} = \frac{\mu_{sim}(x_{i,j},x_{i-1,j-1}) + \mu_{sim}(x_{i,j},x_{i+1,j+1}) + \mu_{sim}(x_{i,j},x_{i-1,j+1}) + \mu_{sim}(x_{i,j},x_{i+1,j-1})}{4} \tag{10}$$

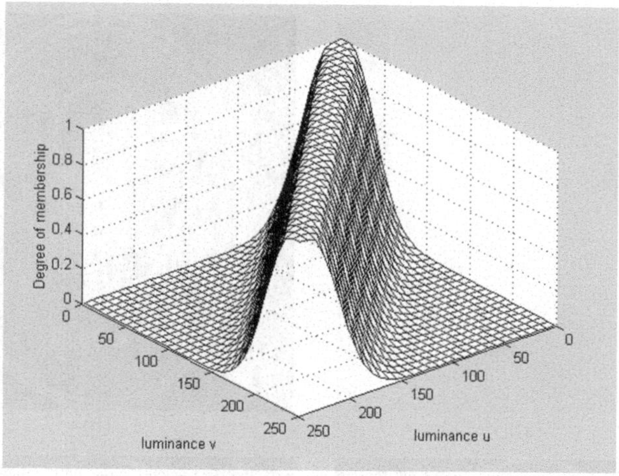

Fig. 5. Example of membership function describing the fuzzy relation "u is similar to v" (k_2=10, k_3=40, L=256)

where $\mu_{sim}(u,v)$ is a parameterized membership function that describes the fuzzy relation "u is similar to v":

$$\mu_{sim}(u,v)=\begin{cases}1 & |u-v|<k_2\\ \exp\left\{-\left(\dfrac{|u-v|-k_2}{k_3}\right)^2\right\} & |u-v|\geq k_2\end{cases} \tag{11}$$

This relation is graphically depicted in Fig.5 (k_2=10, k_3=40). The processing defined by (6-11) is very simple. Clearly, when the value of a neigboring pixel $x_{m,n}$ is similar to the luminance of the central pixel $x_{i,j}$ ($x_{m,n}\approx x_{i,j}$) we have $\mu_{sim}(x_{i,j}, x_{m,n})\approx 1$. On the contrary, when these values are very different ($x_{i,j}>>x_{m,n}$ or $x_{i,j}<<x_{m,n}$), we have $\mu_{sim}(x_{i,j},x_{m,n})\approx 0$. Now, let us focus on eq.(7) that defines a suboperator dealing with the cross neighborhood denoted by $\{x_{i,j-1}, x_{i,j}, x_{i,j+1}, x_{i-1,j}, x_{i+1,j}\}$. When this neighborhood is uniform, we have $a_{i,j}\approx 1$ (see eq.(9)) and then $\delta_{i,j}^{(A)}\approx 0$. Conversely, an object border occurring into this neighborhood gives $a_{i,j}<1$ and thus $\delta_{i,j}^{(A)}>0$. The parameter k_1 is an amplification factor that helps detecting fine details. In any case, the minimum operator (eq.(7)) limits the output to unity, as it should be. The second suboperator defined by eq.(8) similarly deals with the following subset of luminances $\{x_{i-1,j-1}, x_{i,j}, x_{i+1,j+1}, x_{i+1,j-1}, x_{i-1,j+1}\}$. Since the first suboperator better detects diagonal borders, while the second suboperator better addresses horizontal and vertical lines, the final result is achieved by evaluating the maximum of the suboperator's outputs (see eq.(6)) in order to obtain a symmetrical behavior (i.e. the same response to vertical, horizontal and diagonal edges). An example dealing with the well-known Shepp-Logan phantom image is depicted in Fig.6.

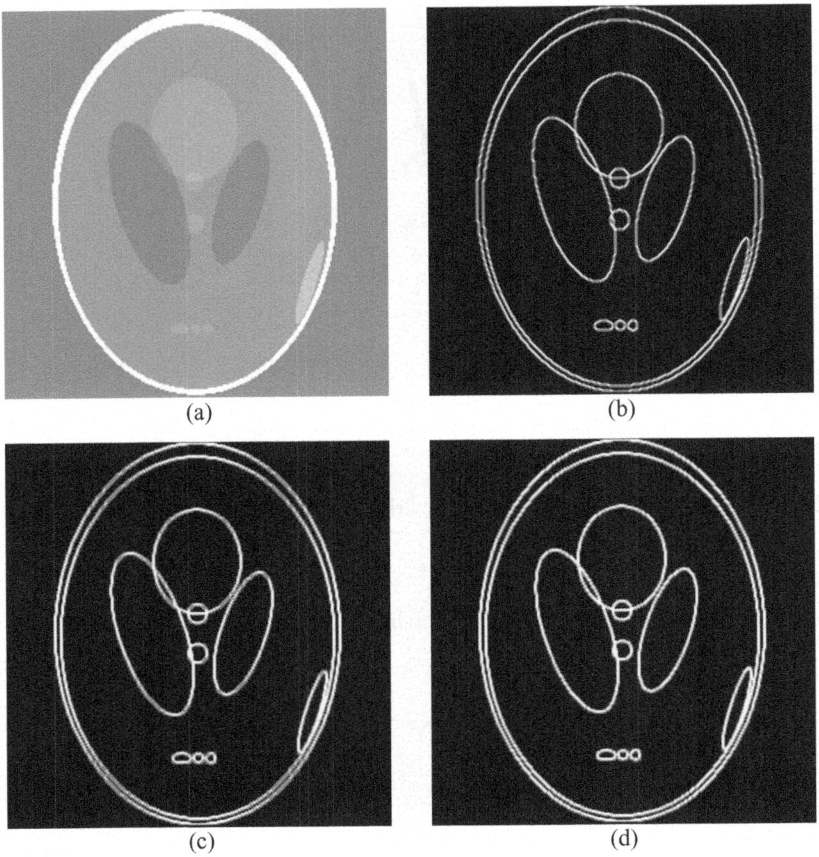

Fig. 6. (a) Test picture, (b) result given by the first suboperator, (c) result given by the second suboperator, (d) final result (k_1=2, k_2=2, k_3=5, L=256)

The behavior of the fuzzy edge detector can be controlled by choosing the values of k_2 and k_3. According to (11), these parameters define the shape of the membership function μ_{sim} and then its ability to detect object contours. Basically the parameter k_2 acts as a threshold.

Pixels yielding luminance differences smaller than k_2 (possibly due to residual noise) are considered uniform regions. In this case, the output of the edge detector is zero. Just as an example, the effects of different choices of k_2 are depicted in Fig.7. It can be seen that all the objects in the picture are detected if k_2=2. By increasing this value, however, the objects whose luminance is close to the background progressively disappear. A small value of k_2 is enough to deal with a reference image that is not exactly noise-free. The parameter k_3 affects the shape of μ_{sim} in a different way.

Small values improve the ability to detect fine details and, generally, increase the weight of the edge regions in the resulting fuzzy segmentation. By choosing the appropriate parameter set, the method can be easily adapted to a specific application.

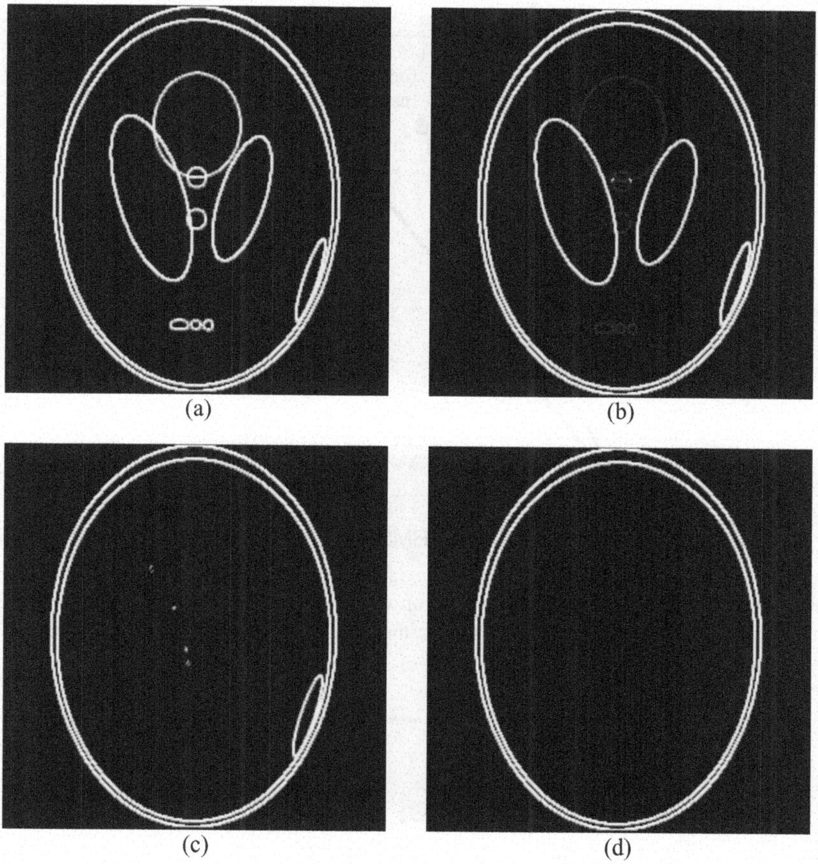

Fig. 7. Effects of different choices of parameter k_2 (k_1=3, k_3=10): (a) k_2=2, (b) k_2=10, (c) k_2=25, (d) k_2=35 (L=256)

4 Some Application Examples

Let us consider again the processed data represented in Fig.2c and 2d. In this example, the well-known effects of different window sizes were exploited in order to achieve two filtered images having the same MSE but different image quality.

By adopting the VRMSE, noise cancellation and detail preservation are now numerically computed, so visual inspection is not necessary anymore. Indeed, the evaluation of the VRMSE for the filtered pictures mentioned in this example is reported in Fig.8 (k_1=3, k_2=2, k_3=8). As aforementioned, the processed images have the same MSE, so both vectors have the same length (|VRMSE|=11). The different orientations of the vectors, however, clearly characterize the different behaviors of these filters.

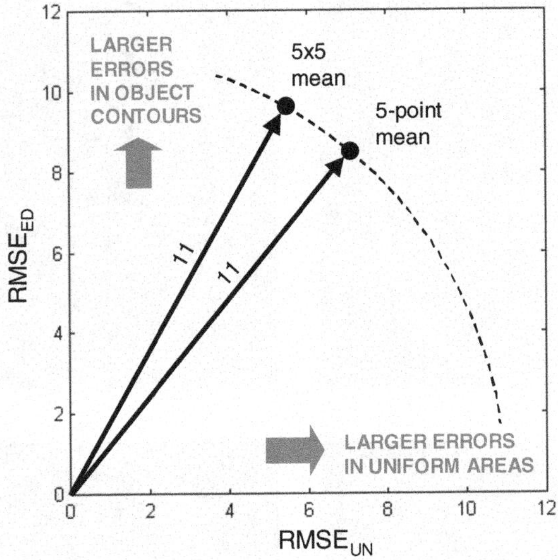

Fig. 8. Example of the graphical representation of the vector errors yielded by the five-point arithmetic mean filter and the 5×5 arithmetic mean filter (The input image is represented in Fig.2b)

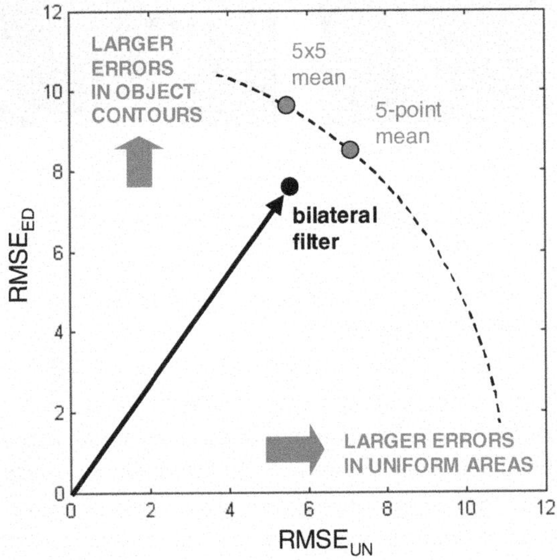

Fig. 9. Example of the graphical representation of the vector error yielded by the bilateral filter (the input image is represented in Fig.2b)

The worse detail preservation of the 5×5 linear operator is denoted by a vector error that is closer to the vertical axis. On the contrary, the worse noise removal performed by the 5-point filter is highlighted by a vector closer to the horizontal axis

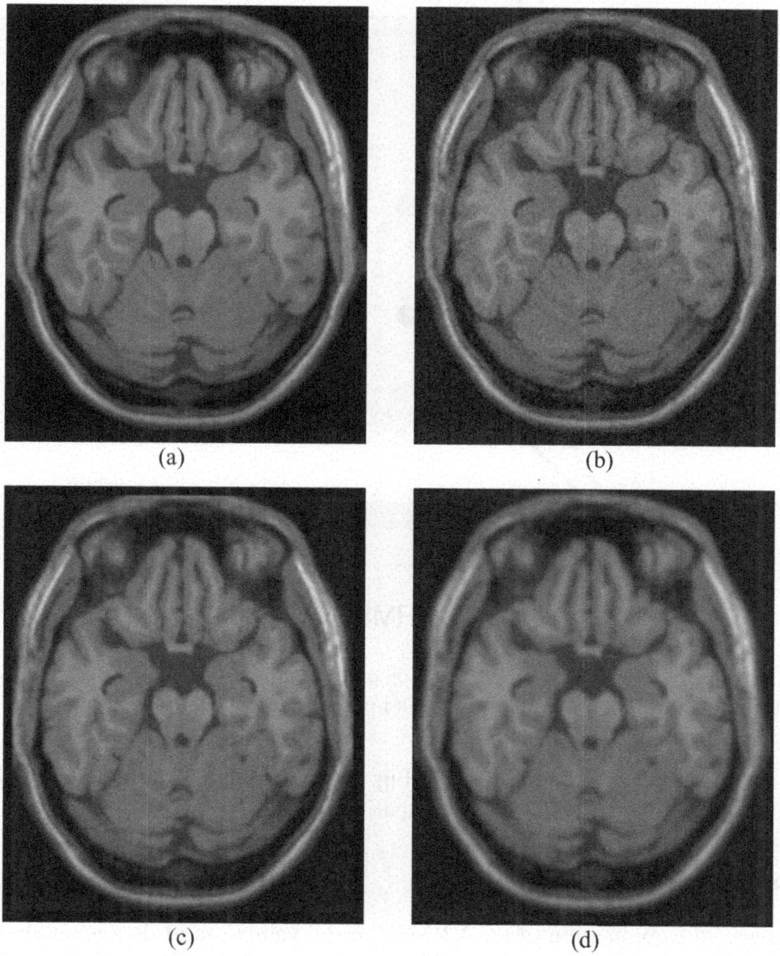

Fig. 10. VRMSE evaluation with T1-weighted simulated BrainWeb data (intensity range 0:100) with Rician noise ($\sigma = 5$) and a 40% intensity inhomogeneity. (a) noise-free image, (b) noisy image, (c) result given by the bilateral filter ($|VRMSE|=5.2$, $RMSE_{UN}=3.67$, $RMSE_{ED}=3.64$), (d) result given by the 3×3 arithmetic mean filter.

(Fig.8). Clearly, nonlinear filters are much more effective. As an example, the VRMSE given by the 11×11 bilateral filter [19-21] applied to the same input data is represented in Fig.9. The lower values of $RMSE_{UN}$ and $RMSE_{ED}$ show the best performance in terms of noise cancellation and edge preservation yielded by this nonlinear filter. A further example dealing with T1-weighted simulated BrainWeb data corrupted by Rician noise is considered in Fig.10. The noise-free picture is shown in Fig.10a while the corrupted data are depicted in Fig.10b. The result given by the bilateral filter is reported in Fig.10c. The VRMSE evaluation of the denoised data shows the well-balanced behavior of this operator: $RMSE_{UN}$ =3.7, $RMSE_{UN}$=3.6. The

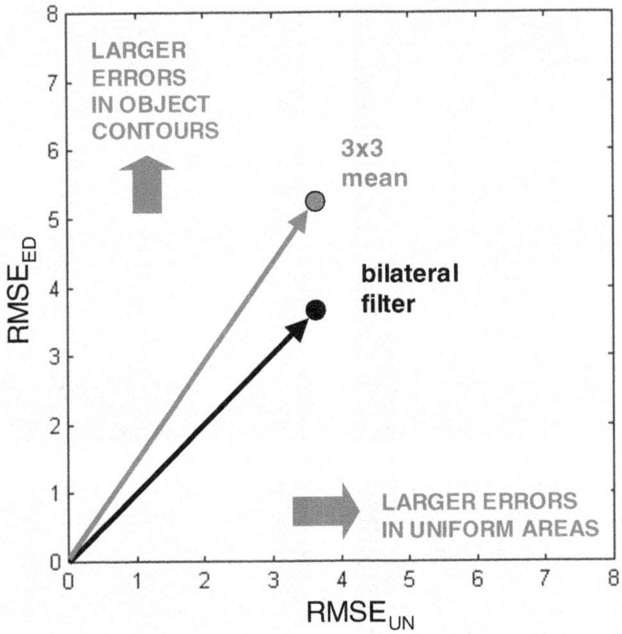

Fig. 11. Example of the graphical representation of the vector errors yielded by the bilateral filter and the 3×3 arithmetic mean filter (the input image is represented in Fig.10b)

vector error is graphically represented in Fig.11 along with the result given by the 3x3 arithmetic mean filter (Fig.10d), as a comparison.

5 Conclusion

In this paper the application of the VRMSE to the validation of denoising algorithms for medical imaging has been discussed. The different components of the VRMSE evaluated for two subsets of the image pixels overcome the limitations of the scalar quality indexes and the need for visual inspection of the filtered data. On the other hand, the method is backward compatible with the scalar MSE that has been widely used in the scientific literature. Moreover, the VRMSE has the advantage of low complexity and very simple implementation. The experiments have shown that the method can numerically evaluate features such as noise cancellation and detail preservation that represent key issues for any denoising technique.

References

1. Pizurica, A., Philips, W., Lemahieu, I., Acheroy, M.: A versatile wavelet domain noise filtration technique for medical imaging. IEEE Transactions on Medical Imaging 22(3), 323–331 (2003)
2. Awate, S.P., Whitaker, R.T.: Feature-Preserving MRI Denoising: A Nonparametric Empirical Bayes Approach. IEEE Transactions on Medical Imaging 26(9), 1242–1255 (2007)

3. Aysal, T.C., Barner, K.E.: Rayleigh-Maximum-Likelihood Filtering for Speckle Reduction of Ultrasound Images. IEEE Transactions on Medical Imaging 26(2), 200–211 (2007)
4. Rodrigues, I., Sanches, J., Bioucas-Dias, J.: Denoising of medical images corrupted by Poisson noise. In: Proc. of 15th IEEE International Conference on Image Processing, ICIP 2008, pp. 1756–1759 (2008)
5. Wang, L., Lu, J., Li, Y., Yahagi, T.: Noise Removal for Medical X-ray images in Multiwavelet Domain. In: Proc. of International Symposium on Intelligent Signal Processing and Communications, ISPACS 2006, pp. 594–597 (2006)
6. Wang, Z., Bovik, A.C.: A Universal Image Quality Index. IEEE Signal Processing Letters 9(3), 81–84 (2002)
7. Wang, Z., Bovik, A.C., Sheikh, H.R., Simoncelli, E.P.: Image Quality Assessment: from Error Visibility to Structural similarity. IEEE Transactions on Image Processing 13(4), 600–612 (2004)
8. Wang, Z., Bovik, A.C.: Modern Image Quality Assessment. Morgan & Claypool Publishers, San Francisco (2006)
9. Shnayderman, A., Gusev, A., Eskicioglu, A.M.: An SVD-Based Grayscale Image Quality Measure for Local and Global Assessment. IEEE Transactions on Image Processing 15(2), 422–429 (2006)
10. Zhang, F., Yang, M.Y., Liang, M.K., Yongmin, K.: Nonlinear diffusion in Laplacian Pyramid Domain for Ultrasonic Speckle Reduction. IEEE Transactions on Medical Imaging 26(2), 200–211 (2007)
11. De Angelis, A., Moschitta, A., Russo, F., Carbone, P.: A Vector Approach for Image Quality Assessment and some Metrological Considerations. IEEE Transactions on Instrumentation and Measurement 58(1), 14–25 (2009)
12. Russo, F., De Angelis, A., Carbone, P.: A Vector Approach to Quality Assessment of Color Images. In: Proc. IEEE I2MTC, International Instrumentation and Measurement Technology Conference, Victoria, Vancouver Island, Canada (2008)
13. `http://www.bic.mni.mcgill.ca/brainweb/`
14. Cocosco, C.A., Kollokian, V., Kwan, R.K.-S., Evans, A.C.: BrainWeb: Online Interface to a 3D MRI Simulated Brain Database. NeuroImage 5(4), part 2/4, S425 (1997)
15. Kwan, R.K.-S., Evans, A.C., Pike, G.B.: MRI simulation-based evaluation of image-processing and classification methods. IEEE Transactions on Medical Imaging 18(11), 1085–1097 (1999)
16. Kwan, R.K.-S., Evans, A.C., Pike, G.B.: An Extensible MRI Simulator for Post-Processing Evaluation. In: Höhne, K.H., Kikinis, R. (eds.) VBC 1996. LNCS, vol. 1131. Springer, Heidelberg (1996)
17. Collins, D.L., Zijdenbos, A.P., Kollokian, V., Sled, J.G., Kabani, N.J., Holmes, C.J., Evans, A.C.: Design and Construction of a Realistic Digital Brain Phantom. IEEE Transactions on Medical Imaging 17(3), 463–468 (1998)
18. Russo, F.: Fuzzy model fundamentals. In: Webster, J.G. (ed.) Wiley Encyclopedia of Electrical and Electronics Engineering. Wiley, Hoboken (1999)
19. Tomasi, C., Manduchi, R.: Bilateral filtering for gray and color images. In: Proc. 6th Int. Conf. Computer Vision, Bombay, India, pp. 839–846 (1998)
20. Elad, M.: On the Origin of the Bilateral Filter and Ways to Improve It. IEEE Transactions on Image Processing 11(10), 1141–1151 (2002)
21. Barash, D.: A Fundamental Relationship between Bilateral Filtering Adaptive Smoothing and the Nonlinear Diffusion Equation. IEEE Transactions on Pattern Analysis and Machine Intelligence 24(6) (2002)

Dielectrophoretic Actuation and Simultaneous Detection of Individual Bioparticles

S.F. Romanuik[1], G.A. Ferrier[1], M.N. Jaric[1], D.J. Thomson[1], G.E. Bridges[1], and M.R. Freeman[2]

[1] Department of Electrical and Computer Engineering, University of Manitoba, Winnipeg, Manitoba, R3T 5V6, Canada
Tel.: +1-204-4748797
thomson@ee.umanitoba.ca

[2] National Institute for Nanotechnology and Department of Physics, University of Alberta, Edmonton, Alberta, T6G 2G7, Canada

Abstract. A microflow cytometer uses a microelectrode array to simultaneously actuate and detect biological particles. Applied microelectrode array potentials in the 10-1000 kHz band generate a dielectrophoretic force which actuates passing bioparticles. Cells are detecting using two independent methods: (1) optical cell assay, in which the change in cellular velocity due to the dielectrophoretic and fluid drag forces is estimated from digital video of the microelectrode array region; and (2) capacitive cytometry, in which a 1.478 GHz coupled capacitance sensor produces a sense signal proportional to the transient capacitance perturbations induced by passing bioparticles. The results obtained from these two independent methods support one another. This work primarily studies baker's yeast (*Saccharomyces cerevisiae*) cells. Additional studies of Chinese Hamster Ovary cells and polystyrene spheres are also presented.

Keywords: single-cell diagnostics, flow cytometry, capacitive cytometry, microwave interferometer, dielectrophoresis, laboratory-on-a-chip, micro-total analysis system, *Saccharomyces cerevisiae*, yeast, Chinese hamster ovary cells, CHO cells, polystyrene spheres, PSS.

1 Motivation: Single-Cell Diagnostics

Microbiology traditionally studies individual biological cells via inference from ensemble averages over heterogeneous bulk samples, from which it is difficult to probe discrete and intrinsic cellular properties [1-5]. Single-cell diagnostics provide an investigation of cellular properties beyond the resolution of bulk sample measurements [6-13]. Moreover, single-cell diagnostics afford quantifying cell population heterogeneity, and the potential to identify, characterize, and separate subpopulations to preferentially select strains [14-15].

Quantifying the heterogeneity of tumour cell populations facilitates determining the cancerous stage of tumours and investigating the response of tumour cells to customized therapies [16]. As an example, certain cancers release tumour cells into blood circulation during their early stages [17]. During the micrometastasis stage of breast

S.C. Mukhopadhyay and A. Lay-Ekuakille (Eds.): Adv. in Biomedical Sensing, LNEE 55, pp. 106–126.
springerlink.com

cancer, it is common to find 1-10 tumourous cells and an excess of 10^7 red blood cells within a millilitre of blood [2]. By detecting these rare tumour cells, cancer can be diagnosed at this early stage [18-19]. Furthermore, the level of circulating tumour cells is a patient prognosis predictor [20].

In vitro stem cell expansion, necessary for generating a sufficient population for transplantation, is accompanied by the differentiation of daughter cell subsets, resulting in a heterogeneous population [15]. Controlling the composition of stem cell transplants is difficult without the isolation of in vitro cell subpopulations afforded by single-cell diagnostics [15].

The above examples demonstrate the motivation for single-cell diagnostics. However, these examples require highly accurate, high-throughput diagnostic devices which are the ultimate goal within the current research and development of single-cell diagnostic tools.

2 Instrumentation I: Microflow Cytometer

Flow cytometry has been widely used for high-throughput cell-by-cell analysis [1,21-24]. However, conventional fluorescence-based flow cytometers are intricate, bulky, and expensive [22-23]. Fortunately, microfabrication technologies have produced miniaturized microflow cytometers well suited for single-cell diagnostics, as microfluidic channel dimensions are comparable with cells [22-23]. Moreover, microfabrication technologies afford low per unit production costs, minute sample consumption, rapid heat and mass transfer rates, high-throughput parallelization, on-site measurements, and on-chip integration with other modules to produce laboratories-on-a-chip (LoCs) and micro-total analysis systems (μTASs) capable of executing complete diagnostic sessions within a single microchip [2,22-24].

Some flow cytometers detect cells using a Coulter counter scheme in which particles suspended in a conductive fluid flow past electrodes, inducing transient pulses within the monitored inter-electrode resistance [22,25-28]. The particles are assumed to be infinitely resistive, true even for conducting particles as the interfacial polarization circumvents internal conduction current [26]. The inter-electrode resistance pulse thus becomes a measure of the detected particle's cross-sectional area alone [26]. Being comprised entirely of electrodes, miniaturized micro-Coulter counters are easier to fabricate and integrate than miniaturized optical systems. However, Coulter counters require a lower density of cells than optical systems, to ensure that multiple cells do not simultaneously occupy the detection zone.

In impedance spectroscopy, the sensing frequency is varied to estimate the probed cell's impedance spectrum. Ayliffe et al. present a microflow cytometer which estimated the impedance spectrum of human polymorphoneuclear leukocytes and teleost fish red blood cells over a 100 Hz to 2 MHz bandwidth [29]. Gawad et al. demonstrated a microflow cytometer that utilizes 1.72-15 MHz impedance spectroscopy to distinguish between normal and ghost human erythrocyte cell subpopulations within a mixed population [30].

Capacitance cytometry measures dielectric particles' polarization as they flow through an electric field [2]. Sohn et al. present a capacitive cytometer that probes the polarization within cells, observing a linear relationship between various eukaryotic cells' DNA content and the transient inter-electrode capacitance perturbations induced as they crossed a 1 kHz electric field [31]. Using their cytometer, Sohn et al. quantified their eukaryotic cells' DNA content and analyzed the cell-cycle kinetics of a eukaryotic population.

This work presents a microflow cytometer which uses a microelectrode array (MEA) to simultaneously actuate and detect individual biological particles (bioparticles) [32-38]. A 10-1000 kHz electric potential ϕ_{DEP} is applied to the MEA, generating a dielectrophoretic (DEP) force $\boldsymbol{F_{DEP}}$ which actuates passing bioparticles. $\boldsymbol{F_{DEP}}$ competes with the fluid drag and gravitational forces to yield trajectories of time-varying elevation h_{cell} and velocity $\mathbf{v}_{cell}$. This cytometer affords the rapid assay of the dielectric character of detected bioparticles.

The cytometer affords two independent methods for simultaneously detecting the response of actuated bioparticles: (1) optical cell assay, in which a microscope fitted with a digital video camera monitors the detection zone; and (2) capacitive cytometry, in which a resonant transmission line (T-line) coupled to the MEA produces a sense signal S proportional to the transient MEA capacitance perturbations ΔC_{MEA} induced by passing bioparticles. The coupled T-line is integrated into a 1.478 GHz interferometer, with narrowband lock-in techniques employed to ensure that ϕ_{DEP} doesn't influence S. Conversely, the superimposed 1.478 GHz interrogation potential ϕ_{RF} is sufficiently low to ensure that its associated $\boldsymbol{F_{DEP}}$ is negligible. In this work, the optical cell assay and the capacitive cytometer shall be used simultaneously for independent experimental confirmation.

This work primarily focuses on the simultaneous actuation and detection of baker's yeast (*Saccharomyces cerevisiae*) cells. However, additional studies of Chinese Hamster Ovary (CHO) cells and polystyrene spheres (PSS) shall also be presented, to illustrate the full range of applications in which this microflow cytometer has been employed thus far.

Our microflow cytometer design was fabricated by Micronit, through the Canadian Microelectronics Corporation (CMC) in a multi-project batch fabrication, using the Sensonit process [39]. A micrograph of our fabricated microflow cytometer is presented in Figure 1.

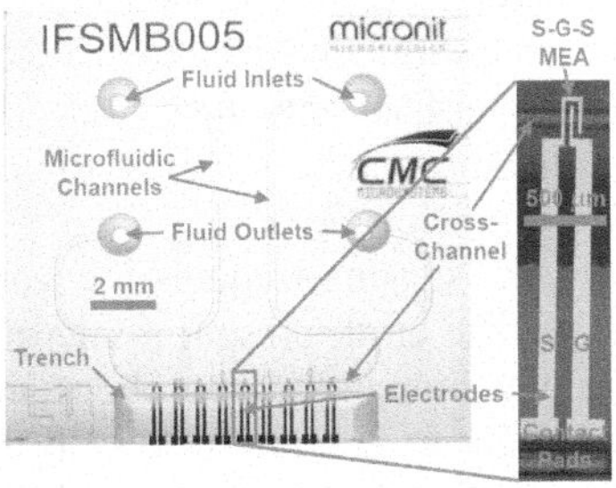

Fig. 1. Microflow cytometer as fabricated by Micronit

Fabrication begins with the wet isotropic etch of 40 μm deep microfluidic channels into a 1.1 mm thick Borofloat wafer's bottom. The design incorporates two 240 μm wide channels, each connected to a fluid inlet and outlet. A 100 μm wide cross-channel is etched between the 240 μm wide channels, passing over the MEA detection zone. This H-channel design affords controlling the cross-channel fluid flow via the flow differential between the two wider channels. The cross-channel cubbies have not been utilized in this work.

The fluid ports and electrode coupling trench were powder-blasted into the 1.1 mm thick wafer's top. The fluid ports have 1400 μm top and 600 μm bottom diameters. The trench is 7.992 mm x 1.700 mm at the top and 7.190 mm x 1.300 mm at the bottom.

200 nm deep trenches were etched into a 0.7 mm thick Borofloat wafer's top, into which 20 nm thick Ta and 180 nm thick Pt layers ere deposited. A micrograph of the principal MEA used in this work is inset within Figure 1. As it spans the cross-channel, this MEA is an interdigitated Signal-Ground-Signal (S-G-S) pattern formed by 25 μm wide microelectrodes with a 25 μm edge-to-edge spacing. These microelectrodes connect to a pair of 100 μm wide electrodes with a 100 μm edge-to-edge separation extending out to the die's periphery. These electrodes are 100 μm wide to reduce the high frequency electrical losses between the periphery and the MEA. At the periphery, these electrodes form 228 μm x 300 μm contact pads, with a 50 μm edge-to-edge separation, within the coupling trench.

The two wafers were fused using a direct heat bond. The Sensonit fabrication process was then concluded by cutting the fused wafers to yield 1.5 cm x 1.5 cm x 1.8 mm dies.

The fabricated die was bonded on top of a 2.6 cm x 3.1 cm x 1.2 cm Plexiglas block. Four 1.1 mm diameter holes, aligned with the fluid ports, were bored into a 1.5 cm x 1.0 cm x 1.1 cm Plexiglas block bonded to the top of the die. The cross-channel and coupling trench were not obscured by Plexiglas, facilitating optical monitoring and MEA coupling. Polyethylene tubing with a 300 μm inner and a 1.09 mm outer diameter was bonded into the holes. Syringe needles were bonded into the inlets' tubes. Syringes connect to the needles for sample injection, and are clamped in a vertical position with their plungers removed. Gravity assisted flow is utilized, where the syringes' relative elevation establishes a pressure differential, inducing a flow differential between the 240 μm wide channels, and thus controlling the 100 μm wide cross-channel's flow.

Figure 2 shows the bonded assembly inside a shielded probe station (SUSS SE1000). The detection zone is illuminated by white-light emitting diodes (Lumex Opto Components SLX-LX5093UWC/G). A microscope (Motic PSM-1000) with a digital camera (Imaging Source DFK 31AF03) monitors the detection zone, capturing UYVY video at 15 frames per second. A FireWire cable connects the camera to a PC running Imaging Source's IC Capture 2.0 software, to display and backlog the captured video (as AVI files) in real-time.

Fig. 2. The bonded microflow cytometer assembly as an optical cell assay

3 Instrumentation II: Capacitive Cytometry Sensor

Capacitive cytometry requires that a sensor be coupled to the MEA, to produce a sense signal S proportional to the transient MEA capacitance perturbations ΔC_{MEA} induced by passing cells. Figure 3 presents the sensor's basis, a coupled-microstrip resonator fabricated by milling Cu T-lines onto a Cu-clad Rogers 5880 substrate [40]. The 90 Ω center T-line is a quarter-wavelength resonator, edge-coupled by 50 Ω short-circuit terminated T-lines, acting as the resonator's input and output ports 1-2 (interchangeable due to symmetry). 150 μm diameter brass wires are soldered at the center T-line's open-circuit edge port. The wires are cut to a point, finely ground, and rinsed with isopropanol and deionized water (DI H_2O). The microstrip structure is mounted into a XYZ positioner, adjusted such that the brass wires firmly contact the pads at the die's periphery, effectively loading the resonator with the MEA. The resonator's other port is terminated with an 85 pF shunt capacitor. Port 3, used for applying the 10-1000 kHz DEP actuation potential ϕ_{DEP}, is coupled to the resonator's end using a 750 Ω resistor. The resulting RC circuit is a lowpass filter (LPF) with a 2.5 MHz cut-off frequency, shorting higher frequencies to ground through the shunt capacitor, thereby isolating the 1.478 GHz interferometer from the ϕ_{DEP} electronics.

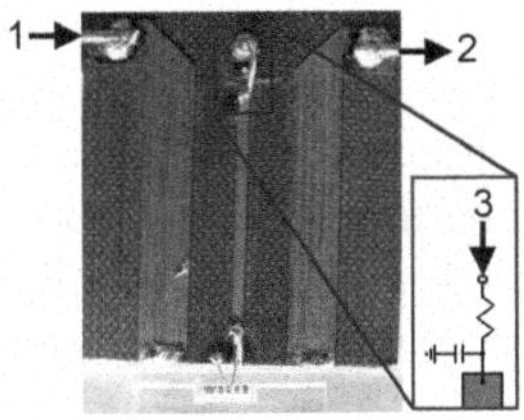

Fig. 3. The coupled-microstrip resonator loaded by the principal MEA

Figure 4 presents the resonator's port 1-2 transmission response, $S_{2,1}$, (measured with a HP 8753E vector network analyzer) when: (1) unloaded and (2) loaded with the MEA. The microfluidic channels are filled with a 33.4 μS/cm diluted methylene blue fluid. The $|S_{2,1}|$ response shows that: (1) the unloaded resonator resonates at 1.826 GHz with a 7.3 dB insertion loss and a 54 Q-factor; and (2) the loaded resonator resonates at 1.478 GHz with a 15.9 dB insertion loss and 20 Q-factor. $|S_{2,1}|$ thus demonstrates that: (i) the MEA's capacitive load shifts f_r by 349 MHz; (ii) an increased electrical loss reduces the resonant magnitude by 8.7 dB; and (iii) this increased electrical loss also diminishes the resonator's energy storage capability, as quantified by the Q-factor. The $\angle S_{2,1}$ response demonstrates that the capacitive MEA load increases the resonator's effective electrical length, thereby increasing the phase shift between ports 1 and 2. Moreover, the loaded $\angle S_{2,1}$ response shows that the rate of change of θ with f_r in the vicinity of 1.478 GHz, $d\theta/df_r$, is approximately -1.56 °/MHz. $d\theta/df_r$ is referred to as the sensor's phase sensitivity.

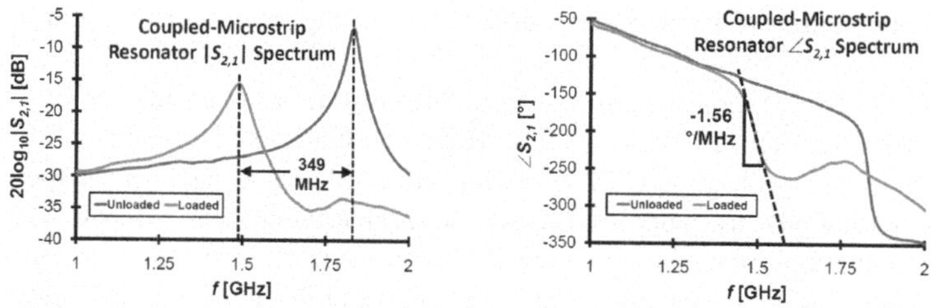

Fig. 4. Resonator's unloaded and MEA loaded $|S_{2,1}|$ (LEFT) and $\angle S_{2,1}$ (RIGHT)

Figure 5 presents the microwave interferometer used to interrogate the coupled-microstrip resonator and generate a sense signal S proportional to ΔC_{MEA}.

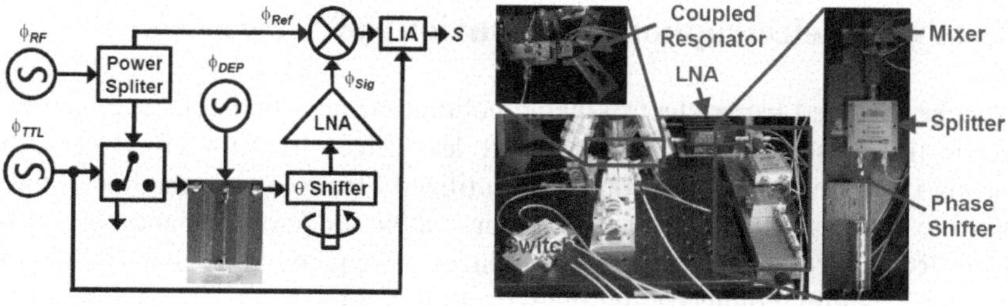

Fig. 5. Schematic (LEFT) and photograph (RIGHT) of the microwave interferometer

The RF source ϕ_{RF} (Agilent E8663B) is split (Mini-Circuits ZAPDJ-2-S) into two paths. The signal path contains a high-isolation switch (Mini-Circuits ZASWA-2-50DR), the coupled-microstrip resonator, an adjustable delay line (Advanced Technical Materials PNR P1213), and a low noise amplifier (LNA; Mini-Circuits ZHL-1724MLN). The LNA output is denoted ϕ_{Sig}. The reference path contains no additional electronics and is denoted ϕ_{Ref}. ϕ_{Sig} and ϕ_{Ref} are multiplied using a mixer (Mini-Circuits ZEM-4300MH+). The switch's control ϕ_{TTL} is toggled on-and-off at 95 kHz, enabling narrow-band detection of the mixer's mean output, S, using a lock-in amplifier (LIA; Stanford SR380). During operation, the driving frequency is set near the 1.478 GHz MEA loaded resonance f_r and the delay line is adjusted such that the reference phase θ between ϕ_{Sig} and ϕ_{Ref} is nominally 90°. A cell's presence within the detection zone perturbs the MEA capacitance by ΔC_{MEA}, thereby shifting the loaded resonance by Δf_r, thus modifying $\angle S_{2,1}$ by $\Delta\theta$, and in turn changing S. When operating near f_r, a sufficiently small ΔC_{MEA} (less than 10 fF) affords approximating S as

$$S \approx G\,|\phi_{Ref}|\,|\phi_{Sig}|\,(d\theta/df_r)\,(df_r/dC_{MEA})\,\Delta C_{MEA} = (dS/dC_{MEA})\,\Delta C_{MEA} \qquad (1)$$

where G is the combined mixer-and-LIA gain and df_r/dC_{MEA} is the frequency sensitivity, being the rate of change of f_r with C_{MEA} in the vicinity of 1.478 GHz. dS/dC_{MEA} is the overall sensitivity, being the rate of change of S with C_{MEA} in the vicinity of 1.478 GHz. dS/dC_{MEA} is estimated in Section 7. The RF electronics are placed inside the shielded probe station to reduce the coupled electromagnetic, thermal, and mechanical noise.

S is sampled by a data acquisition board (Measurement Computing PCI-DAS6034) internal to the PC. The PC runs a GUI written in National Instruments' LabVIEW® 7.1 to display and backlog (as CSV files) the sampled S as a real-time stripchart.

Sensing with GHz frequencies affords several advantages. Below 200 MHz, a cell's electrical properties can vary greatly due to ionic currents [41], electrode polarization [42], interfacial polarization [41], the electrical double layer [43], and large potential gradients across cellular membranes and walls [44]. These effects can be greatly reduced, or even eliminated, by operating above 1 GHz [41]. This assumption is made during the derivation of ΔC_{MEA} presented in Section 6. Moreover, sensing above 1 GHz affords using a single time constant LPF to isolate the interferometer from the 10-1000 kHz ϕ_{DEP} electronics.

4 Theory I: Dielectrophoretic Behaviour of *S. cerevisiae*

Cells can be actuated using electrokinetic techniques, in which cells are manipulated by electric fields established between electrodes. Electrorotation (ROT) and dielectrophoresis (DEP) are the most commonly utilized electrokinetic methods, and have been used to successfully determine if cells are viable, infected, or cancerous [45-47].

During ROT, a rotating electric field induces a frequency-and-material-dependent electric dipole within a dielectric particle, creating a torque as the dipole tries to align itself with or against the field [48-49]. Beyond determining cell viability [50], a cell's ROT spectrum can be used to estimate the various subcellular conductivities and permittivities [51-54].

During DEP, a non-uniform electric field induces and exerts itself upon an electric dipole within a dielectric particle, creating a net translational force [55]. As a measure of a cell's frequency-and-material-dependent polarization, DEP can be used to infer inherent cellular traits [56] and distinguish between cell subpopulations [45,57-60].

There are two approaches to separating subpopulations using DEP: DEP migration (mDEP) and DEP retention (rDEP) [61]. In mDEP, different cell types experience opposing DEP forces, so that a positive DEP (pDEP) force attracts one cell type towards electric field maxima whilst a negative DEP (nDEP) force repels other cells from the maxima [46,62-64]. In rDEP, the DEP force competes with fluid flow forces, so that cells experiencing a strong pDEP force are held against the flow whilst other cells are carried along the flow [61,63,65].

In this work, *S. cerevisiae* cells are suspended in a diluted methylene blue medium (*med*) with a $\sigma_{med} \approx 33.4$ µS/cm conductivity. A 10-1000 kHz MEA potential ϕ_{DEP} actuates the cells with a nDEP or pDEP force $\boldsymbol{F_{DEP}}$ that competes with the viscous fluid drag and gravitational forces to yield trajectories of time-varying elevation h_{cell} and velocity $\mathbf{v}_{cell}$.

Subcellular constitutive electrical parameters have been inferred from experimental ROT spectra [51-54]. The complex heterogeneous *S. cerevisiae* cell necessitates that ROT data be fitted to models consisting of a homogeneous cytoplasmic sphere (*cyt*) surrounded by concentric homogeneous spherical shells, typically representing the cytoplasmic membrane and the cell wall [66-67]. Hölzel obtained excellent agreement with ROT data over a 100 Hz to 1.6 GHz bandwidth using a four-shelled model [54]. These four shells represent the cytoplasmic membrane (*mem*), a permiplasmic space (*pps*), and inner and outer cell walls (*icw* and *ocw*). Each layer within the model is defined by: thickness (or radius), real permittivity, and conductivity. Hölzel used cellular dimensions derived from scanning electron micrographs [68-71], and real permittivities reported elsewhere [43,51-52,66]. Subcellular conductivities were then computed via a combined least squares error (LSE) fit of the ROT spectrum [54]. As subcellular conductivities vary with σ_{med}, Hölzel measured viable *S. cerevisiae* ROT spectra using $\sigma_{med} \approx 0.7$, 20, 90, and 550 μS/cm [54]. Hölzel suggests that only σ_{mem} and σ_{icw} vary with σ_{med} (when $\sigma_{med} \geq 20$ μS/cm). LSE fits to the σ_{mem} and σ_{icw} versus σ_{med} data were produced, from which the constitutive parameters of viable *S. cerevisiae* as used in this work were estimated as presented in Table 1.

Hölzel et al. measured the ROT spectra of heat shocked *S. cerevisiae* using $\sigma_{med} \approx$ 25 μS/cm [53]. A two-shelled model was used, representing the cytoplasmic membrane (*mem*) and cell wall (*cw*). Using dimensional and real permittivity assumptions, subcellular conductivities were computed via a combined LSE fit of the ROT spectrum [53]. The subcellular parameters of nonviable *S. cerevisiae* as used in this work were estimated as presented in Table 1, by assuming Hölzel et al.'s $\sigma_{med} \approx 25$ μS/cm values approximate the subcellular conductivities of nonviable *S. cerevisiae* suspended in our $\sigma_{med} \approx 33.4$ μS/cm medium.

Table 1. Constitute *S. cerevisiae* properties as used in this work

Viable				*Nonviable*			
Layer	***Dimension***	**ε'$_r$**	**σ [μS/cm]**	***Layer***	***Dimension***	**ε'$_r$**	**σ [μS/cm]**
cyt	3.0 μm Radius	51	12 000	*cyt*	3.0 μm Radius	51	100
mem	3.5 nm Thick	3	0.0302	*mem*	3.5 nm Thick	6	0.1
pps	25 nm Thick	14.4	41				
icw	110 nm Thick	60	30.4322	*cw*	110 nm Thick	60	300
ocw	50 nm Thick	5.9	200				
med	N/A	78	33.4	*med*	N/A	78	33.4

Hölzel's multi-shelled models can be replaced by equivalent homogeneous spheres. Consider a sphere with radius r_{sphere} and complex permittivity $\underline{\varepsilon}_{sphere}$ surrounded by a concentric spherical shell with thickness d_{shell} and complex permittivity $\underline{\varepsilon}_{shell}$. This single-shelled sphere can be replaced with an equivalent homogeneous sphere with radius $r_{eq} = r_{sphere} + d_{shell}$ and complex permittivity $\underline{\varepsilon}_{eq}$, given as [72]

$$\underline{\varepsilon}_{eq} = \underline{\varepsilon}_{shell} \left[\left(\frac{r_{sphere} + d_{shell}}{r_{sphere}} \right)^3 + 2 \left(\frac{\underline{\varepsilon}_{sphere} - \underline{\varepsilon}_{shell}}{\underline{\varepsilon}_{sphere} + 2\underline{\varepsilon}_{shell}} \right) \right] \Bigg/ \left[\left(\frac{r_{sphere} + d_{shell}}{r_{sphere}} \right)^3 - \left(\frac{\underline{\varepsilon}_{sphere} - \underline{\varepsilon}_{shell}}{\underline{\varepsilon}_{sphere} + 2\underline{\varepsilon}_{shell}} \right) \right] \quad (2)$$

Equation (2) is first used to replace the cytoplasmic sphere and membrane shell with an equivalent homogeneous sphere. Equation (2) is then used iteratively from the inside out until the entire multi-shelled model is replaced by an equivalent homogeneous sphere.

The time-averaged $\boldsymbol{F_{DEP}}$ experienced by a homogeneous lossy dielectric sphere subjected to a non-uniform time-harmonic electric field $\boldsymbol{E_e}$ is given as [72]

$$F_{DEP} \approx 2\,\pi\,\varepsilon'_{med}\,a^3\,\mathrm{Re}\{\underline{K}\}\,\nabla E_{RMS}{}^2 \tag{3}$$

where E_{RMS} is the RMS value of E_e, a is the sphere's radius, ε'_{med} is the fluid medium's real absolute permittivity, and $\underline{K}$ is the complex Clausius-Mossotti factor. The frequency dependence of $\boldsymbol{F_{DEP}}$ is determined by Re{$\underline{K}$}. $\underline{K}$ is given as [72]

$$\underline{K} = (\,\underline{\varepsilon}_{cell} - \underline{\varepsilon}_{med}\,)\,/\,(\,\underline{\varepsilon}_{cell} + 2\underline{\varepsilon}_{med}\,) \tag{4}$$

Equation (4) implies that the $\omega\rightarrow 0$ DEP behaviour is governed by the conductivities, and that the $\omega\rightarrow\infty$ DEP behaviour is governed by the real permittivities:

$$\mathrm{K}_0 = \lim_{\omega\to 0}\{\underline{\mathrm{K}}\} \approx \frac{\sigma_{\mathrm{cell}} - \sigma_{\mathrm{med}}}{\sigma_{\mathrm{cell}} + 2\sigma_{\mathrm{med}}} \qquad \mathrm{K}_\infty = \lim_{\omega\to\infty}\{\underline{\mathrm{K}}\} \approx \frac{\varepsilon'_{\mathrm{cell}} - \varepsilon'_{\mathrm{med}}}{\varepsilon'_{\mathrm{cell}} + 2\varepsilon'_{\mathrm{med}}} \tag{5}$$

Although the multi-shelled model has been replaced with an equivalent homogeneous sphere, the Maxwell-Wagner interfaces between each layer are embedded within the applications of Equation (2). As such, the real part of Equation (4) can be expressed as

$$\mathrm{Re}\{\underline{\mathrm{K}}\} \approx \mathrm{K}_\infty - \frac{\Delta\mathrm{K}_\alpha}{1+\omega^2\tau_\alpha^2} - \frac{\Delta\mathrm{K}_\beta}{1+\omega^2\tau_\beta^2} - \frac{\Delta\mathrm{K}_\gamma}{1+\omega^2\tau_\gamma^2} - \frac{\Delta\mathrm{K}_\delta}{1+\omega^2\tau_\delta^2} - \frac{\Delta\mathrm{K}_\varepsilon}{1+\omega^2\tau_\varepsilon^2} \tag{6}$$

where each τ is the Maxwell-Wagner charge relaxation time, and each ΔK is the induced change in Re{$\underline{K}$}, of a given interfacial polarization [72]. Equation (6) does not include non-interfacial polarizations, such as the 10-20 GHz dispersion exhibited by DI H_2O [43].

The DEP spectrum for both viable and nonviable *S. cerevisiae*, as used in this work, is presented in Figure 6. Figure 6 was created using MATLAB® to evaluate the real part of Equation (4). Notably, several of the τ values in Equation (6) are similar, resulting in net DEP spectra consisting of two effective dispersions. Figure 6 shows two distinct bands in which the DEP spectra are opposing: (1) the 0-18 kHz viable nDEP and nonviable pDEP band; and (2) the 2.5-282.5 MHz viable pDEP and nonviable nDEP band. A DEP potential ϕ_{DEP} in either band is well suited for distinguishing between viable and nonviable *S. cerevisiae* cells in a 33.4 μS/cm medium.

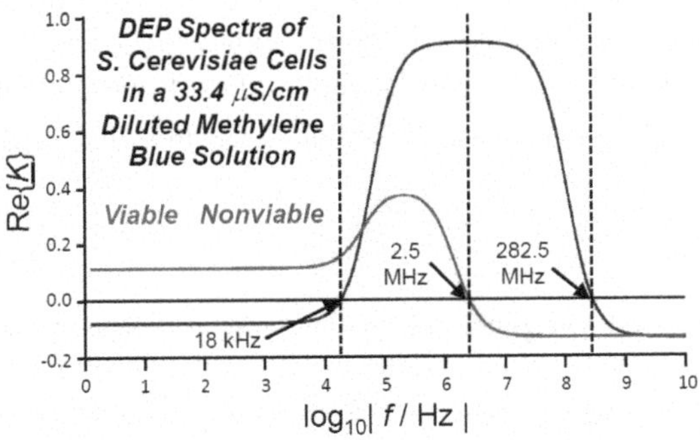

Fig. 6. DEP spectra of viable and nonviable *S. cerevisiae* as used in this work

5 Theory II: Simulated *S. cerevisiae* Trajectories

A two-dimensional finite element model of the detection zone was created in COMSOL® Multiphysics. The cross-channel was modelled as a 40 μm deep and a 1 mm long rectangular subdomain, filled with a $\varepsilon'_{med} = 78\varepsilon_0$ and $\sigma_{med} = 33.4$ μS/cm medium, modelling the diluted methylene blue fluid. The MEA was placed at the subdomain's bottom, centered along its length. Each microelectrode was infinitesimally thin, 25 μm wide, and given a 25 μm edge-to-edge separation. The innermost microelectrode was grounded, and a 1 V_p 1.5 GHz ϕ_{RF} was applied to the outer microelectrodes. Neumann conditions were assumed along the subdomain's boundary. The Electric Currents module solved Laplace's Equation, $\nabla \bullet (\underline{\varepsilon} \nabla \phi) = 0$, and computed the non-uniform electric field $\boldsymbol{E_e}$. Figure 7 presents plots of E_e^2 along various trajectories parallel with the cross-channel's bottom.

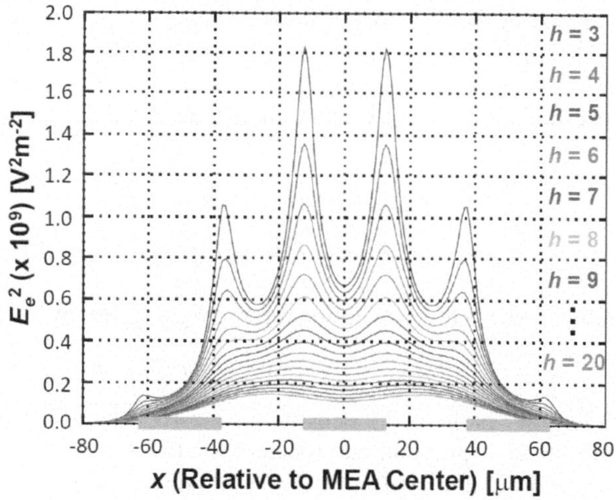

Fig. 7. E_e^2 along trajectories parallel with the cross-channel's bottom. The constant elevation h above the MEA was varied from 3 μm (TOP) to 20 μm (BOTTOM).

Notably, the actuating $\boldsymbol{E_e}$ oscillates at 10-1000 kHz in practice, whereas the simulated $\boldsymbol{E_e}$ oscillates at 1.5 GHz. As the MEA is much smaller than the simulated 1.5 GHz wavelength, Figure 7 is a good approximation of the E_e^2 produced by ϕ_{DEP} in practice (to within a constant scaling factor reflecting the magnitude of ϕ_{DEP}).

Figure 7 shows that: (1) at a given lateral position, E_e^2 decreases as elevation h increases; (2) near the MEA, local maxima in E_e^2 correspond to microelectrode edges; (3) the maxima at the outermost edges are much smaller than the other maxima; (4) near the MEA, local minima in E_e^2 correspond to the microelectrode and inter-microelectrode gap centers; (5) as h increases, the maxima bordering each gap collapse into maxima at each gap's center; (6) as h increases, the maxima at the outermost microelectrode edges vanish; and (7) the location of the minima at the gap centers are independent of h. A nDEP $\boldsymbol{F_{DEP}}$ repels cells from the E_e^2 maxima and attracts cells towards the E_e^2 minima [72]. Conversely, a pDEP $\boldsymbol{F}_{DEP}$ attracts cells toward the E_e^2 maxima and repels cells from the E_e^2 minima [72].

The Particle Tracing post-processing module simulates the trajectory of a *S. cerevisiae* cell passing through the detection zone. In this simulation, the cell is treated as a point particle (corresponding to the cell's center) subjected to user-specified forces.

The cell is initially placed 80 μm to the left of the MEA center and given an initial velocity $\boldsymbol{v}_{cell}$ approximated by the fluid velocity $\boldsymbol{v}_{med}$. As is typical within microfluidics, a Poiseuille $\boldsymbol{v}_{med}$ profile is assumed:

$$\boldsymbol{v}_{med} \approx 6 \langle \boldsymbol{v}_{med} \rangle (h_{cell} / D) [1 - (h_{cell} / D)] \tag{7}$$

where $\langle \boldsymbol{v}_{med} \rangle$ is the mean $\boldsymbol{v}_{med}$, h_{cell} is the cell's elevation, and D is the channel's 40 μm depth [61,73]. $\langle \boldsymbol{v}_{med} \rangle$ and the initial elevation h_{cell0} are matched to experimental observations.

The first specified force is $\boldsymbol{F}_{DEP}$, as given in Equation (3). To use Equation (3), the simulated E_e^2 produced by the 1 V_p MEA must be scaled to match the RMS E_e^2 produced by ϕ_{DEP}. The full DEP spectrum can be simulated by varying the specified $\mathrm{Re}\{\underline{K}\}$.

The second specified force is the viscous fluid drag force $\boldsymbol{F}_{drag}$, given by Stoke's law as

$$\boldsymbol{F}_{drag} \approx -6 \, \pi \, \eta_{med} \, a \, (\boldsymbol{v}_{cell} - \boldsymbol{v}_{med}) \tag{8}$$

where η_{med} = 1 mPa s is the fluid's viscosity and a = 3 μm is the cell's effective radius [61].

The final specified force is the gravitational force $\boldsymbol{F}_{grav}$, given by

$$\boldsymbol{F}_{grav} \approx (4 \, \pi / 3) \, a^3 \, (\rho_{cell} - \rho_{med}) \, \boldsymbol{g} \tag{9}$$

where ρ_{cell} = 1050 kg m^{-3} is the cell's density, ρ_{med} = 1000 kg m^{-3} is the fluid's density, and $\boldsymbol{g} = -9.81\boldsymbol{a}_z$ m^2 s^{-2} is gravitational acceleration [61].

At each time-step, the cell's position P_{cell} and velocity $\boldsymbol{v}_{cell}$ are re-calculated using Equations (3,7-9). The short effective time response of $\boldsymbol{F}_{drag}$ necessitates that time steps no larger than 5 μs be used during the simulation, to ensure stability. The simulation runs until the cell contacts a subdomain boundary.

6 Theory III: MEA Capacitance Perturbation Signatures

Assume that the cross-channel is filled with a real absolute permittivity ε'_{med} and conductivity σ_{med}. Also assume that a time-harmonic potential ϕ_{RF} is applied to the MEA, generating the following non-uniform electric field $\boldsymbol{E}_e$ within the detection zone

$$\boldsymbol{E}_e = \mathrm{Re}\{ \underline{\boldsymbol{E}}_e \exp(j \, \omega \, t) \} \tag{10}$$

where $\underline{\boldsymbol{E}}_e$ is the external electric field phasor, ω is angular frequency, and j is the unit imaginary number.

The electrical energy W_{e0} stored in detection zone follows as

$$W_{e0} = (1/2)\, C_{MEA0}\, \phi_{RF}^{2} \tag{11}$$

where C_{MEA0} is the total MEA capacitance.

A dielectric sphere, with radius a, real absolute permittivity ε'_{cell}, and conductivity σ_{cell}, is inserted into the detection zone. The electrical energy of this region is now W_{ef}, given by

$$\begin{aligned} W_{ef} &= (1/2)\, C_{MEAf}\, \phi_{RF}^{2} \\ &= (1/2)\, C_{MEA0}\, \phi_{RF}^{2} + (1/2)\,\mathrm{Re}\{ \textstyle\int_{Vcell} \underline{\boldsymbol{E}}_{e} \bullet \underline{\boldsymbol{P}}_{cell}\, dV \} \end{aligned} \tag{12}$$

where C_{MEAf} is the new capacitance, $\underline{\boldsymbol{P}}_{cell}$ is the sphere's polarization, and V_{cell} is the sphere's volume.

Equation (12) can be re-arranged so that the change in MEA capacitance ΔC_{MEA} induced by the sphere is proportional to the integral over V_{cell}:

$$\begin{aligned} W_{ef} - W_{e0} &= (1/2)\,(C_{MEAf} - C_{MEA0})\,\phi_{RF}^{2} = (1/2)\,\mathrm{Re}\{ \textstyle\int_{Vcell} \underline{\boldsymbol{E}}_{e} \bullet \underline{\boldsymbol{P}}_{cell}\, dV \} \\ \therefore\ \Delta C_{MEA} &= C_{MEAf} - C_{MEA0} = \phi_{RF}^{-2}\,\mathrm{Re}\{ \textstyle\int_{Vcell} \underline{\boldsymbol{E}}_{e} \bullet \underline{\boldsymbol{P}}_{cell}\, dV \} \end{aligned} \tag{13}$$

The sphere's polarization $\boldsymbol{P}_{cell}$ can be written as

$$\underline{\boldsymbol{P}}_{cell} = (\underline{\varepsilon}_{cell} - \underline{\varepsilon}_{med})\, \underline{\boldsymbol{E}}_{i} \tag{14}$$

where $\underline{\boldsymbol{E}}_{i}$ is the electric field phasor induced within the sphere, $\underline{\varepsilon}_{med}$ is the medium's complex permittivity, and $\underline{\varepsilon}_{cell}$ is the sphere's complex permittivity. If we assume that a is sufficiently small for the non-uniformity of $\underline{\boldsymbol{E}}_{e}$ to be insignificant over V_{cell}, then $\underline{\boldsymbol{E}}_{i}$ follows as

$$\underline{\boldsymbol{E}}_{i} \approx [\, 3\, \underline{\varepsilon}_{med} / (\underline{\varepsilon}_{cell} + 2\underline{\varepsilon}_{med})\,]\, \underline{\boldsymbol{E}}_{e} \tag{15}$$

ΔC_{MEA} then follows by substituting Equations (14-15) into Equation (13):

$$\begin{aligned} \Delta C_{MEA} &\approx \phi_{RF}^{-2}\,\mathrm{Re}\{ 3\, \underline{\varepsilon}_{med}\, [\, (\underline{\varepsilon}_{cell} - \underline{\varepsilon}_{med}) / (\underline{\varepsilon}_{cell} + 2\underline{\varepsilon}_{med})\,]\, \underline{E}_{e}^{2} \}\, (4\pi a^{3}/3) \\ \therefore\ \Delta C_{MEA} &\approx 4\pi a^{3}\,\mathrm{Re}\{ \underline{\varepsilon}_{med}\, \underline{K}\, \underline{E}_{e}^{2} \}\, \phi_{RF}^{-2} \end{aligned} \tag{16}$$

where $\underline{K}$ is the complex Clausius-Mossotti factor as given in Equation (4).

ϕ_{RF} shall oscillate at 1.478 GHz, where the real absolute permittivities dominate the conduction losses, $\varepsilon' \gg \sigma / \omega$. As such, Equation (17) can be approximated as

$$\Delta C_{MEA} \approx 4\pi a^{3}\, \varepsilon'_{med}\, K_{\infty}\, E_{e}^{2}\, \phi_{RF}^{-2} \tag{17}$$

where K_{∞} is the $\omega \rightarrow \infty$ limit of $\underline{K}$ as given in Equation (5b).

Equation (17) demonstrates that a spherical cell produces a ΔC_{MEA} signature proportional to E_{e}^{2} along its trajectory. Assuming that $\varepsilon'_{med} = 78\varepsilon_{0}$, $a = 3\ \mu\text{m}$, and $K_{\infty} = -0.135$: Equation (17) predicts a $-3.16 \times 10^{-26}\ \text{F m}^{2}\ \text{V}^{-2}$ scaling factor for converting the E_{e}^{2} profiles of Figure 7 to the -4.7 to -56.9 aF ΔC_{MEA} signatures induced by *S. cerevisiae* cells.

7 Capacitive Cytometer Calibration

Impedance based cytometers commonly use PSS as model bioparticles [27-28]; capitalizing on their well defined spherical shape, diameter, and homogeneous electrical parameters. Moreover, PSS can be utilized as a coated base for different bioassays (such as biotin-streptavidin and other protein-ligand pair complexes [74-75]).

In this work, the capacitance sensor is calibrated using experimental S signatures generated by unactuated 5.68 ± 0.305 μm diameter PSS (from Polysciences, Inc.) suspended in DI H_2O. These S signatures are compared to theoretical PSS ΔC_{MEA} signatures, thereby estimating the sensor's overall sensitivity dS/dC_{MEA} and resolution $|\delta C_{MEA}|$.

The PSS are modelled as homogeneous spheres with radius a = 2.84 μm, real absolute permittivity $\varepsilon'_{cell} = 2.5\varepsilon_0$, and conductivity σ_{cell} = 2 μS/cm [27]. Assuming that $\varepsilon'_{med} = 78\varepsilon_0$, a = 2.84 μm, and K_∞ = -0.476: Equation (17) predicts a -9.47 x 10^{-26} F m^2 V^{-2} scaling factor for converting the E_e^2 profiles of Figure 7 to the -9.5 to -170.5 aF PSS ΔC_{MEA} signatures.

By matching an experimental S signature's shape to the E_e^2 signatures of Figure 7, h_{cell} is estimated within ±1 μm. The S and ΔC_{MEA} signatures are then compared at the outermost inter-microelectrode gaps, being the features most insensitive to variations in h_{cell}.

Figure 8 presents an experimental S profile (measured using a 3 ms LIA time constant with a 12 dB/octave rolloff) for which h_{cell} was estimated as 6 μm via comparison to Figure 7. The h_{cell} = 6 μm ΔC_{MEA} signature is -45.5 aF at the outermost inter-microelectrode gap, whilst the S signature is -2.64 V at this point. It thus follows that $dS/dC_{MEA} \approx$ 58 mV/aF. The RMS noise V_{RMS} was measured within the 0-1 s window of Figure 8 as 51 mV. The sensor's resolution $|\delta C_{MEA}|$ thus follows as 880 zF. This resolution is acceptable when detecting the -4.7 to -56.9 aF ΔC_{MEA} induced by *S. cerevisiae* cells.

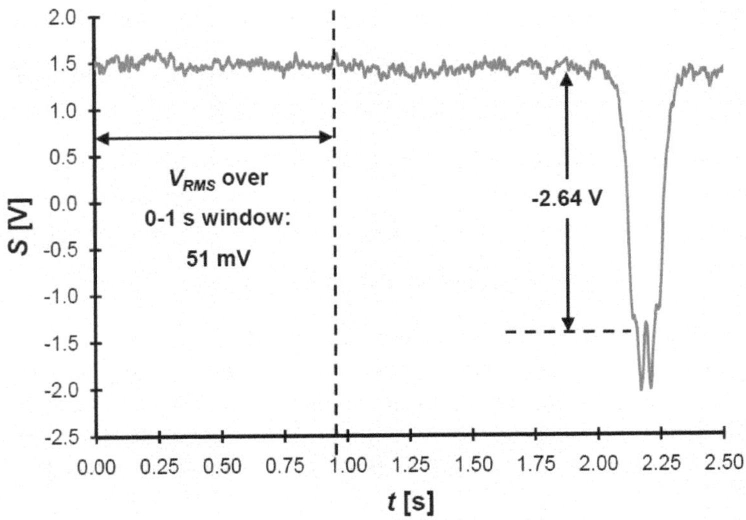

Fig. 8. Experimental PSS S signature used for capacitive cytometry calibration

8 Results I: Nonviable *S. cerevisiae* Optical Assay

In this work, *S. cerevisiae* cells are suspended in a diluted methylene blue solution, to stain nonviable cells with compromised cytoplasmic membranes. To prepare the solution: (1) 10 mg of methylene blue trihydrate was stirred into 10 mL of DI H_2O; (2) the solution was filtered and further diluted with DI H_2O, yielding a total 100 mL volume; (3) 5 mg of sodium citrate dihydrate was mixed into this solution; and (4) this solution was further diluted with 300 mL of DI H_2O. The final solution's conductivity σ_{med} is 33.4 μS/cm (as measured with a Thermo Scientific Orion 3 Star meter).

Dehydrated *S. cerevisiae* cells are stored in sorbitan monostearate granules (Fleischmann's® Traditional). A single 2 mg granule was mixed into 15 mL of the diluted methylene blue solution. The sample was heated by a 70 °C DI H_2O bath (measured with a Fluke 116 multimeter). Once the sample reached 60 °C, it was heated for an additional 4 minutes. This heat shock treatment compromises the cytoplasmic membranes of all cells within the sample, yielding a homogeneous nonviable *S. cerevisiae* population.

The heated shocked sample was injected into the microflow cytometer, which was simultaneously operated as both an optical cell assay and a capacitive cytometer. During this data acquisition session, ϕ_{DEP} was set to 0 V_p, 1 V_p 10 kHz, 0.5 V_p 100 kHz, and 0.5 V_p 1 MHz. ϕ_{DEP} was halved at 100 kHz and 1 MHz to prevent excess cellular-MEA adhesion. Approximately 20 minutes worth of data was collected at each ϕ_{DEP} setting.

Five single-cell crossings were randomly selected from each ϕ_{DEP} dataset for post-processing optical assay video analysis using Tracker 2.60 [76] to yield $x_{cell}(t)$ profiles. These x_{cell} profiles were exported into MATLAB® to: (1) compute v_{cellx} profiles; (2) fit each v_{cellx} profile to a uniformly spaced x_{cell} line, spanning ±75 μm in 0.5 μm steps (with x_{cell} = 0 μm corresponding to the MEA center); (3) estimate the initial cellular velocity v_{cellx0} for each crossing as the mean v_{cellx} within the x_{cell} = -75 to -37.5 μm domain; (4) estimate the final cellular velocity v_{cellxf} for each crossing as the mean v_{cellx} within the x_{cell} = 37.5 to 75 μm domain; and (5) compute the percent change in v_{cellx} for each crossing as

$$\%\Delta v_{cellx} \approx 100\ (\ v_{cellxf} - v_{cellx0}\ )\ /\ v_{cellx0} \tag{18}$$

The $\%\Delta v_{cellx}$ analysis of the single-cell crossings are summarized in Table 2:

Table 2. $\%\Delta v_{cellx}$ analysis of captured single-cell crossing video

ϕ_{DEP}	*0.0 V_p*	*1.0 V_p 10 kHz*	*0.5 V_p 100 kHz*	*0.5 V_p 1 MHz*
Experimental $\%\Delta v_{cellx}^{(1)}$	-0.1	+7.5	-42.0	-49.1
Experimental $\%\Delta v_{cellx}^{(2)}$	+27.0	-29.8	-31.9	-38.7
Experimental $\%\Delta v_{cellx}^{(3)}$	+15.6	-46.8	-5.1	-19.7
Experimental $\%\Delta v_{cellx}^{(4)}$	+1.4	-21.2	-55.0	-29.6
Experimental $\%\Delta v_{cellx}^{(5)}$	-10.5	-13.9	-10.4	-39.3
Mean of Exp. $\%\Delta v_{cellx}^{(1\text{-}5)}$	+6.7	-20.9	-28.9	-35.3
Std. Dev. of Exp. $\%\Delta v_{cellx}^{(1\text{-}5)}$	±14.6	±20.0	±21.0	±11.1

A typical v_{cellx} profile was estimated for each ϕ_{DEP} dataset, by averaging the five fitted v_{cellx} values at each point along x_{cell}. The Particle Tracing simulation of Section 5 was used to reproduce each typical v_{cellx} profile, as shown in Figure 9. In these simulations, v_{cellx0} at x_{cell} = -75 μm was matched to the corresponding typical v_{cellx0} value, Re{$\underline{K}$} was matched to the nonviable *S. cerevisiae* DEP spectrum of Figure 6, h_{cell0} was initially assumed to be 8 μm, and $<v_{cellx}>$ was computed using Equation (7). h_{cell0} and $<v_{cellx}>$ were then adjusted until the simulated v_{cellx} profile approximated the corresponding typical v_{cellx} profile.

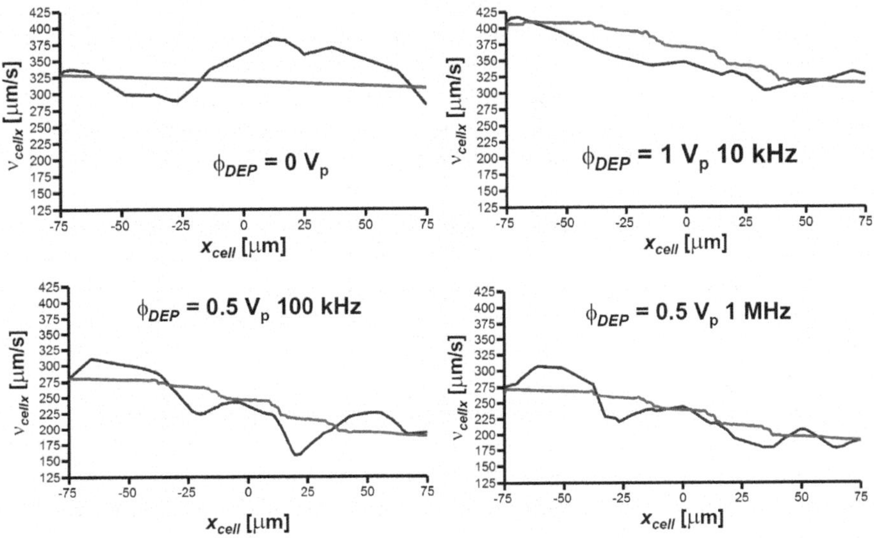

Fig. 9. Simulated and typical experimentally observed v_{cellx} profiles

The $\%\Delta v_{cellx}$ analysis for each typical and simulated crossing are summarized in Table 3:

Table 3. $\%\Delta v_{cellx}$ analysis of typical experimental and simulated single-cell crossings

ϕ_{DEP}	*0.0 V_p*	*1.0 V_p 10 kHz*	*0.5 V_p 100 kHz*	*0.5 V_p 1 MHz*
Computed Re{$\underline{K}$}	0.00	0.13	0.35	0.22
***Estimated v_{cell0}* [μm/s]**	330	410	280	272
***Estimated h_{cell0}* [μm]**	6.0	8.5	8.0	7.0
$\%\Delta v_{cellx}$ of Simulated Crossing	-4.9	-22.2	-31.1	-27.7
$\%\Delta v_{cellx}$ of Typical Exp. Crossing	+6.3	-19.6	-29.4	-35.1

9 Results II: Nonviable *S. cerevisiae* Capacitive Cytometry

Figure 10 presents ΔC_{MEA} signatures corresponding to the experimentally observed and simulated single-cell crossings featured in Section 8. For the sake of compactness, only three experimental ΔC_{MEA} signatures are presented for each ϕ_{DEP} dataset. In each case, the three signatures which best characterized the dataset's diversity are presented.

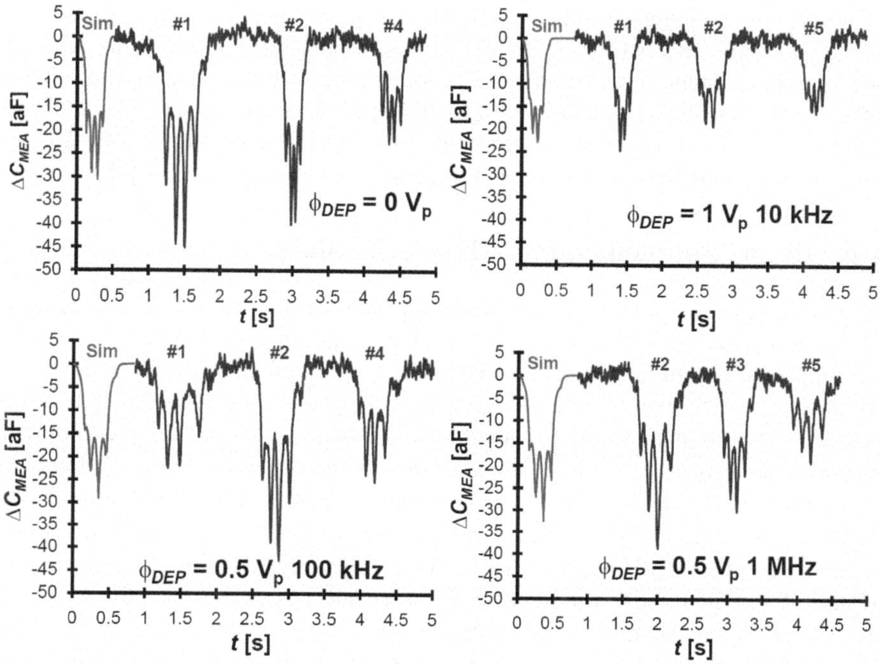

Fig. 10. Simulated and typical experimentally observed ΔC_{MEA} signatures

$\%\Delta v_{cellx}$ are to be estimated from the ΔC_{MEA} signatures alone. As the $x_{cell} = \pm 75$ μm positions within the ΔC_{MEA} signatures are difficult to define, v_{cellx0} and v_{cellxf} must be defined using alternative intervals. v_{cellx0} shall be estimated using transit time from the first microelectrode's innermost edge extrema to the central microelectrode's first edge extrema. v_{cellxf} shall be estimated using the transit time from the central microelectrode's last edge extrema to the last microelectrode's innermost edge extrema. $\%\Delta v_{cellx}$ is then computed via Equation (18). The results of this $\%\Delta v_{cellx}$ analysis are summarized in Table 4. Table 4 supports the DEP behaviour inferred from Table 2, with deviations between the two tables expected as a consequence of defining v_{cellx0} and v_{cellxf} using different intervals.

Table 4. $\%\Delta v_{cellx}$ analysis of single-cell capacitance signatures

ϕ_{DEP}	*0.0 V_p*	*1.0 V_p 10 kHz*	*0.5 V_p 100 kHz*	*0.5 V_p 1 MHz*
Experimental $\%\Delta v_{cellx}^{(1)}$	+1.5	0.0	-49.8	-9.8
Experimental $\%\Delta v_{cellx}^{(2)}$	+14.3	-32.3	-17.9	-30.9
Experimental $\%\Delta v_{cellx}^{(3)}$	+14.3	+27.9	-23.0	-15.0
Experimental $\%\Delta v_{cellx}^{(4)}$	+11.1	-38.3	-27.1	-27.1
Experimental $\%\Delta v_{cellx}^{(5)}$	+14.3	-5.7	-37.8	-18.6
Mean of Exp. $\%\Delta v_{cellx}^{(1\text{-}5)}$	+11.1	-9.7	-31.1	-20.3
Std. Dev. of Exp. $\%\Delta v_{cellx}^{(1\text{-}5)}$	±5.5	±26.7	±12.8	±8.7
Simulated $\%\Delta v_{cellx}$	-2.8	-15.6	-16.7	-15.3

Once the methodology of inferring the DEP response of detected bioparticles from their ΔC_{MEA} signatures is refined, the optical assay subsystem would no longer be required for independent confirmation. The remaining instrumentation could then be miniaturized and integrated on-chip with the microflow cytometer. As such, this work serves as the developmental prototype of a capacitive cytometer with DEP actuation which could be miniaturized and integrated on-chip as a LoC or μTAS.

Appendix A: Supplementary CHO Cell Analysis

If this work is to serve as the developmental prototype of a microflow cytometer for biomedical applications, it is necessary to demonstrate the applicability of our device in the analysis of mammalian cells. CHO cells are mammalian cells commonly used for transfection and expression, and have been adopted as an industrial standard for large-scale recombinant protein production [77]. In this section, we present ΔC_{MEA} signatures induced by CHO cells transfected with the gene for human β-Interferon (IFN-β; provided by the University of Manitoba's Department of Microbiology).

The effective ε'_{cell} of CHO cells ranges from $60.2\varepsilon_0$ to $62.9\varepsilon_0$, comparable to that of *S. cerevisiae* cells [78]. However, CHO cells are typically 15-25 μm in diameter, significantly larger than the 6 μm diameter *S. cerevisiae* cells. Consequently, CHO cells induce larger ΔC_{MEA} signatures than *S. cerevisiae* cells.

To maintain population viability, CHO cells must be suspended in a solution such as phosphate buffered saline (PBS; for which σ_{med} = 10^4 μS/cm). Within such a highly conductive solution, the DEP spectrum exhibited by CHO cells will be entirely nDEP. Figure 11 presents ΔC_{MEA} signatures induced by CHO cells as the magnitude of a 20 kHz ϕ_{DEP} is increased from 0-2 V_p.

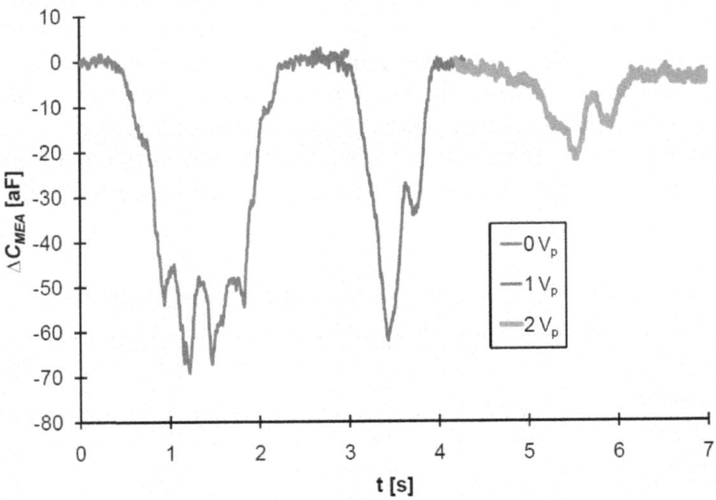

Fig. 11. Typical ΔC_{MEA} signatures induced by CHO cells flowing in PBS (0.4% with volume). A 20 kHz ϕ_{DEP} produces a nDEP force repelling CHO cells away from the MEA, yielding progressively smaller ΔC_{MEA} signatures as $|\phi_{DEP}|$ increases from 0-2 V_p.

When $|\phi_{DEP}| = 0$ V$_p$, CHO cells pass over the MEA at a constant h_{cell}, inducing large symmetric ΔC_{MEA} signatures. When $|\phi_{DEP}| = 1$ V$_p$, $\boldsymbol{F}_{DEP}$ repels CHO cells away from the MEA, yielding asymmetric and decaying ΔC_{MEA} signatures as h_{cell} increases. When $|\phi_{DEP}| = 2$ V$_p$, $\boldsymbol{F}_{DEP}$ is so strong that CHO cells are rapidly pushed to a high h_{cell} by the outermost microelectrode edge extrema (as confirmed by the captured video). Consequently, the overall ΔC_{MEA} signature and change in h_{cell} between the two most significant extrema are less than that observed in the 1 V$_p$ ϕ_{DEP} case.

Appendix B: Supplementary PSS Analysis

The microflow cytometer was simultaneously operated as an optical assay and a capacitive cytometer as unactuated PSS suspended in DI H_2O flowed past the MEA. By matching an experimental S signature to the E_e^2 signatures of Figure 7, the PSS elevation h_{cell} was estimated within ±1 μm. By assuming a Poiseuille fluid flow [73] with a mean velocity $\langle \mathbf{v}_{med} \rangle$ of $870\boldsymbol{a}_x$ μm/s (matching experimental observations), $\mathbf{v}_{med}$ was approximated by substituting h_{cell} into Equation (7). The PSS velocity $\mathbf{v}_{cell}$ was then assumed to approximate $\mathbf{v}_{med}$ at h_{cell}. The full-width-half-maximum (FWHM) of each S signature was then utilized to estimate the transit time Δt_{MEA} required for the said PSS to pass over the MEA. A dataset comprised of 97 single PSS S signatures (confirmed to be associated with individual PSS crossings via the captured video) was then used to produce a predictive model relating the peak value of a given S signature, S_p, to its associated Δt_{MEA}. Figure 12 presents this predictive model along with some of the experimental single PSS S_p versus Δt_{MEA} data used in its construction [71]. The central extrema of this model corresponds to the smallest Δt_{MEA}, which in turn corresponds to the peak of the Poiseuille $\mathbf{v}_{med}$ along the center of the cross-channel. As S_p decreases as h_{cell} is increased: the upper branch of the predictive model corresponds to the lower half of the cross-channel and vice versa. In addition to the individual PSS, several conjoined PSS doublets were observed. As doublets have twice the volume of a single PSS, Equation (17) states that the S_p of a doublet's signature will be twice that of a single PSS signature with the same h_{cell}. As such, multiplets can be identified on the basis of S_p alone. Doublets were observed to conform to a similar predictive model, as shown in Figure 12.

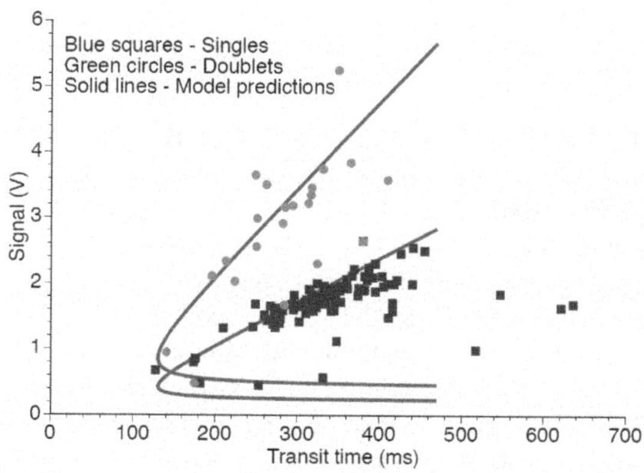

Fig. 12. Experimental S_p versus Δt_{MEA} data points, and the predictive model fitted to them

Under the influence of F_{grav}, the PSS settles during the time required to achieve stationary flow. Hydrodynamic lift forces [61] balance F_{grav}, preventing the PSS from settling to the bottom of the cross-channel. These observations are evidenced by the cluster of data around the top of the upper branch. The almost linear dependence of S_p with Δt_{MEA} in this region affords simultaneously determining $\mathbf{v}_{cell}$ and h_{cell} with reasonable accuracy.

Figure 13 presents histograms of S_p and $\mathbf{v}_{cell}$ [32], which confirm that: (1) over 85% of the PSS passed over the MEA with $h_{cell} \leq 6$ μm or lower, and (2) 72% of the PSS had $400 \leq \mathbf{v}_{cell} \leq 600$ μm/s. The $\mathbf{v}_{cell}$ histogram agrees with the Poiseuille $\mathbf{v}_{med}$, confirming that $\mathbf{v}_{cell}$ is accurately approximated by $\mathbf{v}_{med}$.

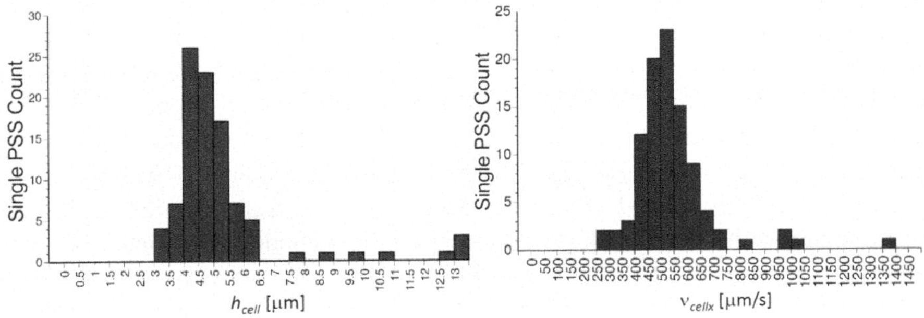

Fig. 13. Single-PSS crossing h_{cell} histogram (LEFT) and v_{cellx} histogram (RIGHT)

This framework is attractive as a prototype of a bioassay in which doublets form only after coated PSS attach themselves to a specific, smaller bioparticle via a biochemical reaction (similar to forming a bond within a biotin-streptavidin complex [74-75]). The presence of sub-microscale bioparticles, such as viruses, could then be detected via the large doublet signatures within a capacitive cytometer's sense signal. This cytometer could then be miniaturized and integrated on-chip as a LoC or μTAS, yielding a device would afford various bead-based bioassays for the on-site detection of sub-microscale bioparticles in very small concentrations.

References

1. Brehm-Stecher, B.F., Johnson, E.A.: Microbio. & Mol. Bio. Rev. 68(3), 538–559 (2004)
2. Witek, M.A., Hupert, M.L., et al.: BioMEMS. In: Wang, W., Soper, S.A. (eds.), pp. 391–441. CRC, Boca Raton (2007)
3. Sims, C.E., Allbritton, N.L.: Lab on a Chip 7(4), 423–440 (2007)
4. Davey, H.M., Kell, D.B.: Microbio. Rev. 60(4), 641–696 (1996)
5. Shapiro, H.M.: J. Microbio. Methods 42(1), 3–16 (2000)
6. Rockabrand, D., Austin, T., et al.: Appl. Environ. Microbio. 65(9), 4181–4188 (1999)
7. Strauß, A., Michel, S., et al.: J. Bacterio. 183(12), 3761–3769 (2001)
8. Bechinger, C., Giebel, K.F., et al.: Science 285(5435), 1896–1899 (1999)
9. Smith, A.E., Zhang, Z., et al.: Proc. Natl. Acad. Sci. 97(18), 9871–9874 (2000)
10. Losick, R., Shapiro, L.: J. Bacterio. 181(14), 4143–4145 (1999)

11. Potma, E.O., de Boeij, W.P., et al.: Proc. Natl. Acad. Sci. 98(4), 1577–1582 (2001)
12. Cai, C., Liu, B., et al.: Anal. Chem. 74(1), 114–119 (2002)
13. Raskin, D.M., de Boer, P.A.J.: Proc. Natl. Acad. Sci. 96(9), 4971–4976 (1999)
14. Katsuragi, T., Tani, Y.: J. Biosci. Bioeng. 89(3), 217–222 (2000)
15. Flanagan, L.A., Lu, J., et al.: Stem Cells 26(3), 656–665 (2008)
16. Han, A., Yang, L., Frazier, A.B.: Clin. Cancer Res. 13(1), 139–143 (2007)
17. Racila, E., Euhus, D., et al.: Proc. Natl. Acad. Sci. 95(8), 4589–4594 (1998)
18. Baker Megan, K., Mikhitarian, K., et al.: Clin. Cancer Res. 9(13), 4865–4871 (2003)
19. Harriette, J.K., Presta, A., et al.: Breast Cancer Res. & Treat. 86(3), 237–247 (2004)
20. Cristofanilli, M., Budd, G.T., et al.: N. Engl. J. Med. 351(8), 781–791 (2004)
21. Shapiro, H.M.: Practical Flow Cytometry. Wiley, Chichester (1995)
22. Yu, C., Shi, L.: BioMEMS. In: Wang, W., Soper, S.A. (eds.), pp. 237–262. CRC, Boca Raton (2007)
23. Huh, D., Gu, W., et al.: Physio. Meas. 26(3), R73–R98(2005)
24. Yi, C., Li, C.W., et al.: Anal. Chim. Acta 560(1-2), 1–23 (2006)
25. Coulter, W.H.: Proc. Natl. Electronics Conf. 12, 1034 (1956)
26. Koch, M., Evans, A.G.R., et al.: J. Micromech. Microeng. 9(2), 159–161 (1999)
27. Cheung, K., Gawad, S., et al.: Proc. IEEE MEMS 17, 343–346 (2004)
28. Wood, D.K., Requa, M.V., et al.: Rev. Sci. Inst. 78(10-104301), 1–6 (2007)
29. Ayliffe, E.H., Frazier, A.B., et al.: J. Microelectromech. Sys. 8(1) (1999)
30. Gawad, S., Schild, L., et al.: Lab on a Chip 1(76) (2001)
31. Sohn, L.L., Saleh, O.A., et al.: Proc. Natl. Acad. Sci. 97(20), 10687–10690 (2000)
32. Nikolic-Jaric, M., Romanuik, S.F., et al.: Biomicrofluidics 3(034103), 1–15 (2009)
33. Ferrier, G.A., Romanuik, S.F., et al.: Lab on a Chip 9(23), 3406–3412 (2009)
34. Ferrier, G.A., Romanuik, S.F., et al.: Proc. MicroTAS 1, 1973–1975 (2009)
35. Ferrier, G.A., Hladio, A.N., et al.: Biomicrofluidics 2(4-044102), 1–13 (2008)
36. Romanuik, S.F., Ferrier, G.A., et al.: Proc. IEEE Sensors 1, 634–637 (2008)
37. Ferrier, G.A., Romanuik, S.F., et al.: Proc. NSTI 2, 589–592 (2008)
38. Ferrier, G.A., Hladio, A.N., et al.: Proc. NSTI 2, 214–217 (2007)
39. `http://www.micronit.com` (2008)
40. Tran, T.D., Oliver, D.R., et al.: Rev. Sci. Inst. 72(6), 2618–2623 (2001)
41. Gascoyne, P.R.C., Vykoukal, J.V.: Proc. IEEE 92(1), 22–41 (2004)
42. Schwan, H.P.: Ann. Biomed. Eng. 20(3), 269–288 (1992)
43. Pethig, R.: Dielectric & Electronic Properties of Biological Materials. Wiley, Chichester (1979)
44. Foster, K.R., Schwan, H.P.: Crit. Rev. Biomed. Eng. 17(1), 25–104 (1989)
45. Becker, F.F., Wang, X.-B., et al.: J. Phys. D: Appl. Phys. 27(12), 2659–2662 (1994)
46. Braschler, T., Demierre, N., et al.: Lab on a Chip 8(2), 280–286 (2008)
47. Gasperis, G.D., Wang, X.B., et al.: Meas. Sci. Technol. 9(3), 518–529 (1998)
48. Goater, A.D., Pethig, R.: Parasitology 117, S117–S189 (1998)
49. Hughes, M.P.: Nanoelectromechanics in Engineering & Biology. CRC, Boca Raton (2003)
50. Goater, A.D., Burt, J.P.H., et al.: J. Phys. D Appl. Phys. 30(18), L65–L69 (1997)
51. Asami, K., Hanai, T., et al.: J. Membr. Bio. 28(2-3), 169–180 (1976)
52. Asami, K., Yonezawa, T.: J. Biophys. 71(4), 2192–2200 (1996)
53. Hölzel, R., Lamprecht, I.: Biochim. Biophys. Acta 1104(1), 195–200 (1992)
54. Hölzel, R.: J. Biophys. 73(2), 1103–1109 (1997)
55. Pohl, H.A.: Dielectrophoresis, Cambridge (1978)
56. Pethig, R., Markx, G.H.: Trends Biotechbol. 15(10), 426–432 (1997)
57. Pethig, R., Bressler, V., et al.: Electrophoresis 23(13), 2057–2063 (2002)

58. Prasad, S., Zhang, X., et al.: J. Neurosci. Methods 135(1-2), 79–88 (2004)
59. Heida, T., Rutten, W.L., et al.: Arch. Physio. Biochem. 110(5), 373–382 (2002)
60. Talary, M.S., Mills, K.I., et al.: Med. Bio. Eng. Comput. 33(3), 235–237 (1995)
61. Huang, Y., Wang, X.B., et al.: Biophys. J. 73(2), 1118–1129 (1997)
62. Gascoyne, P.R.C., Huang, Y., et al.: Meas. Sci. Technol. 3(5), 439–445 (1992)
63. Wang, X.B., Huang, Y., et al.: J. Phys. D Appl. Phys. 26(8), 1278–1285 (1993)
64. Markx, G.H., Huang, Y., et al.: Microbiology 140(3), 585–591 (1994)
65. Markx, G.H., Talary, M., et al.: J. Biotechnol. 32(1), 29–37 (1994)
66. Fuhr, G.: Ph.D. Dissertation. Humbolt University, Berlin (1985)
67. Wang, X.B., Huang, Y., et al.: J. Phys. D Appl. Phys. 26(2), 312–322 (1993)
68. Vitols, E., North, R.J., et al.: J. Biophys. Biochem. Cytol. 9(3), 689–699 (1961)
69. Moore, C.W., Valle, R.D., et al.: Antimicrob. & Chemother. 36(11), 2497–2505 (1992)
70. Mulholland, J., Preuss, D., et al.: J. Cell Bio. 125(2), 381–391 (1994)
71. Ganeva, V., Galutzov, B., et al.: Biochim. Biophys. Acta 1240(2), 229–236 (1995)
72. Jones, T.B.: Electromechanics of Particles, Cambridge (1995)
73. White, F.M.: Fluid Mechanics, 5th edn. McGraw-Hill, New York (2003)
74. Diamandis, E.P., Christopoulos, T.K.: Clin. Chem. 37(5), 625–636 (1991)
75. Pierres, A., Vitte, J., et al.: Biophys. Rev. Lett. 1(3), 231–257 (2006)
76. `http://www.cabrillo.edu/~dbrown/tracker/` (2009)
77. Tharmalingam, T., Sunley, K., et al.: Biotechnol. Prog. 24(4), 832–838 (2008)
78. Saito, H., Suzuki, Y., et al.: URSI GA, Session K01p2 (2008)

Use of Triaxial Accelerometers for Posture and Movement Analysis of Patients

Roman Malarić, Hrvoje Hegeduš, and Petar Mostarac

University of Zagreb, Faculty of Electrical Engineering and Computing

Abstract. Triaxial accelerometers can be used as a low cost solution to wide areas in patient care. This paper describes the measurement system that comprises two triaxial accelerometers together with ZigBee transceivers to measure posture and movement of patients wireless. The system, including calibration of accelerometers and measurement is explained in detail.

Keywords: accelerometers, ZigBee standard, posture and movement measurement.

1 Introduction

The micro machined inertial sensors (accelerometers and gyroscopes) are used for many different applications including navigation, impact detection, position, tilt, inclination, shock, vibration, motion detection, etc... They are also used for human movement tracking. The human body has a posture control that enables humans to move and maintain stability. The measurement of human posture and movement is valuable for many different medical applications and diagnostics [1]. Currently, there are many different techniques that are used to measure it. They can be classified as inside-in, inside-out and outside-in systems [2], depending on the position of the source and sensor. In inside-in systems, like accelerometers, the sensor and a source are located on the body of patient, which may be obtrusive if patients have to wear it all the time. Fall detection of older people is one such example. However, for detecting the range of voluntary motion of different joints in the body, this technique is acceptable. This paper will present an easy and simple method to measure the inclination of different parts of human body using accelerometer together with ZigBee transceiver to communicate with PC wireless. The measurement system is achieved with only two triaxial accelerometers. Other applications sometimes make use of gyroscopes, and also magnetometers [3,4].

2 Range of Human Movement

Human body can make a wide range of movement using its skeletal and muscular systems. This movement differs from person to person, and is usually also affected by the person's age. Joint actions are described in relation to the anatomical position. Movement is defined (Fig. 1.) by referring to the three planes (Sagittal Plane, Frontal Plane and Transverse Plane), and three axis (Frontal, Sagittal and Longitudinal) [5].

S.C. Mukhopadhyay and A. Lay-Ekuakille (Eds.): Adv. in Biomedical Sensing, LNEE 55, pp. 127–143.
springerlink.com

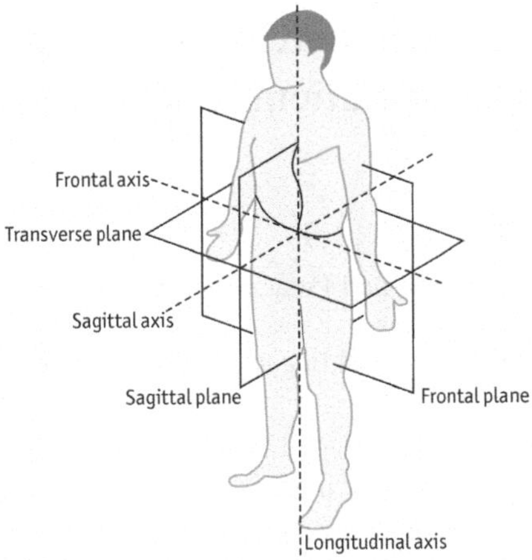

Fig. 1. The three planes and three axis of human movement

Amongst the many possible types of movement the human body can make, this article will address only the movement that can be reliably measured with accelerometers, such as flexion in which body is bending forward, extension, in which body is bending backwards and lateral flexion in which the body is bending to one side. Some movement of body such as rotation of head is difficult to measure with accelerometers, and for such measurements, a gyroscope is better solution. The movement of spine joint can be seen in Fig. 2. Approximate maximum flexion and extension movement of spine joint is shown on the left, figure in the middle shows the maximum lateral flexion movement of spine (bending to one side), and figure on the right shows the head movement.

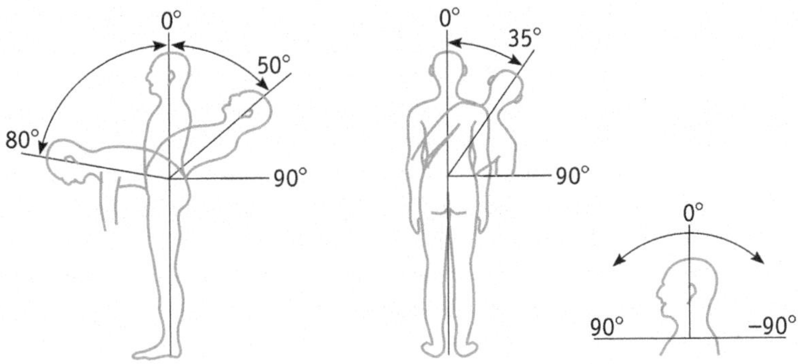

Fig. 2. Spine and head movement

The possible movement of lower extremities (legs and toes) can be seen on Fig. 3. The possible movement of lower extremities include flexion, extension, abduction (spreading toes apart), and adduction (bringing the toes together).

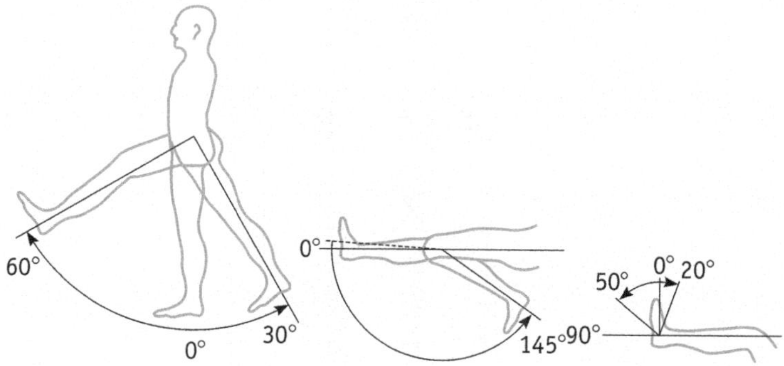

Fig. 3. Movement of lower extremities

The movement of hands and fingers can be seen in Fig. 4. These movements include flexion, extension, abduction and adduction. All movement of spine, legs, arms and hands presented on figures 2 , 3 and 4 can be measured with measurement method presented in this paper.

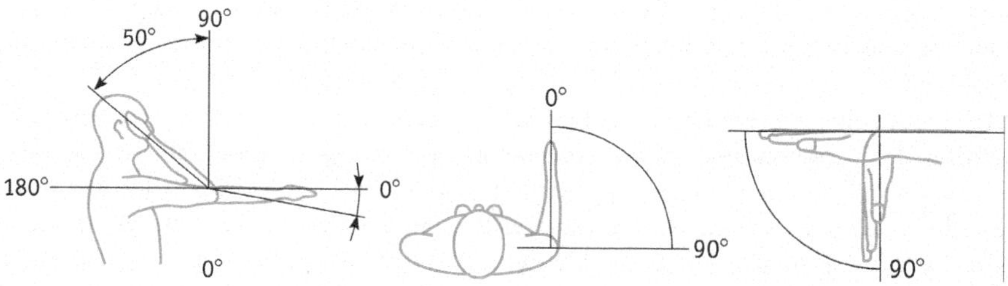

Fig. 4. Arm and hand movement

3 Accelerometers

Accelerometer is a device for measuring acceleration, but is also used to detect free fall and shock, movement, speed and vibration.

There are many different accelerometers defined by way of working, such as: capacitive, piezoelectric, hall-effect or heat transfer. Most common type of accelerometer, and also used in this application is the one detecting internal capacitor's change of capacitance. The change of capacitance is caused by reducing or expanding the plate distance. One plate is still on accelerometer case, and the other is mobile under the influence of acceleration. In one accelerometer there can be several capacitors serially connected or in bridge. Principle of changing capacitance is shown in Fig. 5. The most essential features of accelerometer are range, sensibility, number of axis and output signal's independence of supply voltage.

Range is defined with earth's gravity g (1g = 9.80665 m/s^2), and in today's accelerometers it can be from ±1.5g to more then ±100g. Sensitivity is defined as ratio of output voltage and gravity on sensor. With high sensitivity it is possible to detect very small acceleration fluctuation, small vibration and motion.

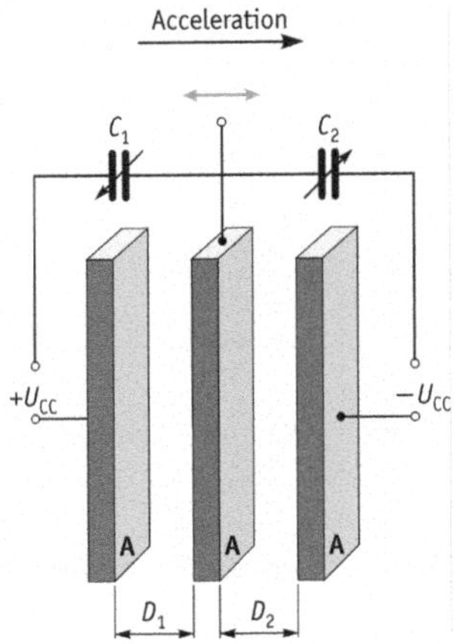

Fig. 5. Change of capacitance in accelerometer

Accelerometers used in this application are micro electromechanical sensors (MEMS). MEMS is the name used for technology which combine mechanical parts, usually sense, and electronics circuits necessary to drive mechanical parts and give output signals dependent of sense activity, all integrated in one device. MEMS technology is widely used, from robotics to agriculture. It is small in dimension, and usually low cost.

MEMS accelerometers can be found with 1 axis, 2 axis or 3-axis detection of acceleration. For applications of three-dimensional (3D) positioning it is necessary to use 3-axis accelerometer. In this application MMA7260Q MEMS accelerometer from Freescale™ is used as a measuring device. It is an accelerometer that can change the sensitivity programmatically to enable wide variety of applications. The sensitivity can be changed from ±1.5g to ±6 g in four steps. It is a low cost, low power consumption device, and also low profile device measuring 6 mm x 6 mm x 1.45 mm. These characteristics make it ideal for high accuracy applications to measure vibration, motion, position, and tilt of objects.

For example, this accelerometer is widely used in applications for protection of electro-mechanical parts and devices such as hard disk drives, mobile phones, or laptop computers. It is also used for human monitoring in health care institutions, for fall detection of old and infirm persons [6]. Free fall detection can be perceived by monitoring all three output signals from accelerometer axis X, Y and Z. If they all give values for zero g acceleration then accelerometer is experiencing free fall.

4 Measuring Tilt with Accelerometers

Measuring the angle with accelerometer is in fact measuring the tilt. Tilt can be measured when accelerometer sense no motion. Thus, unlike applications to measure shock or velocity that require detection of dynamic acceleration, application to detect tilt requires sensitive accelerometer to detect static acceleration. The angle between two axes can be calculated from stationary and final position of accelerometer (Fig. 6.). The sensors position determines the static acceleration in each axis, which can be between -1.0 g to +1.0 g when the angle is tilted from -90° to +90°.

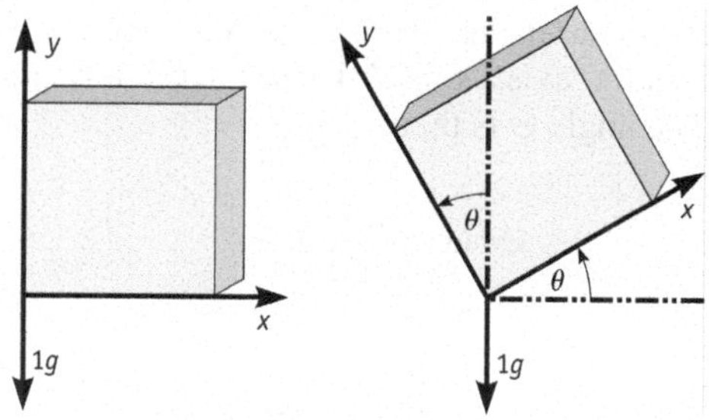

Fig. 6. Initial stationary position of accelerometer is on the left, and final position is on the right

Tilt of an accelerometer can be calculated using the trigonometric rules. Angle of tilt Θ can be calculated as:

$$\Theta = \sin^{-1} A_x \tag{1}$$

where A_x is the acceleration on the X axis due to gravity.

One of the major problems with accelerometer in measuring tilt is that sensitivity of accelerometer is dependent on the angle of tilt of accelerometer. The sensitivity of the accelerometer is defined as V/g. To obtain the most resolution per degree of change, the accelerometer should be mounted with the sensitive axis parallel to the plane of movement where the most sensitivity is desired [7]. On Fig. 7. the acceleration of typical one axis accelerometer in g's is shown as it tilts from -90° and +90°. The sensitivity for this axis is best for angles from -45° to +45°, and is reduced between -90° to -45° and also between +45° to +90°. Therefore, it is necessary to use another axis to expand the range of tilt that can be measured with required precision.

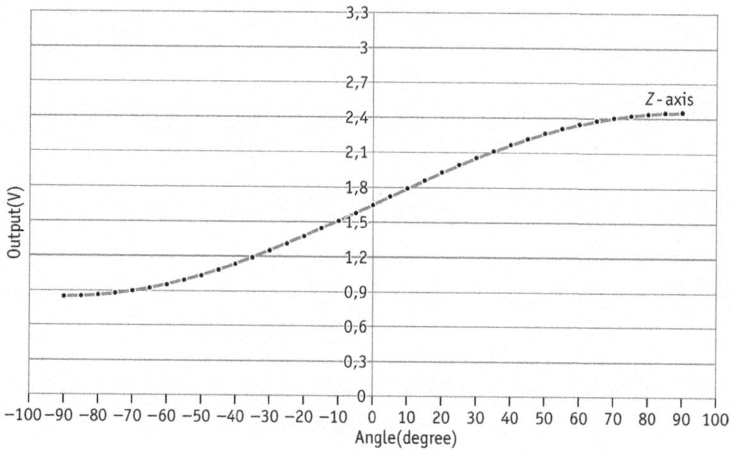

Fig. 7. Nonlinearity of accelerometer's output

The better results are made using two axis accelerometers. On Fig. 8. A_x and A_y are accelerations on X and Y axis. Axis Z doesn't sense any gravitation, as Z axis is vertical to plain X-Y. The angle Θ is then:

$$\Theta = \arctan\left(\frac{A_x}{A_y}\right), \tag{2}$$

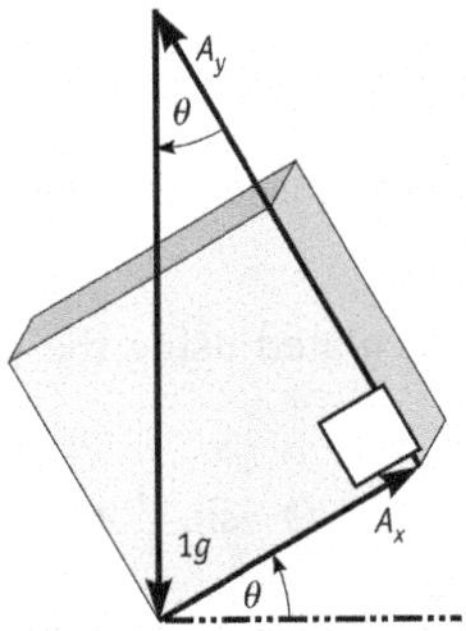

Fig. 8. Tilt defined as angle between X and Y axis

By combining two axis accelerations, and calculating the angle according to (2), a constant sensitivity can be maintained across 360° rotation (Fig. 9.) [7]. Using two axes to calculate angle of tilt also overcomes the problem of knowing if the accelerometer is tilted 30° or 150° according to the 0,5g accelerometer output. With two axes solution this problem is solved by looking at the sign of A_x and A_y [8].

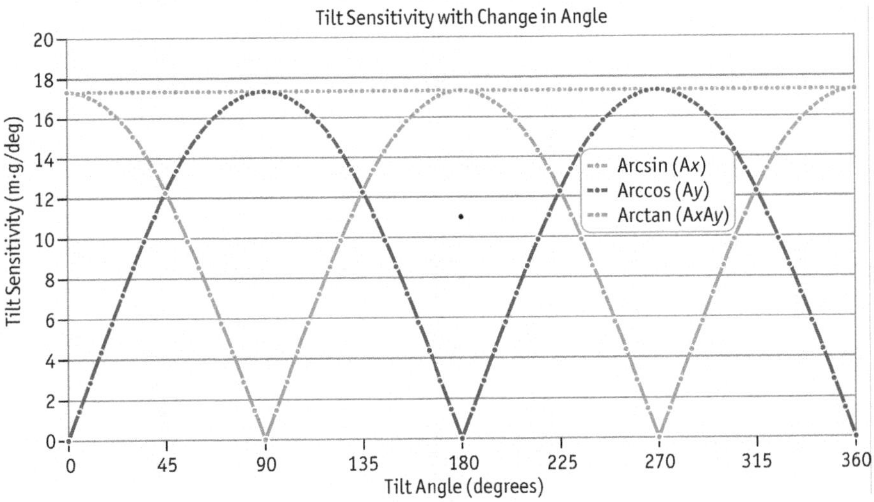

Fig. 9. Tilt sensitivity using two axes

5 Freescale™ ZSTAR Accelerometer Board

The Freescale™ ZSTAR wireless sensing triple axis board is used in our application. It is a low power, small and portable device (board size is 56 mm x 27 mm) using MMA7260Q triple-axis accelerometer together with 8-bit MC9S08QG8 microcontroller unit and MC13191 wireless ZigBee low power transceiver for sending measurement data to USB stick board. USB stick board has MC13191 transceiver, together with HC908JW32 USB enabled microcontroller (Fig. 10.). The transceiver is working on 2.4 GHz industrial, scientific, and medical (ISM) band. It is ideal for low power, long battery life application such as this. Both boards have integrated 2.4 GHz Loop antennas on board [9].

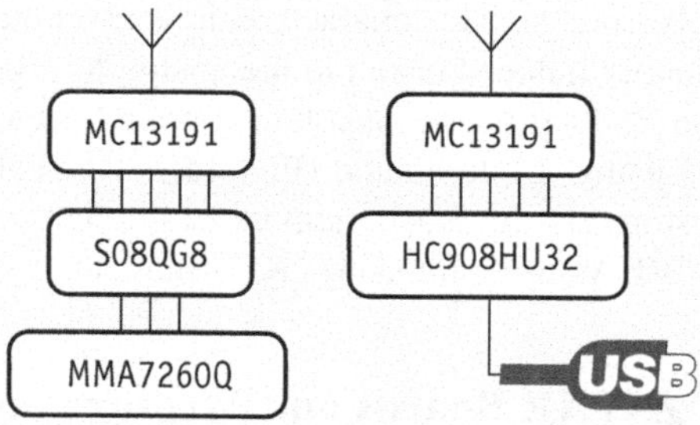

Fig. 10. Wireless ZSTAR accelerometer sensing board

The microcontroller MC9S08QG8 measures the three axis sensor data from the MMA7260QT accelerometer with integrated 8 - channel, 10-bit analog to digital converter, then creates a data packet and sends it with the SMAC (Simple Media Access Controller) to the MC13191 Transceiver. The sensor board needs current in the range of 1 mA, and only 1 µA in standby mode which is provided from the CR2032 Lithium battery located on the bottom of the board. The simple ZSTAR RF protocol also transfers the calibration data. These data are stored in non-volatile Flash memory and are transferred on request [9]. The sensor board has been evaluated and it has been shown that the distance over which the USB stick will receive the data from sensor board is over 20 meters in buildings. The distance is lower if there some obstacles, but the boards successfully communicated even in rooms divided by brick walls. The HC908JW32 microcontroller belongs to the family of 8-bit microcontrollers that offer the Universal Serial Bus (USB) full speed functions. The software in the USB stick board converts the data from the transceiver to the USB connection, and places it in USB stick memory. The data are transferred to the PC through the simple serial protocol communication (virtual serial port). The user can communicate with it from LabVIEW using Virtual Instrument Software Architecture (VISA). It is a standard for configuring, programming, and troubleshooting instrumentation systems comprising GPIB, VXI, PXI, Serial, Ethernet, and/or USB interfaces, and it can provide the programming interface between the hardware and development environments such as LabVIEW [10]. The sensor is capable of measuring all three axes approximately 30 times per second providing nearly a real-time response from the Sensor. The measurement data are packed to a special data frame 10 bytes long which is then decoded in the LabVIEW program. The program communicates with the sensor board using simple set of commands. For example, to receive data, command 'V' is used, to establish handshake command 'Z', and to select g-level command 'G'.

The data frame, besides accelerometer data, also includes information on temperature and bandgap voltage, which is used to calculate battery voltage. The accelerometer MMA7260QT provides three separate analog levels for 3-axis. These outputs are ratiometric which means that the output offset voltage and sensitivity will scale linearly with applied supply voltage. This characteristic is essential as it provides system level cancellation of supply induced errors in the analog to digital conversion process [11]. Accelerometer on ZSTAR board must be calibrated occasionally, for 0g and 1g (earth's gravity) levels using 'k' command (Fig. 14.). This calibration must be done for each axis, and are stored in the Flash memory of the Sensor Board. They can also be retrieved by the LabVIEW program using 'K' command.

6 Calibration of ZSTAR Boards and Measurement Results

To calibrate accelerometers, and to test a measurement system, a special metal bar was constructed to serve as a model of human joints (Fig. 11.). The metal bar has three joints to simulate a human leg or arm. The bar is made of aluminum, with size of 25 mm x 25mm. Each section is 55 cm long, containing one potentiometer used for measuring the angle. The maximum angle that potentiometer can be turned is 270°, which means that each section of bar can bend ± 135°, if angle 0° is presumed when bar is positioned straight. With this model, a human leg can be simulated in all three main joints (knee, hip and ankle). Potentiometers have resistance of 1 kΩ, and are connected to the voltage of 5 V. The potentiometer output voltage is measured using National Instruments USB6008 DAQ card, which has input resistance of 144 kΩ. As one bar is moving apart from other bar, it causes potentiometer's resistance to change. Due to high input resistance of DAQ card, it is practically linear change, providing very robust and accurate method of angle measurement. The maximum measurement error of this system is less than 0,1 %, and translated to the measurement of angle in the working range of potentiometer of 270°, this error can be estimated to +/- 0,2°. This systematic error can be corrected programmatically; therefore, the accuracy of this method is dependent only to the accuracy of A/D converter and the linearity of the potentiometer. As the voltage on potentiometer resistance is measured using DAQ card, this voltage must be converted to degrees. In order to do that, the bar must be calibrated. Each section of bar is calibrated to +/- 90° reference positions using reference angle bars, and also to 0° using flat surface. Between those values, the angles are calculated with linear interpolation.

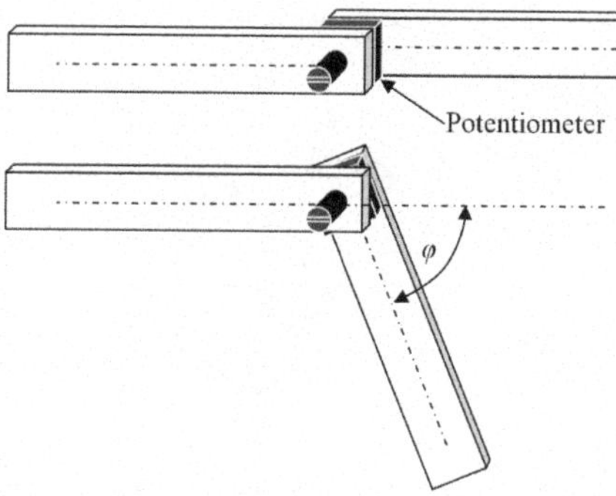

Fig. 11. Placement of potentiometer in joint, and measuring the angle from initial to final position

This model bar was used to calibrate accelerometers and measure the accuracy with which accelerometers can be used to measure angles. Angle of joint is defined with positive definition in Cartesian system from positive part of X axis to positive part of Y axis. The angle presentation is shown in Fig. 12.

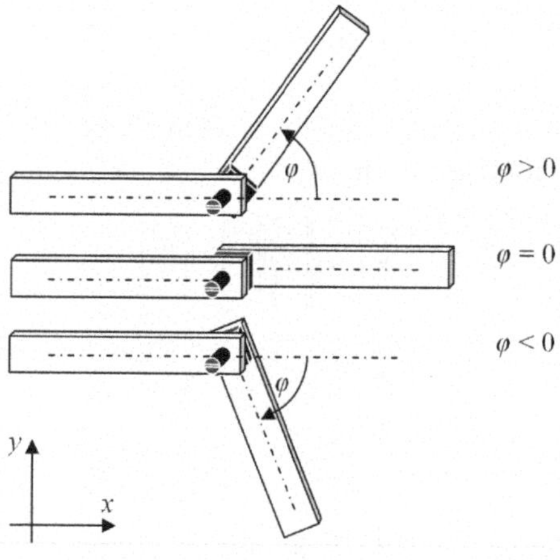

Fig. 12. Definition of angles between two bars

Accelerometers are placed the metal bar as presented on Fig. 13. Initial position of patient is represented on top. Bottom figure shows angle of joint φ_j and angle measured by second accelerometer φ_2. They differ from each other as the whole extremity of interest is moving during the measurement. This is why two accelerometers must be used to remove error caused by moving upper part of extremity.

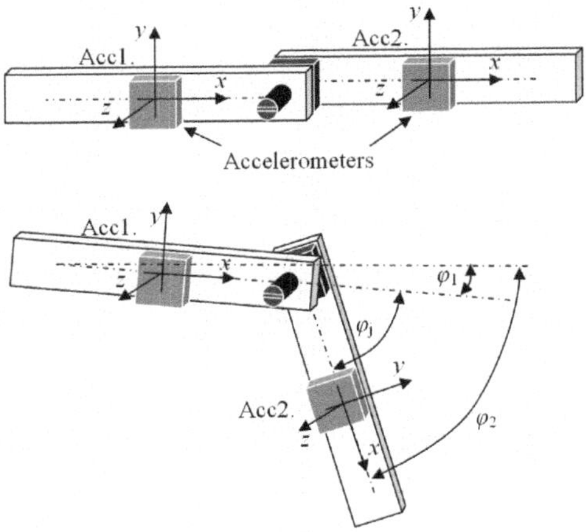

Fig. 13. Representation of angles between two bars, with accelerometers placed on bars

First accelerometer measures the angle of the upper part while the second measures the angle of the lower part. And the real angle of joint movement is difference between the two measured angles:

$$\varphi_j = \varphi_2 - \varphi_1 \tag{3}$$

With data received from accelerometers from initial and final position, maximum angle of joint can be calculated as the difference between the two angles:

$$\varphi = \mathrm{abs}(\varphi_{j2} - \varphi_{j1}) \tag{4}$$

Joint angle is calculated from data given by accelerometers ($\vec{a}$ and $\vec{b}$ on Fig. 14.) by using dot product of vectors, and norm two:

$$\vec{a} \bullet \vec{b} = \left\| \vec{a} \right\| \left\| \vec{b} \right\| \cos(\varphi) \tag{5}$$

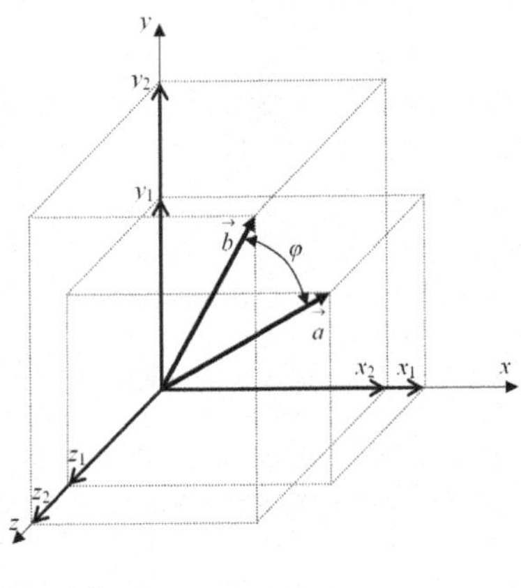

Fig. 14. Vectors $\vec{a}$ and $\vec{b}$ are with their components and angle between them

Dot product between two vectors $\vec{a}$ and $\vec{b}$ is also defined as:

$$\vec{a} \bullet \vec{b} = \left[x_1\ y_1\ z_1\right] \cdot \left[x_2\ y_2\ z_2\right]^T = x_1 x_2 + y_1 y_2 + z_1 z_2\,, \tag{6}$$

and norm is defined as:

$$\left\|\vec{a}\right\| = \sqrt{{x_1}^2 + {y_1}^2 + {z_1}^2}\,. \tag{7}$$

Vectors and angle between them are shown on Fig. 15. Finally angle can be calculated from (5) by using (6) and (7) as:

$$\varphi = \arccos\left(\frac{\vec{a} \bullet \vec{b}}{\left\|\vec{a}\right\|\left\|\vec{b}\right\|}\right) = \arccos\left(\frac{x_1 x_2 + y_1 y_2 + z_1 z_2}{\sqrt{{x_1}^2 + {y_1}^2 + {z_1}^2}\sqrt{{x_2}^2 + {y_2}^2 + {z_2}^2}}\right) \tag{8}$$

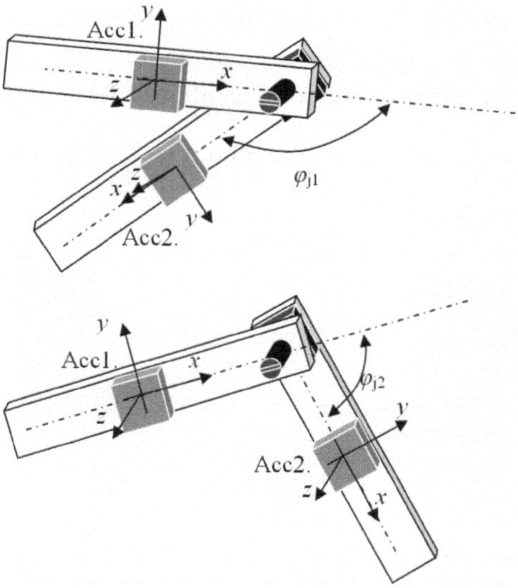

Fig. 15. Representation of joint's angle. Upper part shows the initial position. Lower figure shows the final position. Joint angle is difference between them.

This calibration procedure has been repeated several times in order to investigate the accuracy of the accelerometer to measure inclination of patients. The results have shown that accelerometers can measure angle with uncertainty of around ± 1°. A metal bar has been photographed along with two ZSTAR accelerometer boards attached to it during calibration procedure (Fig. 16.).

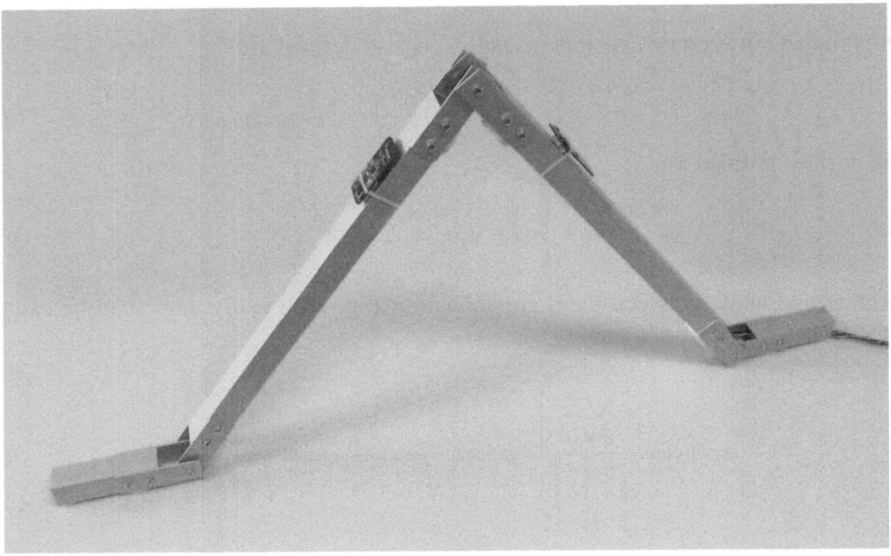

Fig. 16. Metal bar with two ZSTAR accelerometer boards attached

Occasionally, accelerometers have to be calibrated to minimum and maximum of earth's gravity $\vec{g}$ in each direction. Calibrating the accelerometer's Y axis is shown on Fig. 17.

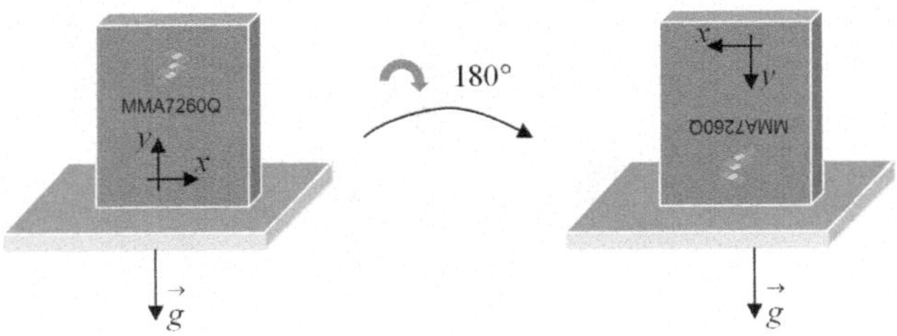

Fig. 17. Accelerometer in calibration positions when Y axis sense maximum and minimum of gravity $\vec{g}$

Voltage output is recorded for –g and +g (U_{-g} and U_{+g}) by using automatized calibration procedure. This procedure must be repeated for all three axes, and a special program written in LabVIEW has been made to accommodate this procedure. With these values measured, accelerometers can be calibrated in $\vec{g}$ units, by interpolating other values (Fig. 18.).

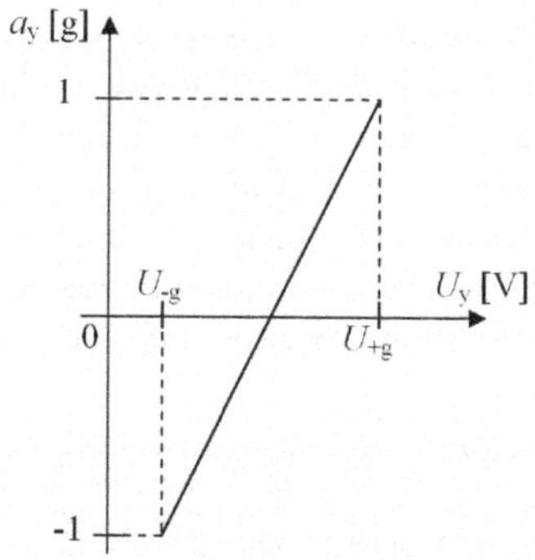

Fig. 18. Calibration of Y axis in $\vec{g}$ units

7 LabVIEW Application

All the software has been programmed using the LabVIEW graphical environment. Operator can choose three different actions from the front panel of main program: Settings, Calibration and Measurement. By choosing one of the buttons, new front panel will be opened. By choosing the 'Settings' (Fig. 19.) tab, the operator can choose the ports for each ZSTAR accelerometer board, as well as number of samples for each measurement. Also, with 'samples per second' the operator can choose how many samples will be taken per second, thus changing the speed of measurement process. With changing these two variables, the operator can choose speed over accuracy, and vice versa. The results on "ZSTAR ANGLE MEASUREMENT" front panel (Fig. 21.) will be the average of the number of samples set on the 'Settings' tab. If there are 100 samples per second, and there are 100 samples per second, then one measurement will last exactly one second.

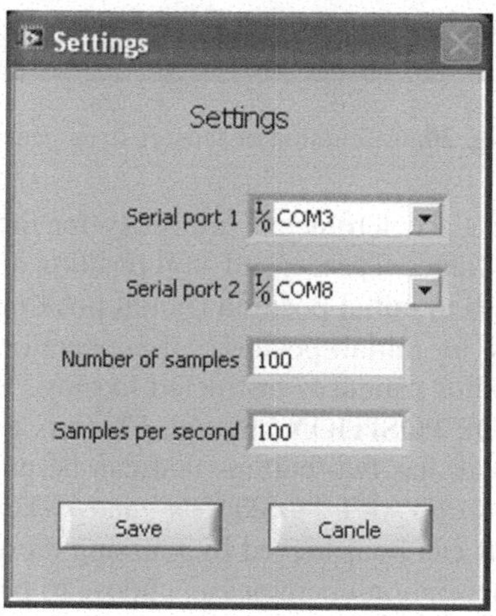

Fig. 19. The ZSTAR accelerometer board calibration procedure

The 'Calibration' tab will bring another front panel (Fig. 20.) where operator can occasionally calibrate ZSTAR sensor boards as already explained on figures 17. and 18. This process is automatized as operator only needs to position each axis to +g and then to –g for several seconds in order for boards to settle down. When standard deviation of measurement drops to predefined experimentally estimated value, program will automatically save the calibration data. Process must be repeated for all three axes and both accelerometers. The whole calibration procedure takes less than five minutes.

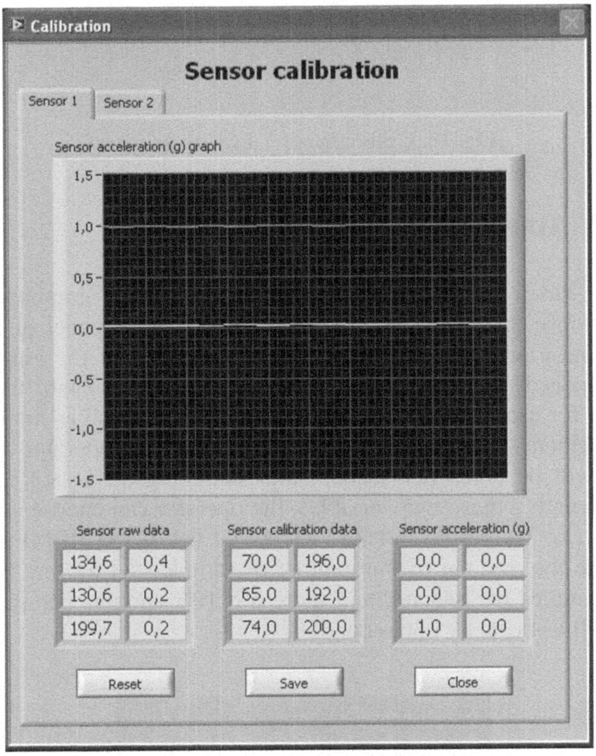

Fig. 20. Calibration of sensors front panel

After calibration process accelerometers are ready for the measurement (Fig. 21). Before the measurement, values for start and stop position are reset. Then a patient is instructed to move the limb to initial position (initial position may differ from case to case). As patient settles to initial position, the operator then presses "START POSITION" button, and after patient is instructed to move the limb to final position, the operator presses "STOP POSITION" button. There is also other option to make measurement. ZSTAR board has two buttons that can be programmed, and therefore it is not necessary to press "START POSITION" and "STOP POSITION" on PC to make measurement, as this can be achieved by pressing these two buttons on ZSTAR board itself. This is another option operator can choose to perform measurement. The operator can choose to measure movement with one of the sensors, or to choose differential measurement, and measure with both sensors for best results.

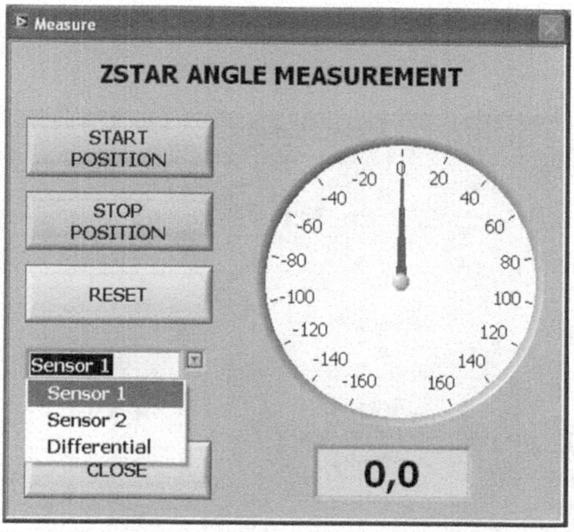

Fig. 21. Front panel of angle measurement application

There is also LabVIEW application used for calibration of metal bar. Each section of bar is calibrated to +/- 90° reference positions using reference angle bars, and also to 0° using flat surface (Fig. 22.).

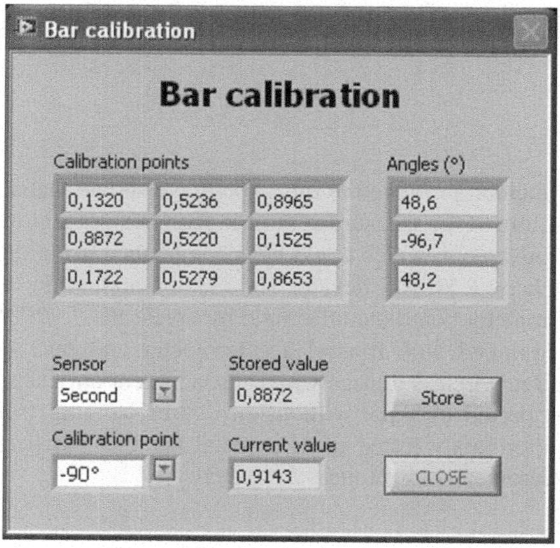

Fig. 22. Calibration data for metal bar

As metal bar has three different joints, the movement of metal bar is also visualized in real time. In this way, the operator can be certain which section of metal bar is actually moving (Fig. 23.).

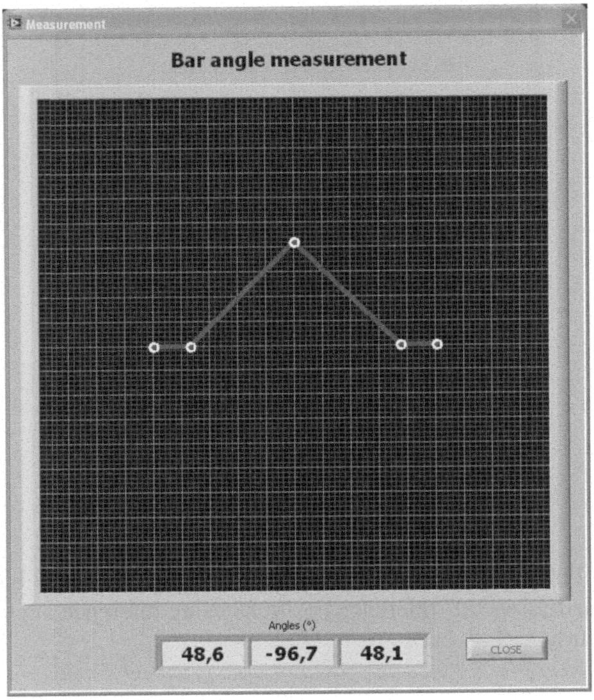

Fig. 23. Metal bar angle measurement and visualization

8 Conclusion

The measurement method presented in this paper have shown that accelerometers can be applied successfully to the measurement of human posture and movement, excluding the rotational movement, which could be accomplished using the additional gyroscope, also available as a MEMS device. The method applied in this paper includes the ZSTAR accelerometer board manufactured by Freescale™. ZSTAR accelerometer board, which is equipped with triaxial accelerometer and microcontroller, is also equipped with ZigBee wireless transceiver. Thus, a patient only needs to wear a small device for a short period of time, without any wires hanging from the body. The method has been thoroughly tested using a metal bar that has been constructed for calibration of accelerometers as tilt measuring device.

References

1. Luinge, H.J.: Inertial Sensing of Human Movement, Doctoral thesis, University of Twente Publications (2002)
2. Mulder, A.: Hand centered Studied of Human Movement Report, Technical Report 94-1, Human movement tracking technology (July 1994)

3. National Institute of Technology and Evaluation of Japan, Human characteristics database, Measurement method of the range of voluntary motion of joint, http://www.tech.nite.go.jp/human/eng/contents/crom/active/troca.html
4. Roetenberg, D., Luinge, H., Veltink, P.: Inertial and magnetic sensing of human movement near ferromagnetic materials. In: Proceedings of the Second IEEE and ACM International Symposium on Mixed and Augmented Reality, October 7-10, pp. 268–269 (2003)
5. Mac, B.: Range of motions, http://www.brianmac.co.uk/musrom.htm
6. Maarit, K., Irene, V., Jimmie, W., Per, L., Lars, N., Timo, J.: Sensitivity and specificity of fall detection in people aged 40 years and over. Gait and Posture 29, 571–574 (2009)
7. Tuck, K.: Tilt Sensing using linear accelerometers, Freescale Application note, Application note 3461, Rev 2 (July 2007)
8. Clifford, M., Gomez, L.: Measuring Tilt with Low-g Accelerometers Sensor Products, Freescale Application note 3107, Rev 0 (May 2005)
9. Lajšner, P., Kozub, R.: Wireless Sensing Triple-Axis Reference Design (ZSTAR), Application note 3152
10. http://www.ni.com/visa/
11. Lajšner, P., Kozub, R.: Wireless Sensing Triple Axis Reference Design. Designer Reference Manual, ZSTARRM, Rev. 3 (January 2007)
12. Tuck, K.: Implementing Auto-Zero Calibration Technique for Accelerometers, Application Note 3447

Instrumentation and Sensors for Human Breath Analysis

Melinda G. Simon[1] and Cristina E. Davis[2,*]

[1] Departments of Biomedical Engineering, [2] Mechanical and Aeronautical Engineering, University of California Davis, One Shields Avenue, Davis CA 95616, USA
cedavis@ucdavis.edu

Abstract. Exhaled breath contains a vast milieu of compounds, both volatile and non-volatile, that appear to correlate with physiological processes on-going in the body. These breath biomarkers hold enormous diagnostic potential when they are adequately measured and monitored. Thus, instrumentation geared towards breath analysis applications has expanded rapidly in the last decade, although challenges for future research still exist. This chapter briefly reviews the history of analytical instrumentation and breath biosensors that have been reported in the literature, and corresponding data analysis approaches that have been attempted to date.

Keywords: breath analysis, biomarker identification, disease diagnostics, chemometrics.

1 Brief History of Breath Diagnostics

Physicians have anecdotally long been aware of the potential for disease diagnosis from breath samples, and both ancient Greek and Chinese medical text allude to breath odor in clinical diagnostics. In more recent times, we now have evidence that the smell of acetone on the breath of a diabetic patient is indicative of blood glucose imbalances, a sweet and musty odor can be indicative of liver disease, and a fishy odor on a patient's breath may represent a symptom of kidney failure [1]. However, it was not until 1969-1971, that instrumentation platforms were used to aid in the precise detection of chemicals in the breath. Dr. Linus Pauling and others were among the first to use mass spectrometry to analyze the content of breath samples [2]. Since that time, thousands of chemicals and small proteins have been detected in breath using a variety of instrumentation platforms and sensors [3]. While much of the current research in the field of breath analysis is focused on identifying breath biomarkers that are statistically related to certain disease outcomes, there are other active areas of research in the field, including: establishing protocols to achieve reproducibility of data, fabrication of more sensitive instruments and sensors, and miniaturization of sensors for portable point-of-care diagnostic applications.

Breath analysis is now starting to be applied for the diagnosis of disease and the monitoring of human health. In recent years, researchers have worked to identify

* Corresponding author.

S.C. Mukhopadhyay and A. Lay-Ekuakille (Eds.): Adv. in Biomedical Sensing, LNEE 55, pp. 144–165.
springerlink.com

biomarkers for a wide range of pulmonary diseases and infections, such as: lung cancer [4], asthma [5], chronic obstructive pulmonary disease [6], gastrointestinal reflux disease [7] , *Pseudomonas* bacterial infections associated with cystic fibrosis [8], diabetes [9], systemic sclerosis [10], *Helicobacter pylori* infection [11] , heart transplant rejection [12] and many others. Once we understand the identity of these biomarkers, we can then monitor for the presence/absence, unique combination or concentration changes of the compounds over time. This can tell us early diagnostic information about the patient, and also can be helpful in tracking the progress of a therapeutic treatment or a response to a pharmaceutical product.

The benefits to breath analysis over more invasive means of diagnosing or monitoring diseases are numerous. In contrast to more traditionally-used diagnostic techniques, such as blood analysis and/or advanced imaging, breath analysis is non-invasive and frequently offers a more comfortable procedure for the patient. This could have positive benefits, such as increasing the likelihood of patient compliance for regular testing and disease monitoring. In addition, breath sample collection can be very rapid and may take as little as a few minutes. Also, the simplicity and non-invasive nature of breath sample collection allows for the possibility of miniaturization of breath collection devices. In the future, micro- or nano-scale breath collection and analysis devices may be portable enabling wider adaptation of the technique in clinical point-of-care medicine.

2 Traditional Instrumentation Platforms Used for Breath Analysis

Due to the wide variety of compounds (volatile and non-volatile) that have been detected in human breath, a number of instrumentation platforms have been employed for the detection and quantification of these chemicals. Depending on the biomarker panel required for diagnosing or monitoring a specific disease, different techniques or a combination of several of these techniques could be used. Detection of chemicals in human breath has traditionally required three steps: isolation or pre-concentration of the biomarker chemical(s), separation of the biomarker chemicals from the complex background of all breath metabolites, and detection/identification of the biomarkers that are deemed important.

Pre-concentration of breath analytes is usually accomplished by two major methods: absorbing breath onto some type of sorbent material, such as a solid-phase micro extraction (SPME) fiber [13]; or collecting exhaled breath condensate, lyophilizing the mixture, and re-suspending it into smaller volumes of matrix for subsequent separation and detection of known biomarker chemicals or proteins [14-16]. Separation or isolation of part of a breath sample for testing can be accomplished using gas chromatography [17], liquid chromatography [18], or even gel electrophoresis [14], depending on the size of the chemicals or proteins to be isolated. These techniques are well established in the biochemistry literature, and so we focus here on illustrating how they are applied specifically for breath analysis.

Techniques to detect and identify chemicals in breath samples are under constant development, but many researchers rely upon more classical instrumentation platforms such as mass spectrometry or visible spectroscopy. In the following sections, typical instrumentation platforms and their utility for breath analysis are presented first, followed by newer methods for chemical separation or identification. Given that so many different types of instrumentation are available for breath analysis, selection of an instrumentation platform for a given application can be difficult. In order to maximize the capability to detect low abundance chemicals, the merits and limitations of each of the following types of instrumentation should be carefully considered for each application. In conjunction with a thorough review of the relevant references below, an instrumentation platform may be selected that will maximize the possibility of detection of new or known breath biomarkers.

2.1 Mass Spectrometry (MS)

Mass spectrometry (MS) has been a popular instrumentation choice for chemical analysis since the 1970s, and is currently used for a variety of applications and in many industries (e.g. petrochemical, pharmaceutical, chemical, consumer products) to detect the presence or concentration of compounds. Mass spectrometer designs vary in the manner by which they ionize samples, and in their detection schemes. Because of its superior ability to separate and detect low concentrations of chemicals and proteins, it has been adopted as the "gold standard" for the detection of biomarkers in human breath. Samples may be injected as vapors, liquids, or through the use of a polymeric solid-phase micro extraction (SPME) fiber, which pre-concentrates the sample. There are several different categories of mass spectrometers, and a few of these are illustrated below along with their application to breath analysis.

Although different types of spectrometers exist, they identify chemicals by ionizing compounds and then exploiting some fundamental principle of the compounds using an appropriate detection method. For example, in traditional MS instruments, we detect the mass to charge ratio of the ionized molecules. The pattern of ions detected in a mass spectrometer is characteristic of a certain chemical, and so the chemicals in an unknown sample can be determined by observation of a mass spectrometer's output by a trained technician or scientist. Experimentally-obtained spectra can also be compared to a library, such as the NIST library to confirm the biomarker identity. In other MS instruments, such as time-of-flight, the other physical parameters of the breath biomarkers are exploited for detection. These are briefly reviewed below. Different designs may be more beneficial for detecting different types of chemicals, so the choice of a mass spectrometer system should include careful consideration of the type of chemicals that a researcher expects to find in a given sample.

2.1.1 Gas Chromatography / Mass Spectrometry (GC/MS)

A gas chromatograph (GC) may be easily connected to a mass spectrometer (MS), and the GC front end allows for pre-separation of chemicals prior to MS chemical identification. After injection into the instrument, the chemicals from the breath sample enter a long, very narrow tube called a capillary column. Molecules that are below their boiling point then attach to the stationary phase that coats the inside of the column, commonly composed of polydimethyl siloxane (PDMS) or another polymer. The temperature of the capillary column is then raised according to a programmed temperature ramp rate, which causes compounds to boil off of the stationary phase as the temperature is increased. In general, low boiling point molecules will elute (emerge) from the end of the column before higher boiling point molecules. Thus, a gas chromatography column effectively separates molecules in a sample by their boiling points.

Since the 1970s, GC/MS has been employed to detect many compounds in breath, some of which are putative biomarkers for lung cancer[19], asthma [20], cystic fibrosis [21], interstitial lung diseases [22], type I diabetes mellitus [9], pulmonary tuberculosis [23], organ transplant rejection [12] and various other biomarkers indicative of oxidative stress [20]. Typically, researchers compare breath samples from a diseased and a healthy group of volunteers in order to find a set of volatile organic compounds common to the breath samples of the diseased group. The alveolar gradient—that is, the difference in concentration of a compound detected in an individual's breath sample compared to the environmental air—is a useful measure for determining the exogenous or endogenous nature of compounds detected in breath samples. Phillips et al. suggest that both negative and positive alveolar gradients can be used for distinction of disease and healthy samples [24]. Many techniques have been employed to develop algorithms that are able to identify possible biomarkers, including forward stepwise discriminant analysis [19] and pattern recognition analysis [23].

Gas chromatography mass spectrometry is a popular instrumentation choice for the analysis of breath samples due to its ability to detect compounds in very low parts-per-billion (ppb) and parts-per-trillion (ppt) abundances. Aside from the benefits of its sensitivity, this instrumentation system requires a relatively large initial monetary investment, the instrumentation is large and non-portable (although miniaturization of GC/MS systems is under development), and analysis can be time-consuming. One challenge of GC/MS analysis of breath is the removal of water vapor from breath samples, which could damage the capillary column of the GC if it were not removed. Sorbent traps may be used to remove some of the water vapor from samples.

One exciting application for breath analysis using a GC/MS has been some pioneering work to enable detection of cancer and organ transplant rejection non-invasively. Philips et al. used breath samples, analyzed using a GC/MS, to elucidate a set of common VOCs occurring in the exhaled breath of patients experiencing rejection of a transplanted organ or afflicted with lung cancer, which were generally not present in healthy subjects [12, 19]. Breath analysis would offer a non-invasive alternative to current diagnostic techniques such as CT scans or biopsies. The implementation of breath analysis for these applications will undoubtedly require further clinical study.

Due to the low (ppb to ppt) concentration of biomarkers in exhaled breath condensate, many researchers have employed techniques to pre-concentrate compounds of interest in the breath sample. The technique of solid-phase microextraction (SPME) was first described in the 1990s by Pawliszyn [13], and is now being employed in breath analysis research [25-27]. A SPME device consists of a fiber covered with a 10-100 μm coating of one or several polymers. Commercial SPMEs are commonly composed of one or more of the following polymers: divinylbenzene, carboxen, and polydimethyl siloxane [25], although custom SPME fibers have been composed of materials such as O-2,3,4,5,6-(pentafluorobenzyl) hydroxylamine (PFBHA) [26, 27]. In several cases, a SPME fiber coating has been employed to derivatize compounds of interest in a breath sample, so that they may be detected more easily. Deng et al. used a coating of PFBHA, which reacted with acetone in the breath sample. The product of this derivatization reaction was then detected in a GC/MS and correlated to the acetone concentration in the sample using standards [26]. Another group recently used the same material to derivatize aldehydes in breath samples, which are products hypothesized to originate from the interaction of reactive oxygen species with lipid peroxidation [27].

Normally, a SPME fiber is exposed to the vapor headspace above a liquid-phase EBC sample for a pre-determined absorption time. Volatile compounds in the vapor phase partition onto the fiber via gas-phase diffusion. Heat and agitation may also be used to volatilize components in the sample and to increase the rate of gaseous diffusion. The mechanism of adsorption of volatile compounds onto the SPME fiber has not been well studied. It is not currently known whether the adsorption mechanism is primarily that of diffusion into pores of the polymer or whether transient weak bond-like interactions form between the solid phase and volatile compounds, although both of these mechanisms likely contribute. After adsorption, the SPME fiber is then introduced to the heated inlet port of a gas chromatography instrument, where thermal desorption of volatile compounds from the fiber occurs. Absorption time and temperature, the composition of the solid phase, and desorption time and temperature must be selected carefully and with experimentation, in order to provide a consistent and reliable methods of detection for a particular set of biomarkers in breath. Interestingly, the SPME headspace extraction technique has found application in the area of environmental and occupational health monitoring [28]. Breath analysis offers a quick and relatively inexpensive technique for monitoring the exposure of employees to harmful chemicals in the workplace.

Although the SPME technique offers the promise of detection of low concentration compounds, there are several concerns associated with its use for breath analysis. The amount of a given compound detected in a SPME sample is proportional to the partitioning properties of that compound onto the solid fiber, which is not necessarily proportional to the concentration of the compound in the sample itself. Some fundamental knowledge of the mechanism of adsorption to SPME fibers, as well as saturation limits, must be determined in order to guarantee accurate quantification of results using this technique.

An alternative technique for the pre-concentration of breath samples is the use of one or multiple sorbent traps. These sorbents are commonly made of porous polymers, graphitized carbon, or carbon molecular sieves and have been used to concentrate many different types of compounds which may occur in breath samples [20, 29]. The mesh size of the sorbent matrix can determine the size and type of molecules that are excluded from the sample after passing through the sorbent material. Sorbent materials are often commonly used for breath analysis to remove water vapor from the sample, in order to prevent degradation of the capillary column in the GC.

2.1.2 Liquid Chromatography / Mass Spectrometry (LC/MS)

Liquid chromatography in breath analysis research has been employed to detect non-volatile breath components that are difficult to detect using gas chromatography. These components range from organic molecules, such as aldehydes [30], to larger molecules such as prostaglandins [31], and even amino acids [15]. In contrast to the open capillary column that is characteristic of gas chromatography, the column used for separation in liquid chromatography, also called the stationary phase, differs greatly in that it is packed with small particles. Sample is driven through the separatory column using a pump, and a mobile phase, composed of one or several solvents, facilitates the separation and movement of compounds through the column. Due to the small particles packed into the LC column, high pressures are required to transport compounds through the column [32]. In contrast to gas chromatography, liquid chromatography (LC) does not separate chemicals in a mixture based upon their boiling points, but rather upon their size and polarity. Compounds in the injected sample are initially associated with stationary phase particles, and are eluted in an order that is based on their polarity. Two configurations that have been used in breath analysis research and biomarker discovery in breath are reverse-phase LC [33, 34] and hydrophilic interaction LC (HILIC) [15, 16].

In reverse-phase chromatography, the stationary phase is composed of a weakly polar or non-polar compound, and the mobile phase consists of a polar chemical. Thus, the gradient of a polar solvent in the mobile phase can be steadily increased throughout the run, in order to elute increasingly polar components of the sample with time [32]. This technique has been used to validate the detection of 8-isoprostane and prostaglandin E_2 in radioimmunoassays [33, 34]. These two compounds are presumed biomarkers for cystic fibrosis, and in clinical trials, researchers found a higher level of 8-isoprostane in cystic fibrosis patients than in healthy control samples.

Another, more recently-developed technique for liquid chromatography separation is termed hydrophilic interaction liquid chromatography (HILIC). In this method, the stationary phase is composed of silica gel or particles, which are modified with chemical groups to make them polar. The mobile phase for this type of chromatography usually consists of an aqueous solvent component, whose fraction is increased with time in order to elute non-polar components first and polar components toward the end of a run. The capability of using aqueous solvents as the mobile phase confers advantage over normal-phase chromatography, in which harsh polar solvents are often used as the mobile phase, leading sometimes to degradation of the column [15]. Conventz et al. chose to use this technique to detect the polar amino acids tyrosine, proline, and amino acid derivatives nitrotyrosine and *trans*-L-4-hydroxyproline in breath samples, since these compounds would not be retained well on a reverse-phase chromatography column [16]. Hydroxyproline and nitrotyrosine are presumed biomarkers for pulmonary fibrosis and inflammation, respectively, while proline and tyrosine were measured as precursors for these compounds. Additionally, the measurement of proline and tyrosine allowed the researchers to correct for dilution effects, which are a large source of inconsistency in breath sampling. The technique was shown to be effective in detecting these compounds in the breath of normal subjects, with a detection limit in the μg/L range for proline and the ng/L range for the other three compounds [16]. In another recent breath analysis study, HILIC was employed for the detection of the polar compounds lysine and N^{ε}-carboxymethyllysine (CML), a product of the reaction of lysine with sugars [15]. This compound was first detected in 2006, using normal-phase liquid chromatography, combined with electrospray ionization and mass spectrometry [18].

One technique to facilitate detection of compounds in breath which metabolize quickly or are otherwise unstable is derivatization. In this technique, a derivatization agent is added to the EBC sample, where it reacts with a specific compound in the EBC sample in order to facilitate detection of that compound or increase its stability and thus, the likelihood of its detection. To enable detection of aldehydes in breath, for example, one group added 2,4-dinitrophenylhydrazine, which reacted with aldehydes in the EBC sample [30]. Aldehydes in exhaled breath originate from lipid peroxidation reactions involving reactive oxygen species in the body. The period during which increased amounts of reactive oxygen species are generated is termed oxidative stress, and thus detection of aldehydes from these reactions provides a means of monitoring the oxidative stress in an individual. In a study, the aldehydes malondialdehyde, hexanal, and heptanal were detected at higher concentrations in breath samples from individuals suffering from COPD than breath samples from healthy (nonsmoking) controls. Additionally, malondialdehyde was detected at a higher level in COPD patients than in smokers without COPD. These results suggest that some aldehydes may eventually serve as diagnostic biomarkers for diseases such as COPD.

Aside from mass spectrometry, fluorescence detection has also been coupled with LC for detection of components in human breath [35]. Thiobarbituric acid was used to derivatize malondialdehyde, a presumed marker for oxidative stress, for detection using an HPLC-fluorescence detection scheme. The use of fluorescence and ultraviolet detectors for analyzing breath will be discussed further in subsequent sections. Finally, liquid chromatography is sometimes simply used as a method for the purification or separation of desired components of a sample prior to analysis methods other than mass spectrometry. In one study, HPLC was used to separate 8-isoprostane and prostaglandin E_2 from the breath matrix before detection of these compounds by radioimmunoassay [36].

The advantage conferred by chromatography coupled to mass spectrometry lies in the ability to detect many compounds in the same sample. Other methods of detection of breath compounds, such as some immunoassays and sensors, are very specific for detection of only one compound. As we have seen, chromatography coupled with mass spectrometry is capable of separation and detection of hundreds or even thousands of compounds which differ greatly in size and polarity. While LC/MS analysis offers accurate detection of many compounds down to concentrations in the ng/L range, the identification of compounds using this technique is largely aided by methods such as isotope dilution. Due to the broad range of compounds present in breath, deuterated standards or isotopes of many biomarkers may not yet be available, and other techniques may provide more definitive chemical identification in these cases.

2.1.3 Other Hyphenated Mass Spectrometry Techniques

Some of the smallest molecules that have been detected in human breath, such as nitric oxide and hydrogen peroxide, have very low boiling points and are not easily detected using traditional chromatography. In contrast, some of the largest molecules detected in human breath are proteins, such as cytokines and leukotrienes, and they can have a molecular weight in the range of 20-60 kDa. Clearly, the detection of these compounds requires different instrumentation than that of the NO and H_2O_2. Between these two extremes in the spectrum of chemicals that may be identified in human breath, many instrumentation platforms and techniques have been employed. Each technique offers advantages and disadvantages for the detection of certain chemicals, and the detection of a range of biomarkers required to diagnose a certain disease will likely employ more than one of these techniques.

After separation of compounds in breath samples using gas or liquid chromatography, compounds are usually introduced into a mass spectrometer, where they are ionized and detected using one of several schemes. Most methods of ionization are most efficient at ionizing a certain type of compound and ionize other compounds weakly, rarely, or not at all. Here, several commonly used ionization and detection schemes for breath analysis research are discussed.

Some metabolites generated by inflammatory cells are short-lived and relatively unstable chemical species. Selected ion flow tube mass spectrometry (SIFT/MS) offers an instrumentation option for the detection of these unstable compounds. This technology is distinguished from other, similar technologies such as ion mobility spectrometry (IMS, discussed in section 2.4), by the introduction of reactive ions in the flow tube reactor of the instrument. These reactive ions interact with certain chemical species in the sample gas and form product ions whose mass-to-charge (m/z) ratios are predictable. The presence and abundance of these chemicals can then be quantified from peaks in the mass spectrum. In these instruments, ionization of the reactant ions occurs by microwave discharge, after which the reactant ions are focused by a quadrupole mass filter into the flow tube reactor. The sample gas is introduced in this chamber and selected ions from the sample gas react with the reactive ions produced earlier. Products of these reactions are guided out of the flow tube reactor and onto a detector via another quadrupole mass filter. Selective ion monitoring (SIM) is often used in SIFT-MS, and allows only ions of certain m/z ratios to pass through the quadrupole and to the particle multiplier detector [37]. This technique has demonstrated success in the ability to measure compounds in breath using only a single breath as a sample, and can detect compounds down to ppb concentration levels. Unlike chromatography systems, SIFT/MS does not require the use of standards to determine the concentration of compounds in a sample—concentration can be determined by knowing the kinetics of the ion reactions as well as the flow rates and pressures in the system [38].

Because of their ability to detect chemicals that quickly react or degrade, SIFT/MS has been used to measure unstable compounds in exhaled human breath. The use of reactive ions H_3O^+ and O_2^+ in one study allowed for the observation of compounds known as haloamines in human breath. Haloamines, such as monobromamine and monochloramine are produced from hypobromus and hypochlorous gases, which are generated upon stimulation of inflammatory neutrophils and eosinophils [39]. Detection of the gases could provide information on the source of inflammation in a patient, which could help distinguish diseases with similar symptoms [37]. Many other biomarkers have been detected using this method to analyze breath, including propanol [40], methanol [41], and acetaldehyde [42].

In proton-transfer reaction mass spectrometry (PTR/MS), the reactive ion that is responsible for the ionization of compounds in the breath sample is a hydronium ion H_3O^+. This type of ionization method has been used to detect various volatile organic compounds in breath samples, including acetone and isoprene [43], benzene and acetonitrile [44], and many others [45]. Similar to the SIFT/MS technology, concentrations of compounds detected using PTR/MS can be determined from kinetic and physical information about the system. PTR/MS is especially useful for breath analysis since the hydronium ion will not react with most compounds present in air, but under the appropriate electric field and pressure conditions, will react with many volatile organic compounds. Finally, the detection of these volatile compounds using

PTR/MS is not affected by the water vapor that is present in breath samples and is a source of concern for some other detection methods [46].

2.2 Ion Mobility Spectrometry (IMS)

The basic physical principles of IMS technology are outlined below, however the interested reader is directed to an excellent overview of the physics behind ion mobility and the instruments which use this technology to characterize chemical mixtures [47]. All ion mobility spectrometers separate and detect chemicals by first ionizing them, manipulating the ions as they pass through different strength electric fields, and then detecting these ions using electrodes for both positive and negative ions. Initially the incoming sample is ionized using a variety of different mechanisms including: photo-UV ionization, electrospray ionization (ESI), or exposure to a radioactive source, such as ^{63}Ni. In most instrument configurations, the resulting ions from the breath sample are then allowed to flow into the detection chamber via an ion shutter. In this chamber, the ions are exposed to an asymmetric waveform, and also to a superimposed applied voltage. The applied voltage is varied to steer certain ions between two electrodes and to the detector surface. Ions which are not exactly balanced by this applied voltage will not travel to the detector, but will collide with either the positive or negative electrode and lose their charge. IMS thus separates ions based on their mobility in an electric field, which is determined not only by the mass of the material, but also by its charge and the connectivity of atoms in the molecule [47]. The strength of IMS for breath analysis lies in its ability to detect very low concentrations of compounds under the right conditions, and the capability for multi-dimensional data acquisition[48].

Instrumentation employing IMS technology has been developed and characterized to allow the detection of common chemicals found in breath, such as toluene, xylene and benzene [49], and other VOCs [49, 50]. These techniques used either high speed capillary columns [49, 50] or multi-capillary columns [51] for pre-separation of components before introduction to an IMS sensor. A multi-capillary column consists of many 50-300 mm diameter capillary tubes packed into a larger column [51], and is useful in IMS to prevent the clustering of ionized compounds while also preventing negative effects of humidity in the sample [48]. Several other studies involving the measurement of compounds in breath using IMS have been reviewed [48]. In these studies, the breath samples of patients with a bacterial lung infections or chronic obstructive pulmonary disease (COPD) were compared to samples obtained from healthy individuals. Several other studies have used IMS technology to identify biomarkers for sarcoidosis [52] and lung cancer [53]. Much development work, including the use of MS in order to construct databases of compounds detected using IMS, remains to be completed before these techniques will be implemented for clinical diagnostic use [48].

Miniature differential mobility spectrometry, a variant of IMS, has emerged as a promising technique for the detection of very low concentrations of compounds. This feature makes DMS technology particularly suited to applications in breath analysis, as compounds in breath are exhaled at very low concentrations that can be difficult to detect. In addition, the miniaturization of this technology offers promise for the possibility of portable and handheld breath analyzers in the future. In a DMS, the ionization method is frequently the same as in traditional IMS, chemical ionization with ^{63}Ni. However, a DMS does not contain an ion shutter and allows for simultaneous detection of both positive and negative ions [47]. Recently, characterization of the DMS for breath analysis has been performed with breath samples spiked with acetone [54]. DMS technology has also been employed for the development of a sensor to measure NO_x compounds, which could conceivably be used to monitor exhaled NO from breath [55]. The development of this technology is expected to proceed rapidly as characteristic spectra for many compounds are obtained, enabling reliable chemical identification of many compounds in breath.

2.3 Electrochemical Sensors

Some of the smallest molecules that have been detected in human breath, such as nitric oxide and hydrogen peroxide, have very low boiling points and are not easily detected using traditional chromatography. Instead, researchers have developed several alternative techniques for the detection of these lighter compounds. Electrochemical sensors take a biological input and convert it to an electrical signal. In the case of electrochemical sensors for breath analysis, the electrical current produced on the sensor electrode is directly proportional to the concentration of the chemical in the sample. These types of sensors offer the possibility of miniature or hand-held sensing capabilities, but so far they are limited to the detection of one or a few compounds of interest in the sample. Despite this limitation, many materials have been identified for electrochemical sensors that provide them excellent selectivity and sensitivity for the analyte of interest over other similar analytes.

One example of how material properties of an electrochemical sensor can be important for breath analysis is illustrated very well by a class of sensors that have been developed for nitric oxide (NO) detection in breath samples. Successful design of a NO biosensor requires that it be sensitive and selective. Typically, the selectivity of NO sensors has been attained by using a selectively permeable membrane specific for the analyte. Permeability is also an important consideration, as a material with excellent NO selectivity but poor permeability would not be capable of detecting low concentrations. While some of the sensors have a high NO selectivity, other sensor qualities, such as permeability, response time, or sensitivity may be low. Finally, many novel sensing matrix materials have been utilized for nitric oxide biosensors. As shown here, no single sensor design proposed yet has demonstrated dominance in all of these qualities; thus, possibilities for innovation and development in the field exist.

Researchers have made use of NO's nonpolar nature, which serves it well in signaling because of its ability to cross the cell membrane easily. Therefore, a nonpolar selective membrane would allow NO to cross more easily to the electrode surface than some of the interfering molecules, such as NO_2^-, which is polar. Nafion is a popular membrane material that is selective for nitric oxide, which has been used in several of these biosensors and was recently combined with single-walled carbon nanotubes (SWNTs) to improve its sensitivity [56]. Nafion is a highly fluorinated polymer with a poly(tetrafluoroethylene) (PTFE) backbone [57]. The fluorine in the chemical structure enhances the material hydrophobicity, and it becomes more permeable to nonpolar chemicals, such as nitric oxide [58]. Nafion is able to provide excellent selectivity for NO over similar ions, such as nitrate, ammonia, and ascorbic acid, because of its hydrophobic nature and the ability of the polymeric chains to pack closely, thus providing a size exclusion mechanism of selectivity [58]. In one study, ultra-sensitive membranes were fabricated by mixing Nafion with single-walled nanotubes (SWNTs), which are renowned for their electrical properties [56]. Other materials have been used as selective membranes for NO sensors, including a fluorinated xerogel [58], and a polymer-metal polymer matrix [59]. Various materials have been used for the electrode surface of these sensors, including a carbon fiber micro disk electrode [56], glassy carbon electrode, gold disk electrode [59], and even platinum and iron nanoparticles, to increase the surface area of the sensor [57]. Nitric oxide sensors have been heavily researched for use in monitoring exhaled breath, with some sensors available commercially. The NIOX MINO is an amperometric NO sensor that is available commercially from Aerocrine (Stockholm, Sweden). This sensor has a low limit of detection of 5 ppb and has been demonstrated to be comparable to the current "gold standard" method of NO detection in breath, chemiluminescence [60]. This sensor has been recently approved by the FDA for clinical monitoring of asthma [61]. Given that NO is a major biomarker of interest in human breath, this category of sensors remains quite interesting.

Electrochemical sensors for other small molecules in the breath, such as hydrogen peroxide [62] and carbon monoxide [63] have also been developed. Hydrogen peroxide sensors often incorporate horseradish peroxidase (HRP), which oxidizes hydrogen peroxide and produces electrons which create a current in the sensor. Other polymers and conductive materials which have been employed for H_2O_2 sensors include polythiolated-β-cyclodextrin [64], chitosan with gold-platinum nanoparticles and polyaniline nanotubes [65], and a chitosan-carbon nanotube-nile blue-HRP complex [66]. Carbon monoxide has also been measured from breath with an electrochemical sensor available commercially (Bedfont E50 Mini-Smokerlyzer; Bedfont Scientific, Kent, UK) [63]. Electrochemical sensors could play an important role in the miniaturization of breath detection technology. Already this has been demonstrated in a hydrogen peroxide sensor that was fabricated based on electrolyte metal oxide semiconductor field-effect transistor (^{E}MOSFET) technology. This sensor uses the polymer Os-polyvinylpyridine with incorporated peroxidase as the sensing matrix. Using this device, researchers were able to detect H_2O_2 in artificial exhaled breath samples [67]. Hopefully this example will lead to research into other miniaturized electrochemical sensors for multiple analytes, and the validation of such technology by testing in healthy and diseased patients.

2.4 Optical Sensors

While there has been an abundance of research on electrochemical sensors for breath compounds, optical sensing methods have also been explored [68]. These techniques are ideal for smaller molecules, such as nitric oxide (NO) and hydrogen peroxide (H_2O_2), and include chemiluminescence, electron paramagnetic resonance spectroscopy (EPRS), UV-Vis spectroscopy, colorimetry, and fluorescence [56]. Typically, these types of sensors do not measure NO directly, but rather detect a metabolic product or derivative of NO [56]. One drawback to these optical and affinity-based biosensors is that dynamic, real-time measurement of NO and H_2O_2 has not been realized [68]. Potential advantages to these types of sensors include ease of use via optical observation.

Chemiluminescence techniques to detect nitric oxide are based on gas-phase sensing, which may have some value in clinical settings. In this sensing method, nitric oxide reacts with ozone (O_3) to produce an excited NO_2 molecule, which emits a photon. This photon can then be detected by a photomultiplier [69]. This technique works because nitric oxide is a free radical, and thus it generates an excited-state molecule upon reacting. Since NO_2 is not a free radical, it will not generate this excited-state molecule and thus will not release a photon. Therefore, a major benefit to this technique is its selectivity for NO over NO_2 because of the inability of NO_2 to react and release a photon. Unfortunately, the detection of NO cannot be accomplished in real-time using this technique [68]. Hydrogen peroxide in EBC has also been measured using chemiluminescence techniques. Luminol (5-amino-2,3-dihydro-1,4 phthalazinedione) and horseradish peroxidase were mixed into an EBC sample and allowed to react. The reaction of hydrogen peroxide with luminal produces chemiluminescence, which can then be detected and quantified with a luminometer [70].

Compounds which become colored upon binding to NO have been discovered, thus allowing for colorimetric sensing of NO. Ferrocyanide, 2,2'-azinobis(3-ethylbenzthiazoline-6-sulfonic acid) is one of these compounds, which has been shown capable of detecting nitric oxide in the micromolar concentration range. Another colorimetric compound known as the Griess reagent, which consists of sulfanilamide and N-(1-naphthyl)ethylenediamine dihydrochloride, measures NO indirectly by measuring nitrite and nitrate ions, but it also has a detection limit in the micromolar range. This reagent has been used to measure nitrite in the exhaled breath of children with cystic fibrosis, and to demonstrate a higher level of nitrite in these patients versus healthy subjects [71]. Although these techniques offer ease of use through visual observation of the presence of NO in a sample, the detection limit on most of the electrochemical sensors discussed in this paper was much lower (in the nM range). Therefore, these sensing strategies would only be useful in situations where the NO concentration is presumed to be relatively high [68]. Colorimetry has also been developed for detection of hydrogen peroxide. The oxidation of tetramethylbenzidine by hydrogen peroxide is catalyzed by horseradish peroxidase and creates a colored product which can be detected using a spectrophotometer. Using this technique, researchers were able to show a significant increase in the level of H_2O_2 in EBC samples of stable COPD patients as compared to healthy controls. Additionally, they also found that the concentration of H_2O_2 in unstable COPD patients was higher than in stable COPD patients [72].

Fluorometry offers another technique with the potential for visual observation of the presence of NO and H_2O_2. Several compounds have been discovered which become more unstable upon binding NO and subsequently fluoresce more strongly. These compounds include 2,3 diaminonaphthalene [73], dihydrorhodamine [74], and cytochrome c [75]. Although the sensitivity of this technique is comparable to many of the electrochemical sensors (nM range), selectivity against nitrite is poor [68]. Hydrogen peroxide has also been detected using fluorescent techniques [76, 77]. It can form a complex with the enzyme horseradish peroxidase, which then forms a fluorescent dimer of para-hydroxyphenyl acetic acid. Using this method, researchers were able to distinguish EBC samples from asthma patients and healthy controls [77].

Electron spin resonance (ESR) spectroscopy is a technique that has been used for detecting NO binding to iron in iron-conjugated molecules such as hemoglobin.

These molecules give a characteristic ESR spectrum when bound to NO. This mechanism of sensing may provide adequate sensitivity for some applications, with a detection limit of 0.5 μM, but requires low temperatures to obtain good resolution. This low temperature requirement may alone prevent the use of ESR spectroscopy in in vitro NO sensing. While ESR offers the possibility of real-time detection of NO, the selectivity of this technique is poor [68].

3 Other Techniques Used in Characterization of Human Breath Samples

In addition to the volatile and fatty acid compounds that can be detected in breath samples, larger molecules such as proteins and fatty acids emerge from the airway lining fluid in aerosol form as we exhale. Although both gas and liquid chromatography coupled with mass spectrometry have been used to detect these heavier compounds [33, 78], these techniques are time consuming, expensive, and the equipment is not portable. As an alternative to traditional chromatography, several research groups have explored the use of fluorescent bead immunoassays[79-81], enzyme immunoassays (EIAs)[78], antibody microarrays[82], gel electrophoresis[14] and PCR[4] for the detection of these heavier breath components.

3.1 Fluorescent Bead Immunoassays

Fluorescent bead immunoassays, such as Immulite® [79] (Seimens Healthcare Diagnostics Inc.) or cytometric bead arrays (Becton Dickinson) [81] have the ability to simultaneously detect a multitude of cytokines and other proteins in breath. These kits contain several sets of small beads, each coated with a different capture antibody. To analyze a breath sample, several beads from each set are mixed together and the sample is added. If the antigen to the antibody bound to a bead is present, the bead will bind this antigen. A solution of fluorescent reporter antibodies is then added. Each type of antibody on the beads responds to a different fluorescent intensity and when analyzed using a flow cytometer, the number of fluorescent events for each fluorescent intensity can be determined. This technology has been employed by many groups in order to discover biomarkers of diseases including COPD, asthma, systemic sclerosis, and acute lung injury/acute respiratory distress syndrome (ALI/ARDS). A fluorescent bead immunoassay was recently used to measure levels of erythropoietin (EPO) and TNF-α in COPD patients and normal subjects, in order to determine if these molecules could be biomarkers of COPD. While there was no observed correlation between EPO and COPD, there was a statistically significant difference in the levels of TNF-α between COPD and non-COPD patients [79]. Another study found levels of leukotriene B4 to be increased in patients with COPD and asthma, as compared to healthy subjects [83]. In a study involving patients with pneumonia and ALI/ARDS, the levels of 5 different interleukins as well as TNF-α were shown to be higher in these patients than in the healthy patients [80]. This result also shows promise that breath testing may prove useful in its ability to detect many types of afflictions.

Edme et al. recently studied the levels of a panel of cytokines in patients with systemic sclerosis, and compared them to levels of those cytokines in normal patients. In this disease, which causes fibrosis of lung tissue and can also manifest itself in pulmonary hypertension, symptoms can be difficult to detect in the early stages of disease. Although many of the cytokine levels were higher in systemic sclerosis patients than in the healthy individuals, the researchers were able to determine that the level of interleukin-4 in an exhaled breath condensate sample could differentiate well between those having the disease and those who did not [81]. Someday, breath testing for systemic sclerosis may provide a reliable, noninvasive means of detection, and prevent suffering due to this disease.

3.2 Enzyme Immunoassay

The assay used in this technology is plate-based rather than bead-based. Instead of a fluorescently-tagged reporter antibody, the reporter antibodies in this assay are conjugated to enzymes which react with a reagent to produce a colored product. The intensity of the color is quantified with a spectrophotometer and is linearly related to the concentration of antigen present in the sample. One study employing this technique investigated the levels of interleukin-6 (IL-6) and leukotriene B_4 in smokers and non-smokers. The researchers for this study observed an increase in IL-6 in smokers which correlated with the number of cigarettes smoker per day, and noted that this molecule could be a biomarker for COPD [78]. A study from the same group the next year demonstrated increased levels of leukotriene B4 in COPD patients and healthy subjects. Additionally, these researchers measured prostaglandin E2 by means of radioimmunoassay and found it to be increased in COPD patients as well, as compared with healthy subjects [36].

3.3 Antibody Microarray

An alternative to the bead-based cytokine detection scheme is the use of a printed antibody array. The array platform offers flexibility in the type and number of antigens which may be screened in the assay. Custom arrays may be easily printed for detection of different panels of antigens, and the assay size is not limited by the requirement of a different fluorescent intensity for each antigen, as in the bead assays. Pre-printed arrays of antibodies to 40 different inflammation markers have been used to measure the difference in cytokine levels between asthmatic and normal subjects. Researchers were able to determine a list of 9 cytokines which were significantly up-regulated in asthmatic patient breath samples, as compared to healthy subject samples [82]. In addition to the use of multiplex cytokine assay technologies in identifying biomarkers of inflammatory diseases, they may also contribute to basic biological research. Observation of the protein products in exhaled breath lends clues as to the gene regulation changes that take place in a disease state as well as the signaling pathways involved in these diseases.

3.4 Polymerase Chain Reaction (PCR)

Providing more direct evidence of the physiological origin of the disease than proteins, some researchers have analyzed the DNA present in exhaled breath condensate samples. Gessner et al. used PCR to amplify DNA segments found in EBC samples and sequenced these segments in order to detect p53 mutations, characteristic of non-small cell lung cancer. The method was able to detect mutations in 4 of the 11 samples which had come from cancer patients, while there were no false positives reported for healthy individuals [4].

Many leukotrienes and other cytokines have been observed in elevated levels in patients with COPD [78, 79], systemic sclerosis [81], ALI/ARDS [80], gastrointestinal reflux disease (GERD) [78], and asthma [7, 82, 84]. In spite of these observations, the difference in concentrations of biomarkers between healthy and disease subjects is often low, and inter-individual variability of biomarker concentrations can be high. Improvements in breath collection devices and detection schemes with higher sensitivity will be required in order to gain clinical approval of these techniques for the diagnosis of disease.

4 Approach to Exhaled Breath Sample Collection

The ease of obtaining exhaled breath samples, compared to other types of biological specimens that must be collected through invasive means, is one of the strongest advantages to the technique of breath analysis. Several commercial devices, including the R-Tube (Respiratory Research, Inc., Charlottesville, VA, USA) [85], EcoScreen (Jaeger, Wuerzburg, Germany) [78], BCA (Breath Meter Technology, Inc., Cleveland, OH, USA) [12], and Bio-VOC sampler (Markes International Ltd., Rhondda Cynon Taff, UK) [28] (have been used to collect samples, although other homemade designs have been used for individual studies. All employ a cooling device to encourage condensation of water and other vapors in the sample, and contain a saliva trap to minimize contamination from this liquid. While the Ecoscreen provides the advantage of keeping the breath sample at a constant temperature during sampling, the instrument is bulky and expensive. The RTube is a more economical and portable method for breath sampling, but some volatiles may be lost during breath collection as the condensate sleeve warms to the environmental temperature. The Bio-VOC device is another simple and economical means of breath sampling, but unlike the RTube, the Bio-VOC captures the end-exhaled breath, which has a different composition than the whole breath [28]. In the BCA device, end-exhaled breath is collected onto an adsorbent trap, which can then be thermally desorbed for analysis [86].

5 Strategies for Biomarker Discovery

A great variety of chemicals have been detected in human breath. While the majority of the chemicals detected in breath samples are products of cells in the respiratory system, other physiological origins produce chemicals that can be detected in the breath, such as the exhaled biomarkers for heart allograft rejection [12] and systemic sclerosis [10]. These biomarkers represent many different signaling pathways in the cell and include both volatile and nonvolatile chemicals.

Of all the volatile breath compounds, nitric oxide and hydrogen peroxide are the most volatile and some of the most common markers measured. The origin of NO in exhaled breath, whether mainly from the circulation or from the alveoli themselves, has not been determined conclusively [87]. Hydrogen peroxide appears to be produced from the action of the enzyme superoxide dismutase on superoxide (O_2^-) ions, which are formed upon inflammation [6, 83, 87]. Due to the small size of both NO and H_2O_2, they can easily traverse the membranes of alveolar cells and thus emerge into the alveolar air, where they are exhaled. Many nonvolatile compounds exhaled in breath can also provide a great deal of information about how signaling processes in the body differ in a diseased state from a healthy state. Inflammatory proteins are frequently measured in exhaled breath condensate. Compounds such as leukotrienes [78] and 8-isoprostane [6] are thought to be metabolite products of the arachidonic acid pathway, and their production is increased during inflammation [87]. Other proteins, such as indole and dimethyl sulfide derive from incomplete metabolism of the amino acids tryptophan or cysteine and methionine, respectively [88]. Yet more protein products detectable in exhaled breath condensate include inflammatory molecules, such as TNF-α and IFN-, and interleukins. These molecules are likely produced by epithelial cells and macrophages in the airways during inflammation [78].

Of the non-protein products that are commonly detected in breath, there are aldehydes, alkanes, and methylated alkanes. Aldehydes are generally believed to result from lipid peroxidation [6], while alkanes and methylated alkanes in breath are believed to result from the action of reactive oxygen species in cells, which inflict oxidative stress and result in the production of these compounds from exogenous sources [24]. While researchers have identified the likely physiological origins of many breath biomarkers, much research remains to connect subcellular-level changes in metabolism to the observable panel of compounds that is provided in a breath sample. Once links are made between breath biomarkers and signal transduction pathways, it will yield new insight into disease processes and offer new opportunities to develop targeted therapeutics.

6 Discussion and Conclusion

Since the GC/MS was first used to analyze breath nearly 40 years ago, instrumentation solutions for the wide variety of chemicals that are present in breath have made substantial progress. While many research groups rely on the use of "gold standard" instrumentation systems, such as mass spectrometry for detecting chemicals in breath samples, the importance of developing new technologies to address problems specific to analyzing breath will be crucial to the continued progress of the field. Specifically, sensors and instrumentation which can adequately detect a wide range of chemicals, even those in very low abundance, will be extremely valuable to the field and ensure its continued development. The convenience of the breath sampling method has also prompted researchers to consider the development of handheld and point-of-care diagnostic and health monitoring devices. Miniaturization of some sensors for use in breath has begun, but many challenges still remain. Challenges inherent to developing these handheld sensors include the need for rapid detection, low power consumption, and ease of use, all while retaining excellent sensitivity and repeatability. As these needs are met, the possibilities of quicker diagnosis and economical and easy-to-use at-home health monitoring equipment may be realized.

Acknowledgements

This work was partially supported by grant UL1 RR024146 from the National Center for Research Resources (NCRR), a component of the National Institutes of Health (NIH), and NIH Roadmap for Medical Research. The authors also acknowledge and thank DARPA (PM Dennis Polla), and California Industry-University Cooperative Research Program for partially supporting this work. The contents of this manuscript are solely the responsibility of the authors and do not necessarily represent the official views of the funding agencies.

References

1. Van den Velde, S., Nevens, F., Van Hee, P., Van Steenberghe, D., Quirynen, M.: GC-MS analysis of breath odor compounds in liver patients. Journal of Chromatography B 875, 344–348 (2008)
2. Pauling, L., Robinson, A.B., Teranishi, R., Cary, P.: Quantitative Analysis of Urine Vapor and Breath by Gas-Liquid Partition Chromatography. Proc. Natl. Acad. Sci. USA 68, 2374–2376 (1971)
3. Phillips, M., Herrera, J., Krishnan, S., Zain, M., Greenberg, J., Cataneo, R.N.: Variation in volatile organic compound in the breath of normal humans. J. Chromatography B 729, 75–88 (1999)
4. Gessner, C., Kuhn, H., Toepfer, K., Hammerschmidt, S., Schauer, J., Wirtz, H.: Detection of p53 gene mutations in exhaled breath condensate of non-small cell lung cancer patients. Lung Cancer 43, 215–222 (2004)
5. Cap, P., Chladek, J., Pehal, F., Maly, M., Petru, V., Barnes, P.J., Montuschi, P.: Gas chromatography/mass spectrometry analysis of exhaled leukotrienes in asthmatic patients. Thorax 59, 465–470 (2004)
6. Borrill, Z.L., Roy, K., Singh, D.: Exhaled breath condensate biomarkers in COPD. European Respiratory Journal 32, 472–486 (2008)
7. Shimizu, Y., Dobashi, K., Mori, M.: Exhaled breath marker in asthma patients with Gastroesophageal reflux disease. Journal of Clinical Biochemistry and Nutrition 41, 147–153 (2007)
8. Balint, B., Kharitonov, S.A., Hanazawa, T., Donnelly, L.E., Shah, P.L., Hodson, M.E., Barnes, P.J.: Increased nitrotyrosine in exhaled breath condensate in cystic fibrosis. European Respiratory Journal 17, 1201–1207 (2001)
9. Novak, B.J., Blake, D.R., Melnardi, S., Rowland, F.S., Pontello, A., Cooper, D.M., Galassetti, P.R.: Exhaled methyl nitrate as a noninvasive marker of hyperglycemia in type 1 diabetes. Proc. Natl. Acad. Sci. USA 104, 15613–15618 (2007)
10. Cope, K.A., Solga, S.F., Hummers, L.K., Wigley, F.M., Diehl, A.M., Risby, T.H.: Abnormal breath ethane concentrations in patients with systemic sclerosis. Arthritis & Rheumatism 52, S591–S592 (2005)
11. Jordaan, M., Laurens, J.B.: Diagnosis of Helicobacter pylori infection with the C-13-urea breath test by means of GC-MS analysis. Journal of Separation Science 31, 329–335 (2008)
12. Phillips, M., Boehmer, J.P., Cataneo, R.N., Cheema, T., Eisen, H.J., Fallon, J.T., Fisher, P.E., Gass, A., Greenberg, J., Kobashigawa, J., et al.: Heart allograft rejection: Detection with breath alkanes in low levels (the HARDBALL study). Journal of Heart and Lung Transplantation 23, 701–708 (2004)

13. Zhang, Z., Yang, M.J., Pawliszyn, J.: Solid-Phase Microextraction. A Solvent-Free Alternative to Sample Preparation. Anal. Chem. 66, 844A–853A (1994)
14. Fumagalli, M., Dolcini, L., Sala, A., Stolk, J., Fregonese, L., Ferrari, F., Viglio, S., Luisetti, M., Iadarola, P.: Proteomic analysis of exhaled breath condensate from single patients with pulmonary emphysema associated to alpha(1)-antitrypsin deficiency. Journal of Proteomics 71, 211–221 (2008)
15. Schettgen, T., Tings, A., Brodowsky, C., Mueller-Lux, A., Musiol, A., Kraus, T.: Simultaneous determination of the advanced glycation end product N-epsilon-carboxymethyllysine and its precursor, lysine, in exhaled breath condensate using isotope-dilution-hydrophilic-interaction liquid chromatography coupled to tandem mass spectrometry. Analytical and Bioanalytical Chemistry 387, 2783–2791 (2007)
16. Conventz, A., Musiol, A., Brodowsky, C., Mueller-Lux, A., Dewes, P., Kraus, T., Schettgen, T.: Simultaneous determination of 3-nitrotyrosine, tyrosine, hydroxyproline and proline in exhaled breath condensate by hydrophilic interaction liquid chromatography/electrospray ionization tandem mass spectrometry. Journal of Chromatography B 860, 78–85 (2007)
17. Prado, C., Marin, P., Periago, J.F.: Application of solid-phase microextraction and gas chromatography-mass spectrometry to the determination of volatile organic compounds in end-exhaled breath samples. Journal of Chromatography A 1011, 125–134 (2003)
18. Gonzalez-Reche, L.M., Kucharczyk, A., Musiol, A.K., Kraus, T.: Determination of N-(epsilon) (carboxymethyl)lysine in exhaled breath condensate using isotope dilution liquid chromatography/electrospray ionization tandem mass spectrometry. Rapid Communications in Mass Spectrometry 20, 2747–2752 (2006)
19. Phillips, M., Cataneo, R.N., Cummin, A.R.C., Gagliardi, A.J., Gleeson, K., Greenberg, J., Maxfield, R.A., Rom, W.N.: Detection of lung cancer with volatile markers in the breath. Chest 123, 2115–2123 (2003)
20. Larstad, M., Soderling, A.-S., Caidahl, K., Olin, A.-C.: Selective quantification of free 3-nitrotyrosine in exhaled breath condensate in asthma using gas chromatography/tandem mass spectrometry. Nitric Oxide 13, 134–144 (2005)
21. Barker, M., Hengst, M., Schmid, J., Buers, H.J., Mittermaier, B., Klemp, D., Koppman, R.: Volatile organic compounds in the exhaled breath of young patients with cystic fibrosis. European Respiratory Journal 27, 929–936 (2006)
22. Kanoh, S., Kobayashi, H., Motoyoshi, K.: Exhaled ethane - An in vivo biomarker of lipid peroxidation in interstitial lung diseases. Chest 128, 2387–2392 (2005)
23. Phillips, M., Cataneo, R.N., Condos, R., Erickson, G.A.R., Greenberg, J., La Bombardi, V., Munawar, M.I., Tietje, O.: Volatile biomarkers of pulmonary tuberculosis in the breath. Tuberculosis (Amsterdam) 87, 44–52 (2007)
24. Phillips, M., Cataneo, R.N., Greenberg, J., Gunawardena, R., Naidu, A., Rahbari-Oskoui, F.: Effect of age on the breath methylated alkane contour, a display of apparent new markers of oxidative stress. Journal of Laboratory and Clinical Medicine 136, 243–249 (2000)
25. Amorim, L.C.A., Carneiro, J.P., Cardeal, Z.L.: An optimized method for determination of benzene in exhaled air by gas chromatography-mass spectrometry using solid phase microextraction as a sampling technique. Journal of Chromatography B 865, 141–146 (2008)
26. Deng, C., Zhang, J., Yu, X., Zhang, W., Zhang, X.: Determination of acetone in human breath by gas chromatography-mass spectrometry and solid-phase microextraction with on-fiber derivatization. Journal of Chromatography B 810, 269–275 (2004)
27. Svensson, S., Larstad, M., Broo, K., Olin, A.-C.: Determination of aldehydes in human breath by on-fibre derivatization, solid-phase microextraction and GC-MS. Journal of Chromatography B 860, 86–91 (2007)

28. Amorim, L.C.A., Cardeal, Z.d.L.: Breath air analysis and its use as a biomarker in biological monitoring of occupational and environmental exposure to chemical agents. Journal of Chromatography B 853, 1–9 (2007)
29. Libardoni, M., Stevens, P.T., Hunter Waite, J., Sacks, R.: Analysis of human breath samples with a multi-bed sorption trap and comprehensive two-dimensional gas chromatography (GC x GC). J. Chromatography B 842, 13–21 (2006)
30. Corradi, M., Rubinstein, I., Andreoli, R., Manini, P., Caglieri, A., Poli, D., Alinovi, R., Mutti, A.: Aldehydes in exhaled breath condensate of patients with chronic obstructive pulmonary disease. American Journal of Respiratory and Critical Care Medicine 167, 1380–1386 (2003)
31. Syslova, K., Kacer, P., Kuzma, M., Klusackova, P., Fenclova, Z., Lebedova, J., Pelclova, D.: Determination of 8-iso-prostaglandin F-2 alpha in exhaled breath condensate using combination of immunoseparation and LC-ESI-MS/MS. Journal of Chromatography B 867, 8–14 (2008)
32. Harris, D.C.: Quantitative Chemical Analysis, 5th edn. W.H. Freeman and Company, New York (1999)
33. Lucidi, V., Ciabattoni, G., Bella, S., Barnes, P.J., Montuschi, P.: Exhaled 8-isoprostane and prostaglandin E-2 in patients with stable and unstable cystic fibrosis. Free Radical Biology & Medicine 45, 913–919 (2008)
34. Montuschi, P., Ragazzoni, E., Valente, S., Corbo, G., Mondino, C., Ciappi, G., Ciabattoni, G.: Validation of 8-isoprostane and prostaglandin E2 measurements in exhaled breath condensate. Inflammation Research 52, 502–507 (2003)
35. Larstad, M., Ljungkvist, G., Olin, A.-C., Toren, K.: Determination of malondialdehyde in breath condensate by high-performance liquid chromatography with fluorescence detection. Journal of Chromatography B 766, 107–114 (2002)
36. Montuschi, P., Kharitonov, S.A., Ciabattoni, G., Barnes, P.J.: Exhaled leukotrienes and prostaglandins in COPD. Thorax 58, 585–588 (2003)
37. Senthilmohan, S.T., Kettle, A.J., McEwan, M.J., Durnmer, J., Edwards, S.J., Wilson, P.F., Epton, M.J.: Detection of monobromamine, monochloramine and dichloramine using selected ion flow tube mass spectrometry and their relevance as breath markers. Rapid Communications in Mass Spectrometry 22, 677–681 (2008)
38. Spanel, P., Smith, D.: Selected ion flow tube mass spectrometry for on-line trace gas analysis in biology and medicine. European Journal of Mass Spectrometry 13, 77–82 (2007)
39. Klebanoff, S.J.: Myeloperoxidase: friend and foe. Journal of Leukocyte Biology 77, 598–625 (2005)
40. Turner, C., Spanel, P., Smith, D.: A longitudinal study of ammonia, acetone and propanol in the exhaled breath of 30 subjects using selected ion flow tube mass spectrometry, SIFT-MS. Physiological Measurement 27, 321–337 (2006)
41. Turner, C., Spanel, P., Smith, D.: A longitudinal study of methanol in the exhaled breath of 30 healthy volunteers using selected ion flow tube mass spectrometry, SIFT-MS. Physiological Measurement 27, 637–648 (2006)
42. Turner, C., Spanel, P., Smith, D.: A longitudinal study of ethanol and acetaldehyde in the exhaled breath of healthy volunteers using selected-ion flow-tube mass spectrometry. Rapid Communications in Mass Spectrometry 20, 61–68 (2006)
43. O'Hara, M.E., O'Hehir, S., Green, S., Mayhew, C.A.: Development of a protocol to measure volatile organic compounds in human breath: a comparison of rebreathing and on-line single exhalations using proton transfer reaction mass spectrometry. Physiological Measurement 29, 309–330 (2008)

44. Karl, T., Jordan, A., Hansel, A., Holzinger, R., Lindinger, W.: Benzene and acetonitrile in smokers and nonsmokers. Berichte des Naturwissenschaftlich-Medizinischen Vereins in Innsbruck 85, 7–15 (1998)
45. Moser, B., Bodrogi, F., Eibl, G., Lechner, M., Rieder, J., Lirk, P.: Mass spectrometric profile of exhaled breath - field study by PTR-MS. Respiratory Physiology & Neurobiology 145, 295–300 (2005)
46. Hansel, A., Jordan, A., Holzinger, R., Prazeller, P., Vogel, W., Lindinger, W.: Proton transfer reaction mass spectrometry: on-line trace gas analysis at the ppb level. International Journal of Mass Spectrometry and Ion Processes 149-150, 609–619 (1995)
47. Borsdorf, H., Eiceman, G.A.: Ion Mobility Spectrometry: Principles and Applications. Applied Spectroscopy Reviews 41, 323–375 (2006)
48. Baumbach, J.I.: Process analysis using ion mobility spectrometry. Analytical and Bioanalytical Chemistry 384, 1059–1070 (2006)
49. Xie, Z., Sielemann, S., Schmidt, H., Li, F., Baumbach, J.I.: Determination of acetone, 2-butanone, diethyl ketone and BTX using HSCC-UV-IMS. Analytical and Bioanalytical Chemistry 372, 606–610 (2002)
50. Ulanowska, A., Ligor, M., Amann, A., Buszewski, B.: Determination of Volatile Organic Compounds in Exhaled Breath by Ion Mobility Spectrometry. Chemia Analityczna 53, 953–965 (2008)
51. Ruzsanyi, V., Baumbach, J.I., Sielemann, S., Litterst, P., Westhoff, M., Freitag, L.: Detection of human metabolites using multi-capillary columns coupled to ion mobility spectrometers. Journal of Chromatography A 1084, 145–151 (2005)
52. Westhoff, M., Litterst, P., Freitag, L., Baumbach, J.I.: Ion mobility spectrometry in the diagnosis of sarcoidosis: Results of a feasibility study. Journal of Physiology and Pharmacology 58, 739–751 (2007)
53. Westhoff, M., Litterst, P., Freitag, L., Urfer, W., Bader, S., Baumbach, J.I.: Ion mobility spectrometry for the detection of volatile organic compounds in exhaled breath of lung cancer patients. Thorax (2009)
54. Molina, M.A., Zhao, W., Sankaran, S., Schivo, M., Kenyon, N.J., Davis, C.E.: Design-of-experiment optimization of exhaled breath condensate analysis using a miniature differential mobility spectrometer (DMS). Analytica Chimica Acta 628, 155–161 (2008)
55. Miller, R.A., Nazarov, E.G., Zhong, M.: NOx monitor using differential mobility spectrometry. Sionex Corporation (2006)
56. Du, F., Huang, W., Shi, Y., Wang, Z., Cheng, J.: Real-time monitoring of NO release from single cells using carbon fiber microdisk electrodes modified with single-walled carbon nanotubes. Biosensors & Bioelectronics 24, 415–421 (2008)
57. Wang, S., Lin, X.: Electrodeposition of Pt-Fe(III) nanoparticle on glassy carbon electrode for electrochemical nitric oxide sensor. Electrochimica Acta 50, 2887–2891 (2005)
58. Shin, J.H., Privett, B.J., Kita, J.M., Wightman, R.M., Schoenfisch, M.H.: Fluorinated Xerogel-Derived Microelectrodes for Amperometric Nitric Oxide Sensing. Analytical Chemistry 80, 6850–6859 (2008)
59. Asakawa, H., Ikeno, S., Haruyama, T.: The molecular design of a PMP complex and its application in a molecular transducer for cellular NO sensing. Sensors and Actuators B 108, 646–650 (2005)
60. Maniscalco, M., de Laurentiis, G., Weitzberg, E., Lundberg, J.O., Sofia, M.: Validation study of nasal nitric oxide measurements using a hand-held electrochemical analyser. European Journal of Clinical Investigation 38, 197–200 (2008)

61. Silkoff, P.E., Carlson, M., Bourke, T., Katial, R., Ogren, E., Szefler, S.J.: The Aerocrine exhaled nitric oxide monitoring system NIOX is cleared by the US Food and Drug Administration for monitoring therapy in asthma. Journal of Allergy and Clinical Immunology 114, 1241–1256 (2004)
62. Thanachasai, S., Rokutanzono, S., Yoshida, S., Watanabe, T.: Novel Hydrogen Peroxide Sensors Based on Peroxidase-Carrying Poly{pyrrole-co-[4-(3-pyrrolyl)butanesulfonate]} Copolymer Films. Analytical Sciences 18, 773–777 (2002)
63. Zetterquist, W., Marteus, H., Johannesson, M., Nordvall, S.L., Ihre, E., Lundberg, J.O.N., Alving, K.: Exhaled carbon monoxide is not elevated in patients with asthma or cystic fibrosis. European Respiratory Journal 20, 92–99 (2002)
64. Camacho, C., Chico, B., Cao, R., Matias, J.C., Hernandez, J., Palchetti, I., Simpsond, B.K., Mascinic, M., Villalonga, R.: Novel enzyme biosensor for hydrogen peroxide via supramolecular associations. Biosensors & Bioelectronics 24, 2028–2033 (2009)
65. Wang, X., Yang, T., Feng, Y., Jiao, K., Li, G.: A Novel Hydrogen Peroxide Biosensor Based on the Synergistic Effect of Gold-Platinum Alloy Nanoparticles/Polyaniline Nanotube/Chitosan Nanocomposite Membrane. Electroanalysis 21, 819–825 (2009)
66. Xi, F., Liu, L., Chen, Z., Lin, X.: One-step construction of reagentless biosensor based on chitosan-carbon nanotubes-nile blue-horseradish peroxidase biocomposite formed by electrodeposition. Talanta 78, 1077–1082 (2009)
67. Anh, D.T.V., Olthius, W., Bergveld, P.: A hydrogen peroxide sensor for exhaled breath measurement. Sensors and Actuators B 111-112, 494–499 (2005)
68. Taha, Z.H.: Nitric oxide measurements in biological samples. Talanta 61, 3–10 (2003)
69. Hadjikoumi, I., Hassan, A., Milner, A.D.: Exhaled nitric oxide measurements in childhood asthma: Comparison of two sampling techniques. Pediatric Research 52, 745–749 (2002)
70. Zappacosta, B., Persichilli, S., Mormile, F., Minucci, A., Russo, A., Giardina, B., De Sole, P.: A fast chemiluminescent method for H2O2 measurement in exhaled breath condensate. Clinica Chimica Acta 310, 187–191 (2001)
71. Cunningham, S., McColm, J.R., Pei Ho, L., Greening, A.P., Marshall, T.G.: Measurement of inflammatory markers in the breath condensate of children with cystic fibrosis. Eur. Respir. J. 15, 955–957 (2000)
72. Dekhuijzen, P.N.R., Aben, K.K.H., Dekker, I., Aarts, L.P.H.J., Wielders, P.L.M.L., Van Herwaarden, C.L.A., Bast, A.: Increased exhalation of hydrogen peroxide in patients with stable and unstable chronic obstructive pulmonary disease. American Journal of Respiratory and Critical Care Medicine 154, 813–816 (1996)
73. Damiani, P., Burini, G.: Fluorometric Determination of Nitrite. Talanta 33, 649–652 (1986)
74. Kojima, H., Hirotani, M., Urano, Y., Kikuchi, K., Higuchi, T., Nagano, T.: Fluorescent indicators for nitric oxide based on rhodamine chromophore. Tetrahedron Letters 41, 69–72 (2000)
75. Barker, S.L.R., Clark, H.A., Swallen, S.F., Kopelman, R.: Ratiometric and Fluorescence-Lifetime-Based Biosensors Incorporating Cytochrome c' and the Detection of Extra- and Intracellular Macrophage Nitric Oxide. Analytical Chemistry 71, 1767–1772 (1999)
76. Szkudlarek, U., Maria, L., Kasielski, M., Kaucka, S., Nowak, D.: Exhaled hydrogen peroxide correlates with the release of reactive oxygen species by blood phagocytes in healthy subjects. Respiratory Medicine 97, 718–725 (2003)
77. Svensson, S., Olin, A.-C., Larstad, M., Ljungkvist, G., Toren, K.: Determination of hydrogen peroxide in exhaled breath condensate by flow injection analysis with fluorescence detection. Journal of Chromatography B 809, 199–203 (2004)

78. Carpagnano, G.E., Kharitonov, S.A., Foschino-Barbaro, M.P., Resta, O., Gramiccioni, E., Barnes, P.J.: Increased inflammatory markers in the exhaled breath condensate of cigarette smokers. European Respiratory Journal 21, 589–593 (2003)
79. Schumann, C., Triantafilou, K., Krueger, S., Hombach, V., Triantafilou, M., Becher, G., Lepper, P.M.: Detection of erythropoietin in exhaled breath condensate of nonhypoxic subjects using a multiplex bead array. Mediators of Inflammation (2006)
80. Sack, U., Scheibe, R., Woetzel, M., Hammerschmidt, S., Kuhn, H., Emmrich, F., Hoheisel, G., Wirtz, H., Gessner, C.: Multiplex analysis of cytokines in exhaled breath condensate. Cytometry 69A, 169–172 (2006)
81. Edme, J.L., Tellart, A.S., Launay, D., Neviere, R., Grutzmacher, C., Boulenguez, C., Labalette, M., Hachulla, E., Hatron, P.Y., Dessaint, J.P., et al.: Cytokine concentrations in exhaled breath condensates in systemic sclerosis. Inflammation Research 57, 151–156 (2008)
82. Matsunaga, K., Yanagisawa, S., Ichikawa, T., Ueshima, K., Akamatsu, K., Hirano, T., Nakanishi, M., Yamagata, T., Minakata, Y., Ichinose, M.: Airway cytokine expression measured by means of protein array in exhaled breath condensate: Correlation with physiologic properties in asthmatic patients. Journal of Allergy and Clinical Immunology 118, 84–90 (2006)
83. Kostikas, K., Gaga, M., Papatheodorou, G., Karamanis, T., Orphanidou, D., Loukides, S.: Leukotriene B-4 in exhaled breath condensate and sputum supernatant in patients with COPD and asthma. Chest 127, 1553–1559 (2005)
84. Kao, P.F., Liebeler, C.L., Blumenthal, M.N.: Measurement of pro-inflammatory and anti-inflammatory cytokines in exhaled breath condensates. Journal of Allergy and Clinical Immunology 113, S289 (2004)
85. Esther Jr., C.R., Jasin, H.M., Collins, L.B., Swenberg, J.A., Boysen, G.: A mass spectrometric method to simultaneously measure a biomarker and dilution marker in exhaled breath condensate. Rapid Communications in Mass Spectrometry 22, 701–705 (2008)
86. Phillips, M.: Method for the Collection and Assay of Volatile Organic Compounds in Breath. Analytical Biochemistry 247, 272–278 (1997)
87. Kharitonov, S.A., Barnes, P.J.: Exhaled Markers of Pulmonary Disease. American Journal of Respiratory and Critical Care Medicine 163, 1693–1722 (2001)
88. Van den Velde, S., Nevens, F., Van Hee, P., van Steenberghe, D., Quirynen, M.: GC-MS analysis of breath odor compounds in liver patients. Journal of Chromatography B 875, 344–348 (2008)
89. Barnes, P.J.: Immunology of asthma and chronic obstructive pulmonary disease. Nature Reviews Immunology 8, 183–192 (2008)

Decomposition of Photoplethysmographical Arterial Pulse Waves by Independent Component Analysis: Possibilities and Limitations

Laila Gbaoui and Eugenijus Kaniusas

Institute of Electrodynamics, Microwaves and Circuit Engineering,
Vienna University of Technology

Abstract. The analysis of the arterial pulse wave becomes an important tool to assess the cardiovascular activity and arterial properties because it contains useful information about the left ventricle activity, the autonomic nervous system dynamics and the heart-brain interaction. The goal of this contribution is to illustrate the ability of the independent component analysis to solve several clinically significant problems including extraction of the reflected and the forward waves from the photoplethysmographical signal as well simultaneous separation of the respiratory and Mayer waves. While the former approach is aimed to increase the reliability of the arterial pulse parameters, the latter one may help to assess more accurately the state of the autonomic nervous system. In particular, the analysis uses dynamical embedding of the photoplethysmographical signal to extract the hidden sources within. In contrast to the standard approaches in time and frequency domain, the applied methodology allows the assessment of the sympathetic and parasympathetic activities based on the independency of the sources contributing to the change in system dynamics.

Keywords: independent component analysis, arterial pulse wave, volume pulse wave, pressure pulse wave, photoplethysmography, pulse wave reflection, blood pressure variability, Mayer wave, respiratory wave.

1 Introduction

The arterial pulse wave in all their forms, pressure, volume or flow represent a promising, most non-invasively, tool for reflecting the status of the cardiovascular system in both clinical and experimental setting.

The volume pulse wave, using the photoplethysmographic (PPG) technique [3], is widely used for measuring the blood pulsations, for monitoring of blood oxygenation saturation and heart rate [23] in intensive care wards and during operative procedures. Its potential to asses the vascular aging [5,45], stiffness [16], endothelial dysfunction [9,14], arteriosclerosis [5] and others disease was recognized in many works. Although, more attention should be paid to its form in the clinical applications because it hides information of great importance in many clinical setting including the respiration monitoring [3,7,21,22,23,28] and the autonomic function assessment such cardiovascular variability assessment, thermoregulation and others. In contrary, the pressure pulse wave

S.C. Mukhopadhyay and A. Lay-Ekuakille (Eds.): Adv. in Biomedical Sensing, LNEE 55, pp. 166–185.
springerlink.com

analysis take more the pulse morphology change in account to assess the left ventricle function and the vascular diseases.

Furthermore, the similarity between the volume and pressure waveform in different body sites and the possibility to derive the central blood pressure from the peripheral pressure using only a constant transfer function encourage the investigation of the low cost and easy handling photoplethysmographical pulse wave.

Particularly important, is the fact that total flow can be derived from the volume pulse wave [11,50]. In [50], the flow pulse wave was extracted from the oximetry pulse wave and compared to the measured flow pulse wave using simple mathematical approach. The results show high correlation between the extracted and measured flow waves. This seems to be very promising to investigate the interaction between the three coherent pulse waves: pressure, flow and volume waves using only the volume pulse wave.

The clinicians usually extract the performance parameters such as augmentation index, reflection index direct from pulse wave [2,35] or using their derivatives to detect the sudden changes [5,16,45]. The approaches are simple and can be used in the real time applications but limited when the waveform is noisy or damped.

A powerful investigation of the pulse wave dynamic provides the non-linear times series analysis [26]. The approach provides information about the time behaviour, the recurrence of the erratic pattern and several hallmarks of the complexity of the underlying system dynamics. However, the non-linear approaches require mostly long time recording and the extracted parameters are more difficult to interpret as the classical time or frequency domain parameters.

Additionally, the most used approaches in the pulse wave analysis are based on the detection of change in the amplitude and time characteristic rather than the analysis of the wave components that cause this change. Thus in the last decades several approaches such as the Fourier decomposition, Gaussian fitting [36], the wave intensity [36,38,39,43] and impedance analysis [32] was developed to decompose the pulse wave into their forward wave generated by the left ventricle ejection and the reflected waves at different sites of the arterial tree system. The approaches are based on different theories [48,43]. Consequently, the number and interpretation of the extracted components depend strongly on the assumed hypothesis which leads to controversial results [43]. However, most of them are not yet entirely introduced in the clinical applications.

Most of these methods, in particularly the wave intensity and impedance analysis, are in fact based on complex mathematical background, give quantitative and temporal changes in the waveform but requires a simultaneous recording of pressure and velocity waves, which is not trivial. Moreover, the most commonly decomposition in the literature is the decomposition of pressure and flow waves.

In order to decompose the pulse wave by a strategic way utilizing the statistical independency of the hidden components, we propose in this contribution a linear decomposition using the Independent Component Analysis (ICA) using only a single channel recording of the PPG pulse wave.

A secondary goal in this contribution is to illustrate the ability of ICA to decompose the PPG pulse wave time variability to assess the autonomic nerve system, particularly the extraction of the Mayer and the respiratory waves. The Mayer wave is generally considered to be a representation of the pressure reflex control system, whereas the respiratory activity represents monitoring parameters of great importance in the sleep medicine and children care units.

Although the successful developments of the ICA in solving the problem of the hidden independent components in the biomedical signal processing[18,19,20,29,47] there are a few works of pulse wave decomposition with ICA [13,28,42], in particularly the pulse wave time variability. The most ICA research is concentred on the artefact elimination [27,40,41].

2 Single Channel Independent Component Analysis

ICA is a statistical technique for decomposing multivariate signals into their underlying source components assuming linear mixing of the sources at the sensors, generally using techniques involving higher-order statistics or temporal decorrelation. Several ICA algorithms can be found in the literature.

As shown in Fig.1 the classical, noise free ICA model assumes that a set of n recorded signals $\boldsymbol{x}(t) = [x_1(t), x_2(t),\ldots, x_n(t)]^T$ for $t = 1\ldots T$ is modelled as a linear combination of $k \leq n$ unknown and statistically independent sources $\boldsymbol{s}(t) = [s_1(t), s_2(t),\ldots, s_l(t)]^T$, i.e.,

$$x_i(t) = \sum_{j=1}^{k} a_{ij} s_j(t) \qquad \text{for } i = 1,\ldots,n \tag{1}$$

The coefficients a_{ij} determine the weight of the sources s_j in the observed signal x_i and form the full rang n x k mixing matrix $\boldsymbol{A}$. The ICA model can be succinctly in vector-matrix form written as

$$x(t) = \boldsymbol{A} \cdot s(t)\,. \tag{2}$$

The ICA algorithms attempt to solve the blind source problem by finding the demixing matrix $\boldsymbol{W}$ separating the sources by only given of a mixture signals x such that

$$\hat{s}(t) = \boldsymbol{W} \cdot x(t), \tag{3}$$

where $\hat{s}$ is the ICA estimate of $\boldsymbol{s}$.

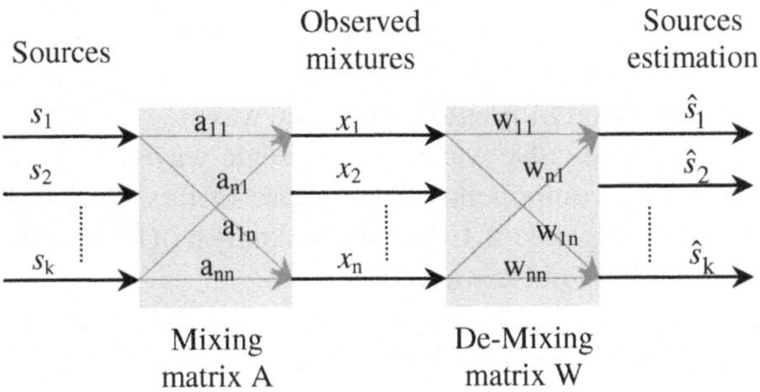

Fig. 1. Classical model of Independent Component Analysis

In the ICA model neither the mixing nor the sources are known, however it is possible to recover the source signals only by analyzing x. The key of the de-mixing algorithm is the spatially statistically independence of the source signals s. It is important to note that the spatial independence differs from the temporal independence, which denote that $s(t)$ is independent from $s(t + 1)$. If at most one of the sources has a Gaussian distribution, or the sources have different spectra, the ICA algorithm can extract the true sources by de-mixing signals up to permutation, scaling and power indeterminacy, by an appropriate estimation of the separating matrix.

Despite the success of the standard ICA model in the real-word applications, the classical setting has several restrictions, which motivate to different extensions of ICA approaches such as convolutive ICA, non-linear ICA or overcomplete ICA, which assumes more sources as sensors. For Instance, by the multidimensional ICA [8] the assumption of having statistically independent component, which lie in one-dimensional subspaces is relaxed to having statistically independent multidimensional subspaces. The approach allows independence between multidimensional signals.

When only a very few or one channel of recording is available the difficulty of separating the signals of interest is increased and necessitates computationally intensive procedures. In this contribution, we used the single channel ICA (SCICA) to extract the underlying components from the single channel recording of the photoplethysmographical pulse wave.

Since the SCICA is an extreme case of the overcomplete ICA (one sensor), thus we introduce the processing steps as shown in Fig.2. First, we break up the single channel recording (1) into a sequence of time delayed m-dimensional state vectors (2) using the Method of Delay and consider these as multi-channel mixing data set for ICA. Than we extract a multiple components by applying the standard ICA algorithm to the

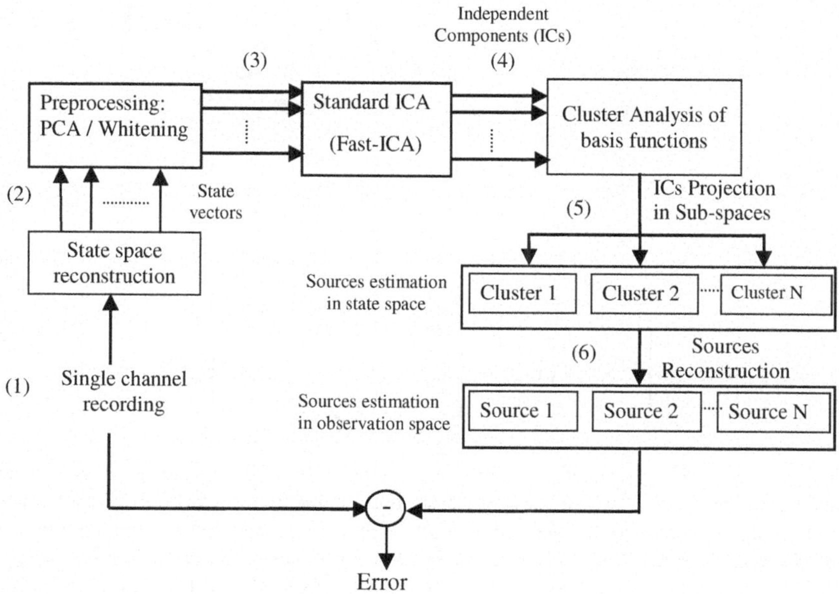

Fig. 2. Procedure steps to apply the independent component analysis to a single channel recording

whitened vectors of delayed samples (3). This implies that multiple ICA outputs (4) may be associated with a single independent component. Thus in last step we cluster the different outputs in state space (5) and reconstruct the independent components (ICs) in the observation space (6).

2.1 Dynamical Embedding

As only a single channel recording of the PPG pulse wave is available and ICA usually requires more observations as sources, thus it is first necessary to build the multichannel data set before applying the ICA algorithms. This is done by breaking up the single channel recording of the pulse wave in a strategic way that the underlying temporal dynamics in a recorded signal is captured.

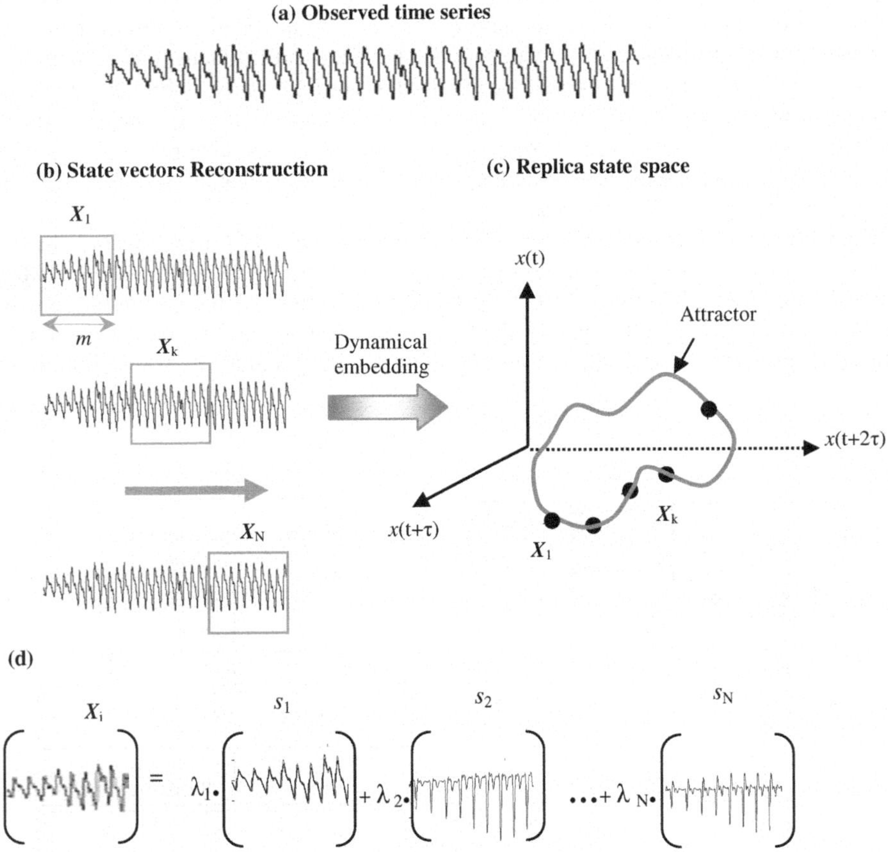

Fig. 3. Example of the dynamical embedding: (a) the recorded PPG waveform is assumed to be generated by a few number of generator processes that interact non-linearly. (b) The reconstruction of the state vectors using the method of delay by setting the time lag τ to one sampling time and the embedding dimension m to a multiple of the cardiac cycle duration. (c) The state vectors evolve an attractor in a manifold embedded in the state space. Its dimension corresponds to the degree of freedom of the dynamical system (d) each state vectors represent a linear summation of the underlying sources.

The non-linear dynamical approach provides much information about the underlying process generating the system dynamics based only on the recorded scalar signal and allows an explanation of the temporal behaviour of the different pattern that appear in irregular and aperiodic signal. To capture this underlying dynamic it is necessary to build an appropriate state space from the scalar time series. This consists of viewing the signal in a high dimensional Euclidean space. The most common state space reconstruction technique in the analysis of non-linear and chaotic systems is the Method of Delay that was first introduced by Takens [10,44]. The basic idea is to construct a matrix of delay $\boldsymbol{M}$ by simply decomposing the scalar recorded signal x into m-dimensional time delayed and overlapped states vectors $\boldsymbol{X}_i = [x_i, x_{i+\tau},\ldots, x_{i+(m-1)\tau}]$ as

$$\boldsymbol{M} = \begin{bmatrix} x_t & x_{t+\tau} & \cdots & x_{t+N.\tau} \\ x_{t+\tau} & x_{t+2\tau} & \cdots & x_{t+(N+1).\tau} \\ \vdots & \vdots & \ddots & \vdots \\ x_{t+(m-1).\tau} & x_{t+m.\tau} & \cdots & x_{t+(m+N-1).\tau} \end{bmatrix} \tag{4}$$

where i is the vector index, τ the time lag and N the number of the consecutive state vectors.

The dynamical embedding assumes that the data are generated by a nonlinear system with a few degrees D of freedom, whose dynamics are generated by a small number of sources forming an unobserved attractor in the original state space. The reconstructed states vectors describe the underlying states of the D-dynamical system and evolve an attractor in the replica state space that preserves the same topological properties of the unobserved attractor. The embedding theory states that m should be at least twice as D to ensure that the mapping provides a faithful representation of the system's attractor. Unfortunately, the value of D is unknown a priori and the ability to estimate it accurately has proven difficult, especially when the data are noisy. This implies that for real word application the value of m should be big enough to capture the information content.

In the practice, if the dynamical embedding is a one stage of an obvious application, it is advisable to estimate the embedding parameters by optimizing the accuracy of the method. If the recorded signal is sampled using an appropriate sampling frequency f_s, one possible technique for estimating the minimal value of m [20] can be based on the lowest frequency of interest f_L and the time lag can be set to one sampling time τ_s, i.e.,

$$m \geq \frac{f_S}{f_L} \tag{5}$$

Once the embedding parameters m, τ and N are adequately chosen, the embedding matrix is rich in information about the underlying temporal dynamics in the recorded signal. In this contribution the PPG pulse wave S_{PPG} was assessed from the left index finger with a sampling frequency of 2 kHz. The measurements were performed in supine position. The SCICA algorithm was performed on S_{PPG} with $\tau = \tau_s$ corresponding to 0.5 ms and m at least 2000 that corresponds to a duration of 1s because the lowest frequency of the underlying sources is unknown a priori. The value of m should be big enough to capture the information (=cardiac) content of pulse wave. To derive the time variability of the pulse wave, the prominent minima in S_{PPG}, which represent the onset of the systolic, were detected as fiducial points for calculation of the beat-to-beat interval signal S_{BBI}. The SCICA algorithm was performed on S_{BBI} with $\tau = \tau_s$ corresponding to 0.3 s because S_{BBI} was interpolated with a frequency of 3 Hz and $m \geq 1000$ because f_L corresponds to the lowest frequency (≈ 0.003 Hz) of the very low oscillations (see section 4.2).

2.2 Extraction and Selection of the Independent Components

At this stage we perform the standard ICA on the whitened states vectors as multi-channel data set. There are several algorithm but we used the Fast-ICA because of its speed and easy implementation. The algorithm uses the fixed point scheme [15,52] for finding the local extrema of the kurtosis

$$Kurt(x)=E\{x^4\}-3(E\{x^2\})^2 \qquad (6)$$

of a zero mean linear observation. It is important to note that finding the kurtosis extremea is equivalent to find the non-Gaussian component. Thus the Fast-ICA algorithm solves the blind source problem by maximizing the non-Gaussianity of sources.

Applying ICA on the embedding matrix we made implicitly the followed assumptions: (1) the underlying sources in state vectors are statistically independent and at least one component has non-Gaussian distributions; (2) the states vectors represent a linear summation of the underlying sources and (3) the sources have a disjoint spectrum.

The first assumptions is fundamental for the blind source separation while the second is not necessarily true for the underlying sources in the pulse wave, but simplify the optimization problem. Generally, the non-linear waves don't interact additively when they meet, however several researchers conclude that the pulse wave decomposition with other methods yield only small difference between the linear and non-linear separation. The third assumption is fundamental for the successful of the separation, however the spectrum of the Mayer and respiratory waves occur in overlapping range. Thus it is expected that the separation of the sources is strongly dependent on the overlapping degree. In contrary, the spectrum of the underlying sources in the raw PPG signal are not a-priori known thus no assumptions can be made.

Applying the ICA on $\boldsymbol{M}$ with a large number of mixing as the expected sources implies that ICA output are still correlated since multiple components are associated with a single channel components. Thus a postprocessing is needed to group the components of interest together. In this case the assumption of the mutually independency of the components is relaxed to the cluster independency. In this step we cluster the basic functions spanning the multidimensional independent subspace where the multidimensional sources lie using the K-means algorithm and project the ICA outputs $\boldsymbol{X}$ into this spaces. This is done by

$$\boldsymbol{M}_{si} = \boldsymbol{A}(:,C_i) \cdot \boldsymbol{W}(C_i,:) \cdot \boldsymbol{X} \qquad (7)$$

Where $\boldsymbol{M}_{si}$ is the matrix of delay of the ith source and C_i the cluster indexes of the ith cluster.

Thus the extraction of the state space is an extreme case of the multidimensional ICA [8]. In the last stage we reconstruct the source in the observation space using $\boldsymbol{M}_{si}$.

3 Pulse Wave Theory

The pulse wave propagation is a complex physiological phenomena observed in the haemodynamic. In the course of the heart systole the blood is ejected and transmitted to the large arteries. The increased blood pressure causes a local widening of the arteries and local accumulation of the blood because of the arteries elasticity. The tension of the widened elastic section causes a contraction and pushes the blood to the next section of the arteries.

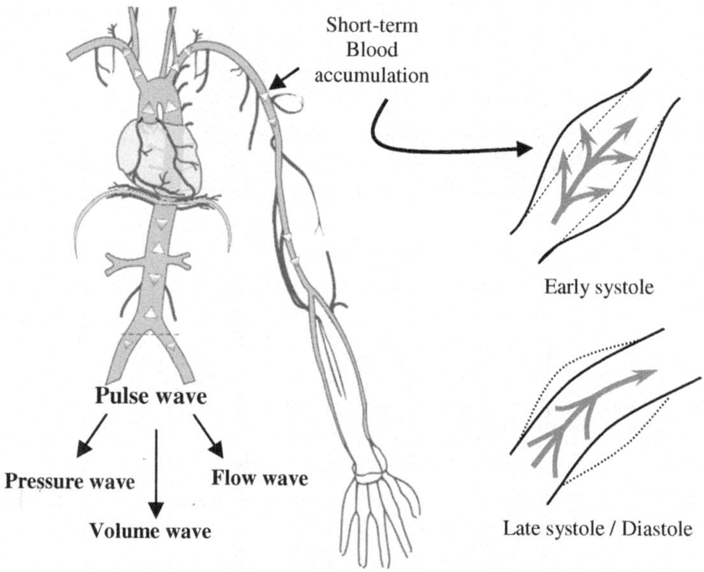

Fig. 4. Propagation mechanisms of the pulse wave based on the continuous transformation of the potential energy into kinetic energy duo to the elasticity of the arteries. In the late systole and during the diastolic the blood flows only into the periphery.

Thus the blood circulation is based on the transformation of the potential or deformation energy into kinetic energy. This process recurs continuously as a propagating pulse along the arterial tree system. So during the pulse wave propagation three coherent waves are observed: the pressure wave duo to the blood pressure changes, the volume wave caused by the arteries cross section change and the flow wave corresponding to the blood flow.

There are several methods to measure the blood volume pulse wave. The most widespread used method is the photoplethysmography. It is often measured non-invasively at the skin surface and is generated by the change in the absorption of infra-red light by blood in the trans-illuminated tissue bed, as shown in Fig.5a. It is similar to the blood pressure pulse wave and has similar change including the damping and loss of pulsation.

PPG contains a varying waveform, which generally corresponds to change duo to the blood pulsation and has a basic frequency corresponding to that of the heartbeat.

This component as depicted in Fig. 5b is superimposed to a slowly varying component that corresponds to the non-pulsatile arterial blood volume, the venous blood and tissues. A typical PPG pulse wave at the periphery is shown in Fig. 5c. The sharp upstroke represents mainly the rapid filling of the arteriolar bed by the stroke volume delivered in course of the heart systole followed by the slow down-stroke that corresponds to volume drain to the venous bed.

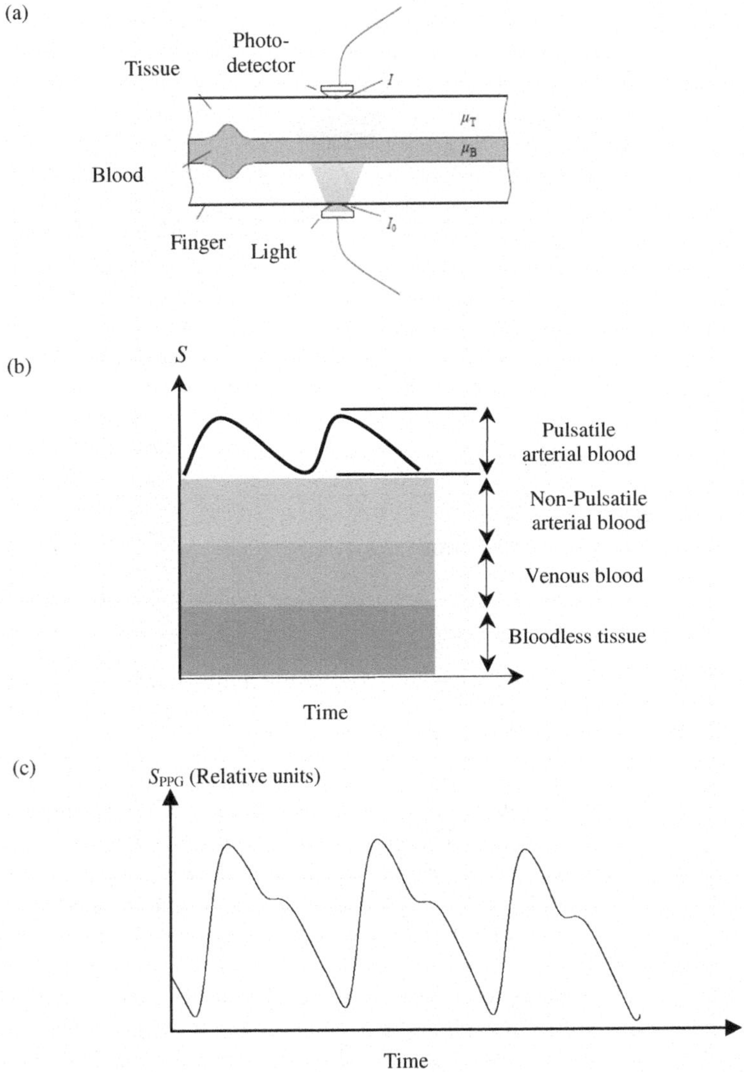

Fig. 5. Photoplethysmography technique: (a) measurement principia, (b) Compartment model of tissue: the photoplethysmographic signal *S* is a composite of the light absorption by pulsatile arterial blood, non-pulsatile arterial blood, venous blood and bloodless tissue and (c) typical PPG pulse wave S_{PPG} at the periphery

The pulse wave spreads to the periphery as forward wave S_{FW}, where it doesn't dissipate but reflected at multiple sites back up to the aorta root and then forwarded back to the periphery. The reflections occur at different branching of the arterial tree system because of resistance mismatch duo to radius and tonus changes.

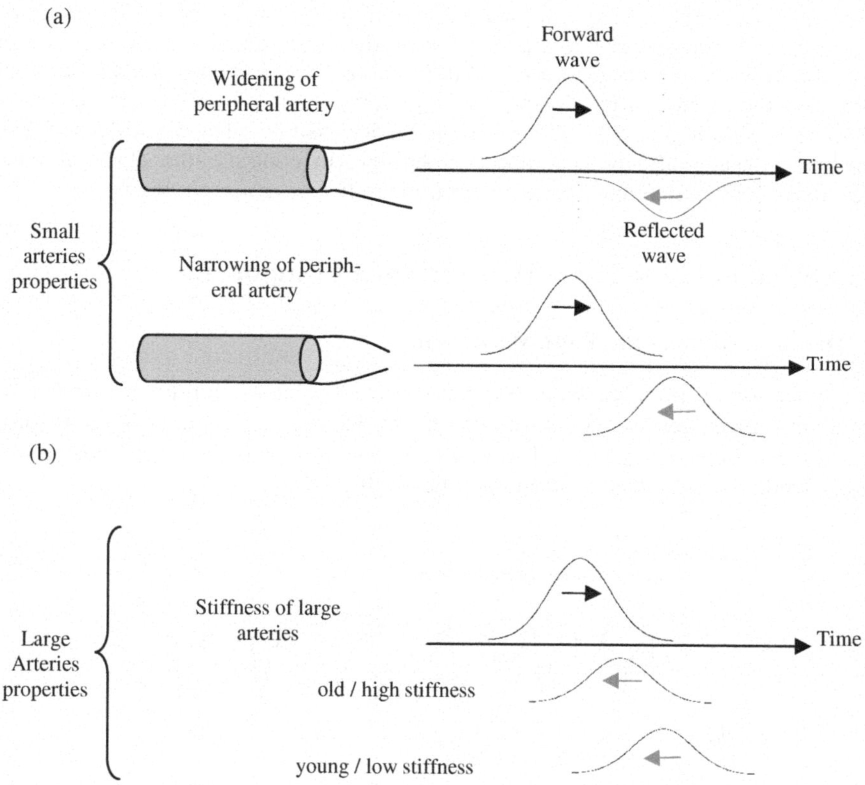

Fig. 6. The properties of the reflected wave depend on the: (a) small arteries properties that determine the degree of reflection and (b) large arteries properties that determine mainly the propagation rate

As demonstrated in Fig.6 the size of the reflected waves S_{RW} depends on the small arteries properties, particularly the resistance mismatch which determines the degree of reflection. The time arrival depends on the path length from the left ventricle to the reflection site and the large arteries properties, in particularly on the stiffness which determines the propagation rate (=speed).

The pulse wave changes in shape while moves towards the periphery and undergoes amplification and different alterations in its form and time characteristics. The classical pulse wave theory ascribes these changes to tapering and the reflection of the pulse wave at different sites of the arterial tree system. A specially, the size and time arrival of S_{RW} have a large contribution in the resulting pulse wave morphology, mainly based on the occurrence of the notch as an inflection point (Class A, B and C) arising from the interaction of the forward wave and the reflected wave. The classical pulse wave analysis approaches are based on the reflection theory and use the time arrival to assess the stiffness of the arteries or its size to assess the vascular disease such as endothelial dysfunction and tonus.

An alternative theory is based on the reservoir model that combines the reflection theory with the windkessel theory [48]. The theory assumes that the compound arterial pulse wave is a sum of a reservoir pressure, duo to the expansion of the arteries during the systole and their passive contraction during the diastole and a wave pressure that drives the forward and reflected waves. Furthermore, the reservoir hypothesis accounts that during the diastole after closing the aorta valve there are no pressure wave with prevailing reservoir pressure. The reservoir theory assumes that the reflection don't contribute considerably to the pulse morphology. In contrary, the classical theory states, that the reflection has a large contribution to the wave morphology.

4 Extraction of the Pulse Wave Hidden Components

4.1 Decomposition of the Pulse Waveform

The application of the SCICA on PPG pulse wave S_{PPG} shows a presence of three independent sources generators contributing to the underlying dynamics in the recorded pulse wave as depicted in Fig.7. The first component S_{IC1} has the largest contribution to S_{PPG}, while the second component S_{IC2} the smallest.

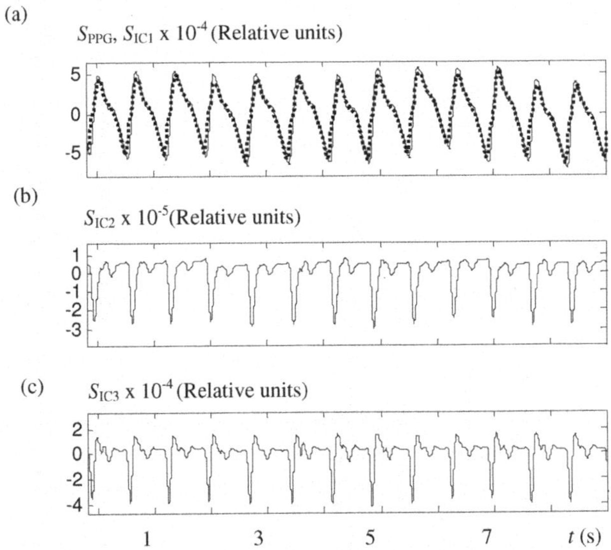

Fig. 7. Decomposition of the phothoplethysmographical signal S_{PPG} (top solid line) into: (a) the first independent component as a forwards wave S_{FW} (dached line) and (b) the second independent component S_{IC2} as local flow wave and (c) the third component S_{IC3} as reflected wave. The interpretation is based on the classical theory that states that the blood pressure waveform comprises out of forward and reflected waves.

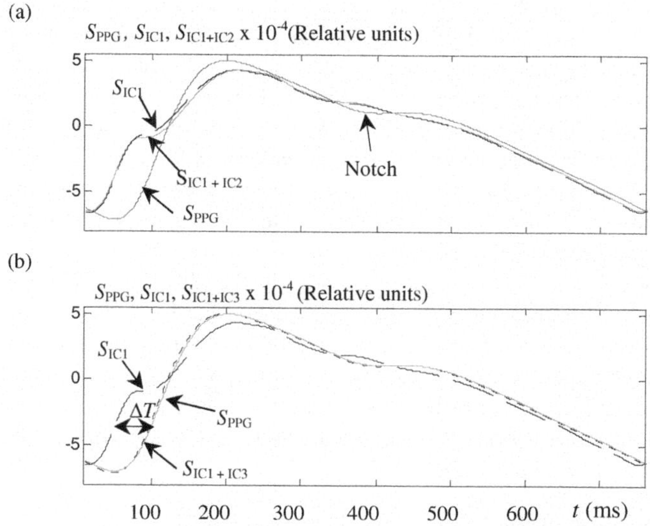

Fig. 8. The contribution of the extracted component into the pulse wave morphology: (a) the component S_{IC2} don't contribute largely to the morphology of S_{PPG} and (b) the component S_{IC3} causes the notch observed in S_{PPG} and leads to delayed recording of S_{PPG} compared to S_{IC1}

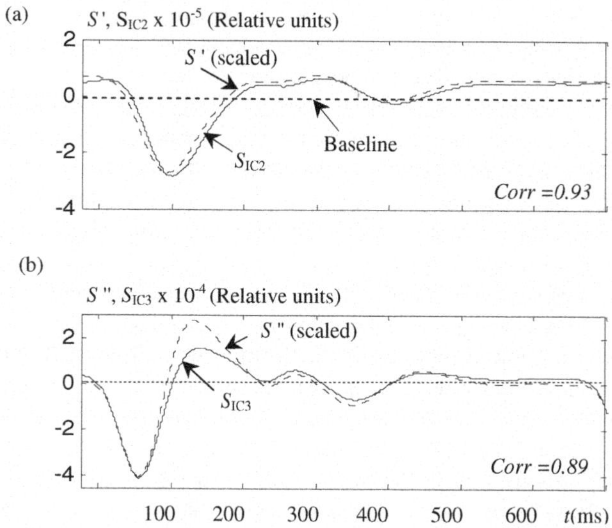

Fig. 9. The correlation between (a) the first derivative S' (inverted and scaled) of pulse wave S_{PPG} and the second extracted independent component S_{IC2}. (b) The second derivative S'' (inverted and scaled) of S_{PPG} versus the third estimated component S_{IC3}.

Of particular important by ICA as discussed in our previous paper [13] is the determination of the time arrival of the extracted components and the delay in between. Furthermore, the analysis shows that S_{IC3} plays a decisive role in the pulse wave morphology and time arrival of the pulse wave, as shown in Fig. 8. It leads to a delayed recording of S_{PPG} (40-60 ms) compared to S_{IC1} and causes the observed notch in S_{PPG} duo to the occurrence of the negative wave at the end of S_{IC3}.

As depicted in Fig. 9, S_{IC2} and S_{IC3} show a large correlation with the first S' and second derivatives S'' of S_{PPG} respectively. This confirms the aforementioned approach based on the derivative of S_{PPG}.

The PPG waveform is generated by change in the absorption of infra-red light by blood in the trans-illuminated tissue bed and reflects the local blood volume changes. Thus its first derivative can be generally accepted as a parameter that describes the local blood flow. On the other hand, as descript above, S' shows a high correlation with S_{IC2}. Consequently, we can attribute this component to the local flow wave.

Ignoring the reservoir pressure, we can attribute S_{IC1} to the forward waveform and S_{IC3} to the integral reflected wave.

4.2 Decomposition of the Pulse Wave Time Variability

The analysis of the heart rate variability (HRV) and blood pressure variability (BPV) provides a powerful, non invasive measure of neurocardiac function that reflects heart-brain interactions and autonomic nervous system dynamics. The spectral analysis of HRV and BPV indicates that such spectra characteristically include three rhythmic oscillations that appear as spectral peaks in predefined frequency ranges [10]. That is, the very low frequency peak which might be associated with humoral activity, while a low frequency and high frequency peaks might be associated with the sympathetic and parasympathetic activities of the autonomic nervous system, respectively.

The low frequency wave, sometimes termed Mayer Wave (MW), has particular significance for diagnostic and patient monitoring proposes. In particular its amplitude and frequency seem to change in connection with hypertension, sudden cardiac death, ventricular tachycardia, coronary artery disease, myocardial infarction, heart failure and diabetes. A difficulty associated with obtaining physiological parameter information based on the MW relates to distinguishing the effects associated with the Mayer wave from effects associated with the respiration wave, particularly in view of the fact that the Mayer wave varies on a frequency similar to that of the respiration, which makes the isolation of these waves with filtering methods difficult.

The MW is largely contributed to the baroreceptor reflex (i.e., a neural receptor as in the arterial walls, sensitive to change in pressure) and is associated with the autonomic nervous system. The MW leads generally fluctuation in the arterial blood pressure which causes a variation in the blood volume of the tissue. These effects can be observed in the PPG pulse wave. Thus it is acceptable to assess effect related to the MW from PPG signal.

As a standard method, the peak detection approach within a specific frequency range assumes that the power of the peak can be used to assess the cardiovascular mechanisms. However, other approaches suggest that a BPV includes rhythmic oscillations that appear in form of peaks and non-rhythmic oscillations that do not appear as a peak but as smeared broad frequency range. In addition, recent studies show that a single cardiovascular control may contribute to different peaks or several controls may contribute to a single peak. Moreover, some studies explain that the very low oscillations are nearly constant, and subdivide the spectrum of the BPV into two domains: low frequency and high frequency domains and analyze the BPV regardless of the VLF oscillations.

Thus the traditional spectral approach on peak detection has drawbacks, generally related to selecting the regulation domains in the spectrum of the beat-to-beat intervals. The application of the well know methods such as factor analysis and principal component analysis that use a linear decorrelation of the components, are limited because of the non-linearity of the pulse wave time variability components.

The nonlinear methods such as Lyapunov exponents [26], fractal correlation dimension [10] are powerful tools for analysis of the BPV dynamics but the physiological interpretation of the extracted parameters is more complex as the parameters of the existing linear methods in time and frequency domain and need long recordings. A solution of the current problem can be attained by application of more complex statistical methods, which allow a separation of the beat-to-beat signal S_{BBI} that represent the pulse wave time variability into independent components.

Moreover, the joint monitoring of the respiratory and cardiac activity is of great importance in clinical use such as in patient monitoring in the emergency, intensive care and children; and the use of methods for indirect extraction of the respiratory activity without any acknowledge of the respiratory frequency estimation is particularly attractive. In order to decompose S_{BBI} by a strategic approach utilizing the statistical independence of the sub-signals and to allow a simultaneous monitoring of the respiratory activity and Mayer wave from the low cost photoplethysmographical signal, we propose again the linear decomposition using the single channel ICA.

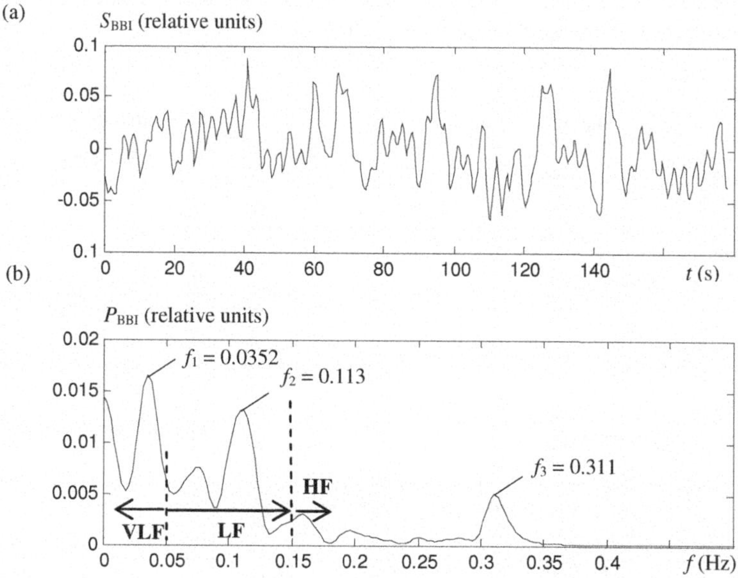

Fig. 10. (a) The beat-to-beat intervals signal S_{BBI} (mean and variance free) derived from S_{PPG} signal and (b) its power spectrum density P_{BBI} that show three dominant peaks in very low (VL), low (LF) and high frequency (HF) range, respectively

The application of the ICA on S_{BBI} (Fig. 10 a) shows generally a presence of three subsignals. As shown in Fig. 11 the first independent component S_1 has a dominant peak in the very low frequency range as indicated in the traditional power spectral analysis, while the dominant peaks of the second S_2 and third component S_3 correspond to the low frequency and high frequency ranges, respectively. The peaks in the spectrum of the three components agree with the peaks in the observed mixture S_{BBI} as shown in Fig. 10. The ICA seems to decompose S_{BBI} into a low frequency component S_{IC1} that corresponds to the averaged blood pressure variability (Fig. 11a) and two higher frequency signals, the S_2 (Fig. 11b) and S_3 (Fig. 11c), that cause the cardiovascular fluctuations occurring around this average.

Furthermore, as depicted in Fig. 11b the component S_2 was found in most sessions within a narrow frequency band around 0.1 Hz. This agreed with the band pass used to isolate the Mayer wave effect from the BPV or the HRV.

Additionally, the analysis shows as depicted in Fig. 12 that the frequency of S_3 is synchronous with the respiratory frequency of the respiratory signal S_R recorded with the skin curvature sensor [24], used as a reference. However the amplitudes of both signals differ at particular times. This was expected, because S_R was recorded at the thorax and S_{PPG} at the left index finger. Consequently, aside the autonomic nerve system modulation and the used algorithm for minima detection in optical signal S_{PPG}, we can ascribe this change to the amplification and alteration in shape and time characteristic that undergoes the pulse wave while moves towards the finger.

As depicted in Fig. 12a S_3 contributes mainly to the high frequency fluctuation in S_{BBI} and seems to affect it in additive way. This is of a particular importance in the isolation of the respiratory activities from pressure signals.

Furthermore, the SCICA seems to give a good decomposition despite the overlapping of the Mayer, respirator and very low waves. The reconstruction of the signal from the extracted components is shown in Fig. 13.

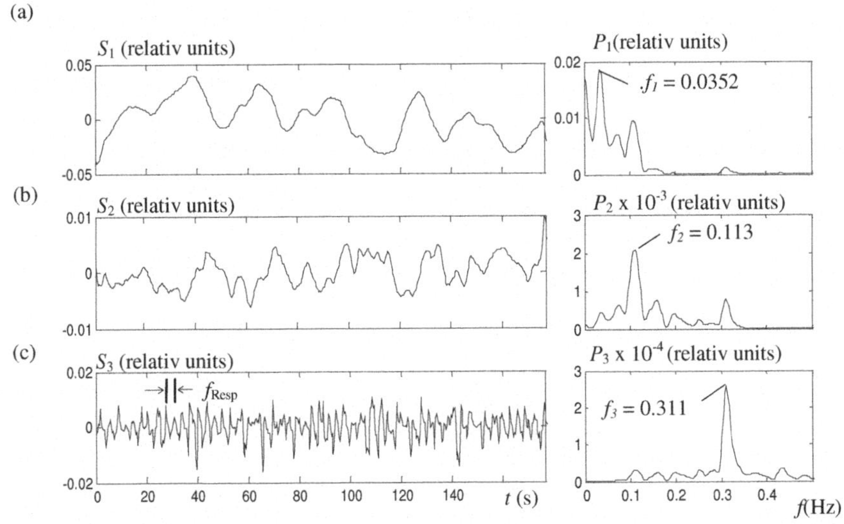

Fig. 11. Decomposition of the PPG pulse wave time variability into: (a) very low oscillation S_I (smoothed), (b) low frequency oscillation S_2 and (c) high frequency oscillation S_3 and the corresponding power spectrum density

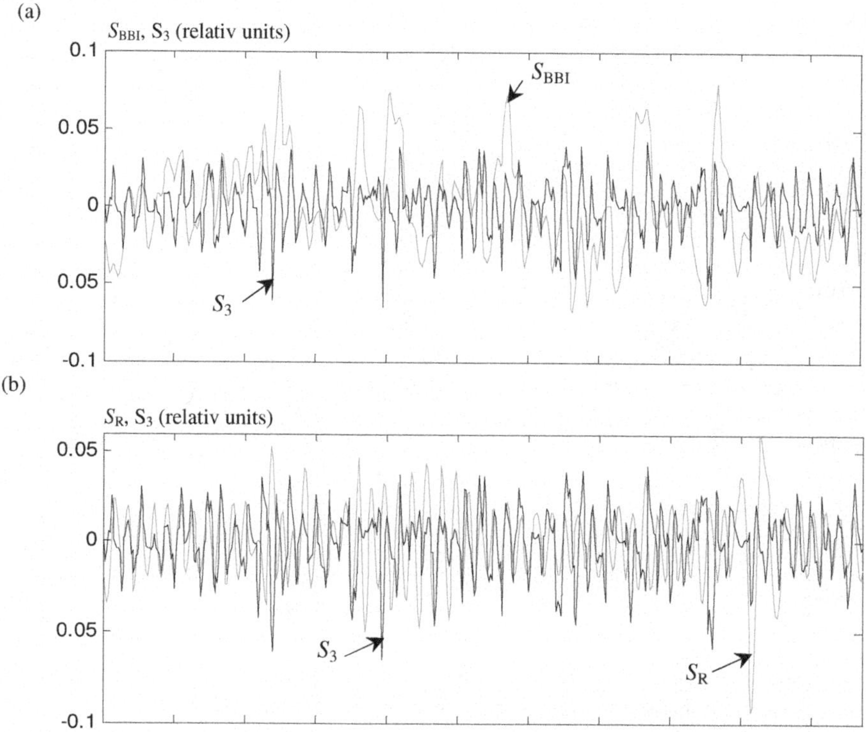

Fig. 12. Comparison between: (a) S_{BBI} (gray) and the extracted high frequency component S_3 (black) shows that S_3 contributes mainly to the high frequency oscillations in the beat-to-beat signal S_{BBI} and represent an additive component, (b) skin curvature signal S_R (gray) as respiratory reference and S_3

5 Discussion

The SCICA shows a powerful tool to decompose the raw pulse wave form S_{PPG} and the derived pulse wave time variability S_{BBI}. However the interpretation of the extracted components from S_{PPG} remains difficult because of the different theories involved. The classical theory ascribes the main change to the reflected waves while the reservoir theory combines the reflection theory with the windkessel theory stating that the reflected wave does not contribute largely to wave morphology. The same controversy can be found in the discussion of the decomposition of the pressure and flow waves using the wave intensity analysis [43].

In the works [11,50], the total flow wave was extracted from the pulse oximetry wave using the first derivative and compared with the Doppler flow wave. The results show high correlation between the both waves. Furthermore, the comparison between the measured and extracted flow waves in [50] shows that the amplitude of the derivative is less than the measured flow wave and a presence of a small time delay (10-30 ms). Similar results were shown between the S_{IC2} and the inverted and scaled first derivative S'(Fig.9). Thus considering this we may attribute S_{IC2} component to the local flow wave at the finger.

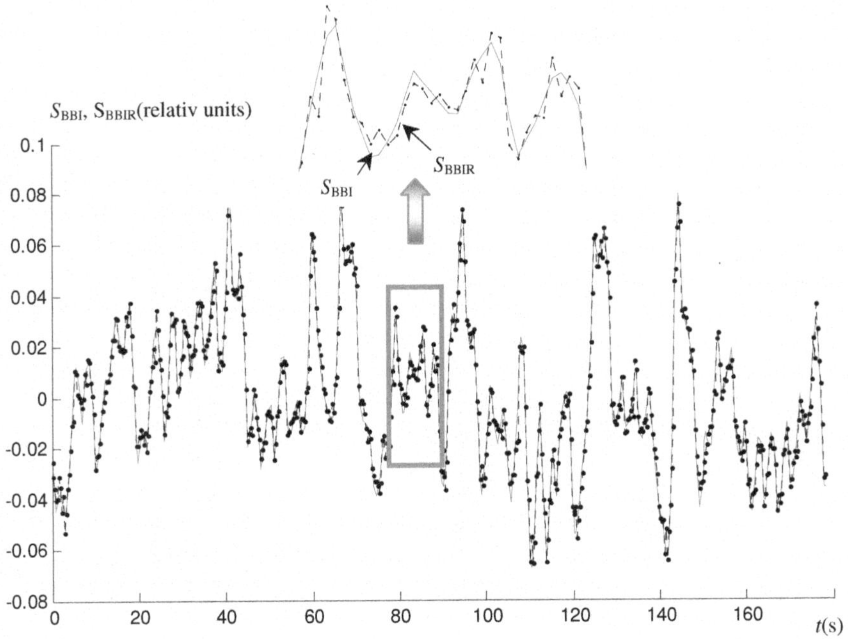

Fig. 13. Comparison between S_{BBI} (gray) before applying SCICA and the reconstructed beat-to-beat signal S_{BBIR} (point) from the estimated independent components with SCICA

On the other hand, SCICA shows a potential tool to decompose the pulse wave time variability regardless the peak position or peak number in the different frequency range of the regulating controls. Thus, in contrary to the standard approaches in time and frequency domain, the proposed approach allows an assessment of sympathetic and parasympathetic activities based on sources contributing to the dynamic change in the observed S_{BBI}. However, it is sensitive to the determination of the embedding parameters, a specially the embedding dimension. Small embedding dimensions corresponding to 2 or 3 min (but no shorter) don't allow a good separation of the very low oscillation, while window larger as 5 min allow generally a good extraction of the very low oscillations. An optimal determination of the embedding dimension seems to be difficult but can be chosen based on the component of interest. In contrary, the decomposition of the pulse wave S_{PPG} is less sensitive to the chose of the embedding dimension.

Furthermore, despite the weak contribution of the high frequency oscillation in the pulse wave time variability in comparison with the heart rate variability, the SCICA shows the possibility to extract the respiratory wave and the Mayer wave. The difference between the extracted and measured respiratory waves could be attributed to many factors such as amplitude alteration, used algorithm and autonomic modulation. Although, the result remain very promising to estimate the respiratory frequency and could be used in the sleep medicine and children units care to detect apnoea.

References

1. Avolio, A.: Input impedance of distributed arterial structures as used in investigations of underlying concepts in arterial haemodynamics. Med. Bio. Eng. Comput. 47, 143–151 (2009)
2. Avolio, A., Westerhof, B.E., Siebes, M., Tyberg, J.V.: Arterial hemodinanics and wave analysis in the frequency and time domains: an evaluation of the paradigms. Med. Bio. Eng. Comput. 47, 107–110 (2009)
3. Allen, J.: Photoplethysmography and its application in clinical physiological measurement. Physiol. Meas. 28, R1–R39 (2007)
4. Allen, J., Murray, A.: Age-related changes in peripheral pulse timing characteristics at the ears, fingers and toes. J. of Human Hypertension 16, 711–717 (2002)
5. Bortolotto, L.A., Blacher, J., Kondo, T., et al.: Assessment of vascular aging and atherosclerosis in hypertensive subjects: second derivative of photoplethysmogram versus pulse wave velocity. American J. of Hypertension 13, 165–171 (2000)
6. Bhattacharya, J., Kanjilal, P.P.: Analysis and characterization of Photo-plethysmographic signal. IEEE Trans. on Biomedical Eng. 48(1), 5–11 (2001)
7. Bartschdorff, D., Zhang, W.: Respiratory rhythm detection with photoplethysmographic methods. In: Proc. 16th IEEE EMBS, vol. 2, pp. 912–913 (1994)
8. Cardoso, J.F.: Multidimensional independent component analysis. In: Proc. of ICASSP 1998, Seattle, WA, pp. 1941–1944 (1998)
9. Chowienczyk, P.J., Kelly, R.P., MacCallum, H., Millasseau, S.C., et al.: Photoplethysmographic assessment of pulse wave reflection: bunted response to endothelium-dependent beta 2-adrenergic vasodilation in type II diabetes mellitus. J. Am. Coll. Cardiol. 34, 2007–2014 (1999)
10. Clifford, G.D., Azuaje, F., McSharry, P.E.: Advanced methods and tools for ECG data analysis. Library of Congress (2006)
11. Cook, L.B.: Extracting arterial flow waveforms from pulse oximeter waveforms. Anaesthesia 56, 551–555 (2001)
12. Davies, M.E., James, C.J.: Source separation using single channel ICA. Signal Proc. 87, 1734–1741 (2007)
13. Gbaoui, L., Kaniusas, E.: Arterial pulse wave decomposition by independent component analysis. In: Proc.4th IEEE Workshop on Medical and Measurement Application, Italy, pp. 111–115 (2009)
14. Hayward, C.S., Kraidly, M., Webb, C.M., Collins, P.: Assessment of endothelial function using peripheral waveform analysis: A clinical application. J. Am. Coll. Cardiol. 40, 521–528 (2002)
15. Hyvärien, A., Oja, E.: A fast fixed point algorithm for independent component analysis. Neural computation 9, 1483–1492 (1997)
16. Imanaga, I., Hara, H., Koyanagi, S., Tanaka, K.: Correlation between wave components of the second derivative of plethysmogram and arterial distensibility. Jpn. Heart Journal 39(5), 775–784 (1998)
17. Iamratanakul, S., McNames, J., Goldstein, B.: Estimation of respiration from physiologic pressure signals. In: Proc. 25th IEEE EMBS conference, Cancun, Mexico, pp. 2734–2737 (2003)
18. Jimenez-Gonzalez, A., James, C.J.: Source separation of foetal heart sounds and maternal activity from single-channel phonograms: a temporal independent component analysis approach. Computers in Cardiology 35, 949–952 (2008)

19. Jimenez-Gonzales, A., James, C.J.: Extracting sources from noisy abdominal phonograms: a single channel blind source separation method. Med. Bio. Eng. Comput. 47, 655–664 (2009)
20. James, C.J., Lowe, D.: Extracting multisource brain activity from a single electromagnetic channel. Artificial Intelligence in Med. 28, 89–104 (2003)
21. Johansson, A., Öberg, P.A.: Estimation of respiratory volumes from the phothoplethysmographic signal. Part I: experimental results. Med. Biol. Eng. Comput. 37, 42–47 (1999)
22. Johansson, A., Öberg, P.A.: Estimation of respiratory volumes from the phothoplethysmographic signal. Part II: a model study. Med. Biol. Eng. Comput. 37, 42–47 (1999)
23. Johansson, A., Öberg, P.A., Sedin, G.: Monitoring of heart and respiratory rates in newborn infants using a new photoplethysmographic technique. J. Clin. Monitoring 15, 461–467 (1999)
24. Kaniusas, E., Pfützner, H., Mehnen, L., et al.: Optimisation of magnetoelstics bilayer sensors for medical applications. Int. J. of Appl. Electromagnetics and Mechanics 28, 193–199 (2008)
25. Kanijilal, P.P., Bhattacharya, J., Saha, G.: Robust method for periodicity detection and characterization of irregular cyclical series inters of embedded periodic components. Am. Physical Society 59, 4013–4025 (1999)
26. Kranz, H., Schreiber, T.: Nonlinear time series analysis. University Press, Cambridge (2004)
27. Krishnan, R., Natarajan, B., Warren, S.: Analysis and detection of motion artifact in photoplethysmographic data using higher order statistics. In: ICASSP 2008. IEEE, Los Alamitos (2008); 1-4244-1484-9/08
28. Leonard, P., Grubb, N.R., Clifton, D., James, N.: An algorithm for the detection of individual breaths from the pulse oximeter waveform. J. of Clinical Monitoring and Computing 18, 309–312 (2004)
29. Li, Z.Y., Liu, S.R., Wang, W.: Using independent component analysis to research heart rate variability. In: Proc. of 27th conference EMB, pp. 5532–5535 (2005)
30. Mendelson, Y.: Pulse Oximetry:Theory and applications for noninvasisve monitoring. Clin. Chem. 39(9), 1601–1607 (1992)
31. Melis, M.D., Moribiducci, U., Rietzschel, E.R., et al.: Blood pressure wave analysis by means of wavelet transform. Med. Bio. Eng. Comput. 47, 165–173 (2009)
32. Mitchell, G.F.: Clinical achievements of impedance analysis. Med. Bio. Eng. Comput. 47, 153–163 (2009)
33. Millasseau, S.C., Guigui, F.G., Kelly, R.P., et al.: Noninvasive assessment of the digital volume pulse comparison with the peripheral pressure pulse. J. Hypertension 36, 952–956 (2000)
34. Millasseau, S.C., Ritter, J.M., Takazawa, K., Chowienczyk, P.J.: Contour analysis of the photoplethysmographic pulse measured at the finger. Application J. Hypertens 24, 1449–1456 (2008)
35. O'Rourke, M.F.: Time domaine analysis of the arterial pulse in clinical medicine. Med. Bio. Eng. Comput. 47, 119–129 (2009)
36. Parker, K.M.: An Introduction to wave intensity analysis. Med. Bio. Eng. Comput. 47, 175–188 (2009)
37. Rubins, U.: Finger and ear photoplethysmogram waveform analysis by fitting with Gaussians. Med. Biol. Eng. Comput. 46, 1271–1276 (2008)
38. Siebes, M., Kolyva, C., Verhoeff, B.J., et al.: Potential and limitations of wave intensity analysis in coronary arteries. Med. Bio. Eng. Comput. 47, 233–239 (2009)

39. Sugawara, M., Niki, K., Ohte, N., et al.: Clinical usefulness of wave intensity analysis. Med. Bio. Eng. Comput. 47, 197–206 (2009)
40. Stetson, P.: Independent component analysis of pulse oximetry signals based on derivative skew. In: Puntonet, C.G., Prieto, A.G. (eds.) ICA 2004. LNCS, vol. 3195, pp. 1072–1078. Springer, Heidelberg (2004)
41. Steton, P.F.: Independent component analysis of pulse oximetry signals. In: Proc. of 26th IEEE EMBS, San Francisco, CA, USA, pp. 231–234 (2004)
42. Shannon, T.T., McNames, J., Ellenby, M.S., Goldstein, B.: Estimation of the additive respiration effect in blood pressure signals: an independent component approach. In: Proc. EMBEC 2002, vol. 3, pp. 1580–1581 (2002)
43. Tyberg, J.V., Davies, J.E., et al.: Wave intensity analysis and the development of the reservoir-wave approach. Med. Bio. Eng. Comput. 47, 221–232 (2009)
44. Takens, F.: Detecting strange attractors in turbulence. In: Rand, D.A., Young, L.S. (eds.) Dynamical systems and turbulence. Lecture Notes in Mathematics, vol. 898, pp. 366–381 (1981)
45. Takazawa, K., Tanaka, N., Fujita, M., Matsuoka, O., et al.: Assessment of vasoactive agents and vascular aging by the second derivative of the photoplethysmogram waveform. J. Hypertension 32, 365–370 (1998)
46. Thayer, J.F., Sollers, J.J., Ruiz-Padial, E., Vila, J.: Estimating respiratory frequency from autoregressive spectral analysis of heart period. In: IEEE EMB, pp. 41–45 (2002)
47. Vetter, R., Celka, P., Vesin, J.M., Scherrer: Sub-Signal extraction of RR time series using independent component analysis. In: Proc. of 20th conference of the IEEE EMBS, vol. 20(1), pp. 286–289 (1998)
48. Westerhof, N., Lankhaar, J.W., Westerhof, B.E.: The arterial windkessel. Med. Bio. Eng. Comput. 47, 131–141 (2009)
49. Wickramasinghe, Y., Spencer, S.A.: Photoplethysmography: a futuristic view for neonatal care. ACTA Peadiatr. 89, 769–770 (2000)
50. Wisely, N.A., Cook, L.B.: Arterial flow waveforms from pulse oximetry compared with measured dopller flow waveforms. Anaesthesia 56, 556–561 (2001)
51. Yao, J., Warren, S.: A short study to assess the potential of the independent component analysis for motion artefact separation in wearable pulse oximeter signals. In: Proc. of 27th IEEE EMBS conference, Shanghai, China, pp. 3585–3588 (2005)
52. FastICA Matlab Package, `http://cis.hut.fi/projects/ica/fastica`

Digital Processing of Diagnostic Images

Domenico Capriglione[1], Luigi Ferrigno[1], Consolatina Liguori[2], Alfredo Paolillo[2], Paolo Sommella[2], and Francesco Tortorella[1]

[1] DAEIMI Univeristy of Cassino – via G. Di Biasio 43, 03043 Cassino (FR), Italy
[2] DIIIE Univeristy of Salerno via ponte don Melillo, 84084, Fisciano (SA), Italy

Abstract. In the Chapter different measurement systems for medical diagnosis are described. Different kinds of diagnostic images are exploited: ultrasound images for carotid analysis, epiluminescence microscopy (ELM) images for skin lesion diagnosis, and mammograms for breast cancer diagnosis. Thanks to the difference in the nature of images and in the investigated quantities the obtainable suggestions can be useful for a wide field of image processing for medical parameter evaluation.

Keywords: digital image processing, medical diagnosis, measurement uncertainty in digital processing.

1 Introduction

Electronic devices for the analysis of health of human organs, tissues, health status have become more and more common in medical science and have introduced enormous benefits to everyday life of everyone of us. In last decades, devices such as Computed Tomography (CT), Magnetic Resonance Inspection (MRI), Ultrasound Imaging, X-ray Radiography, and various kinds of visible and infrared imaging, have become common tools for physicians which use them in order to support the diagnostic activities and to monitor the effectiveness of medical care. The data yielded by all these instruments (often structured or represented as images) allow diagnosticians to "see" inside human body, often with a high level of details (*e.g.* ultrasound imagers can map the speed of blood flowing through arteries and veins). However, these images require a relevant effort and medical experience in order to be effectively interpreted, due to physical phenomena exploited for the acquisition and to the specific nature of resulting images, and have played the role of powerful yet passive tools in the hands of diagnosticians for many years. Medical scientists realized that the rapid progress in computer science and in its various areas could have brought these sophisticated transducers to take a more active part as Computer-Aided Diagnosis tools. The idea is to process the raw images yielded by diagnostic devices in order to highlight structures, phenomena, values of parameters (*e.g.* the speed of blood, the heartbeat frequency) with respect to a background made of useless details non relevant for the diagnosis or due to some kind of superimposed noise. The development of processing procedures can draw from many scientific areas, such as image processing, object recognition, image analysis and restoration, artificial intelligence, to name a few. In fact, researchers involved in the development of computer-aided diagnostic tools have

S.C. Mukhopadhyay and A. Lay-Ekuakille (Eds.): Adv. in Biomedical Sensing, LNEE 55, pp. 186–209.
springerlink.com

to exploit their knowledge in different fields, since the nature of medical images can be very different from one case to another, due to many different reasons. Computer-aided diagnostic tools are designed in order to help physician to make a diagnosis, or to evaluate the effectiveness of a medical treatment. To this purpose, a real challenge arises: the results of the processing should be quantitative, objective and comparable to other results, namely from the same patient. In other words, the digital processing of diagnostic images should be regarded as a measurement process. The quantity to be measured, *i.e.* the measurand, is very particular, often not exactly defined, and this is the reason of several medical image processing difficulties. For instance, in many applications of medical image processing, it is very difficult to have an alternative reference measurement to be used in calibrating a processing procedure, or to have a reference measurand, because measurements have to be performed *in vivo*. In other cases it is difficult to evaluate statistics on results because the number of occurrences of a specific disease is, fortunately, limited. For all these reasons, the interest of researchers is devoted to the definition of methodologies and to the planning of data collection and elaboration in order to assess the effectiveness of the image processing procedures. In the following, applications of medical image processing will be described for different diagnostic problems, in the area of ultrasound imaging, epiluminescence imaging and mammography.

2 Measurement Systems for Carotid Analysis Using Ultrasound Imaging

The atherosclerosis is a degenerative process of the artery status and whose consequences (cerebral infarction, embolus, ictus, ischemia) are the main dead causes in the occidental word. The atherosclerotic process is strongly correlated with carotid thickening, whose presence can be clearly detected in the images of the artery longitudinal section provided by ultrasound techniques [1], [2]. Many non-invasive methods, as the ultrasound techniques, such as B-mode, Color Doppler, Color Power Angio (CPA) can clearly detect these phenomena. In fact, the analysis of the obtained images allows a diagnostician to find out IMT thickenings and plaques and to evaluate its nature and shape. Images are taken by ultrasound equipment and featured by a probe which has to be positioned on the patient neck. The human operator suitably handles the ultrasound probe until a position is reached that gives a noiseless and clear image of the carotid part of interest.

Thickness measurements are then executed by the operator, which uses a track-ball to select measurement points on leading edges of the captured ultrasound image. Because of the human eye resolution in the detection of leading edges, the measurements are characterized by low reproducibility and low accuracy, so that some techniques were developed for the leading edge detection and for carotid Intima Media Thickness (IMT) measurement [3]-[5], but they are still characterized by high operator interactivity and execution time.

Furthermore with 2-D image analysis, some spatial information, in particular about the carotid volume occlusion, could be lost. Consequently, there is a growing clinical interest about 3-D ultrasound imaging, since a 3-D visualization of the carotid artery helps in the accurate definition of the 3-D size and geometry of atherosclerosis plaques [6]-[9].

In the following, two measurement systems will be described. The former allows 2-D carotid analysis, and the latter is able to reconstruct the 3-D profile of the carotid in order to determine the volume of plaques and the percentage of carotid occlusion (stenosis). Both systems are characterized by short execution time, user friendly interface, and overall high accuracy. They allow also that measurement results, concerning images taken in successive times on the same patient, can be used for a precise disease monitoring.

2.1 The Carotid Analysis

Each carotid artery is characterized by a longitudinal tract (common carotid) that, after an enlargement (carotid sinus) containing a flow divider, bifurcates in two arteries the internal and external on the basis of their position respect to the neck skin. Artery walls are made of three layers (tunicae): intima, media and adventitia. In ultrasound impulse based analysis, blood (arterial lumen) and wall layers exhibit difference in wave reflection capability caused by their different density and elasticity. As a result, arterial lumen and tunica media do not reflect ultrasound waves thus allowing the intima-lumen and adventitia-media interfaces to be identified. In Fig. 1 a monochromatic carotid ultrasonic image is reproduced in negative evidencing the carotid interfaces of the two walls (near wall and far wall).

The main symptom of atherosclerosis (found in different age and race people) is the carotid intima layer thickening in proximity of the endotelial lumen surface. This thickening can be also focused on a short artery segment, and in this case it is called plaque. Its detection and evaluation is made carrying out the intima-media thickness measurement, which is stated to be the distance between (2) and (3) for the near wall, and between (4) and (5) for the far wall. Reference values of the intima-media thickness (IMT) are the following:

- Normal: IMT < 1mm;
- Thickening: 1mm < IMT < 1.3 mm;
- Plaque: IMT > 1.3 mm.

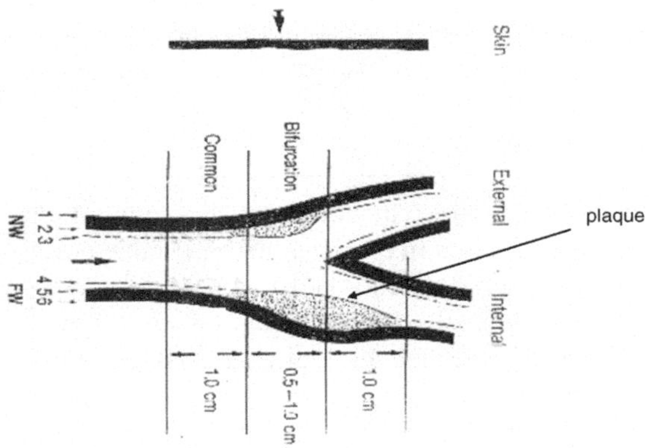

Fig. 1. Negative ultrasound image of the carotidy artery, with the interfaces: 1 Periadventitia – adventitia (NW), 2 Adventitia – media (NW), 3 Intima – lumen (NW), 4 Lumen – intima (FW), 5 Media – adventitia (FW), 6 Adventitia – periadventitia (FW)

2.2 The System for IMT Measurement: User Interface

The procedure for IMT measurement is semi-automatic, so some actions are required to the human operator. At first he moves the ultrasound probe to achieve a noiseless and clear image of the carotid, then on the stopped image he is required to draw and measure with the track-ball of the ultrasound equipment a vertical segment, and finally the user selects the area of interest for the image analysis.

The system user-friendly interface entrusts to display, keyboard and, mouse the interaction with the user. A suitable virtual button allows the image acquisition to be triggered by the user. Once the image has been captured, it appears on the measurement system virtual panel (Fig.2 A). Here the user draws with the computer mouse a rectangle, which selects the area of interest for the image analysis. Having focussed the attention on this area, the processing software finds and underlines on the image the adventitia-media and intima-lumen interfaces of both near and far wall. Also the one wall (near or fall) analysis is allowed to be made. The rest of the panel includes several diagrams reporting for each column from left to right the intima-media thickness for each wall, the lumen-lumen diameter and the adventitia-adventitia diameter. Finally a scroll-bar allows the user to select any transversal section of the carotid along the tract included in the rectangle with the resolution of the pixel and, for each position of the cursor, the corresponding values of the above mentioned quantities are reported in suitable windows (Fig.2).

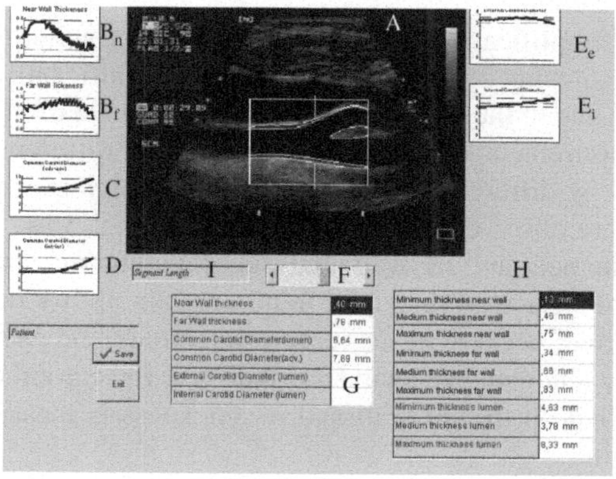

Fig. 2. Output panel: **A.** The interfaces found in the user selected area of the captured image; **B.** IMT trend for near (Bn) and far (Bf) wall; **C.** adventitia-adventitia common carotid diameter trend; **D.** intima-intima common carotid diameter trend; **E.** intima-intima external carotid diameter trend (Ee) and intima-intima internal carotid diameter trend (Ei); **F.** scrollbar for the x value at which the measurement values refer. **G.** measured value table; **H.** statistical representation of data concerning the analyzed area; **I.** cursor for user defined measurement.

2.3 The System for IMT Measurement: Processing Software

The image processing software implements original algorithms, which were ad hoc designed and set up. The core of the measurement software is constituted by a Pattern Recognition and Edge Detection (PRED) algorithm [10] and a Measurement algorithm.

The PRED algorithm has the task of finding the two lumen-intima and media adventitia interfaces for each wall. Using the obtained results, the Measurement algorithm measures lumen diameter, tunica media diameter, intima-media thickness (of near wall and/or far wall). The PRED algorithm with the aim of finding, in the area selected by the user, all the pixels belonging to the two interfaces of interest, analyzes the intensity gradient in order to identify the relative minimums and maxima corresponding to the interfaces. Because of the noise due to the blood turbulences, ultrasound equipment operation and so on, suitable closing and median filtering algorithms are applied to the images allowing to better analyze the gradient [10]. Finally on all these edge points a smoothing filter is applied to obtain a noiseless edge.

The Measurement algorithm, for each column of the selected area, evaluates distances between the corresponding edge points are defined as number of pixels and so as difference between row indices. As a result, diameters and thickness are evaluated as product between a number of pixels and the conversion constant factor obtained in the calibration phase.

2.4 The System for IMT Measurement: Experimental Results

First at all the uncertainty introduced by the measurement software is evaluated using a priori analytical evaluation [11], based on the application of the ISO-GUM [12] to the relationships used for the thickness and diameter measurements. In this evaluation, the contributes of the spatial quantization, of the ultrasound equipment measurement capability in the calibration segment evaluation and of the detection algorithm were considered.

Making realistic simplifications [10], an uncertainty of about 0.02 mm for a segment of 1.00 mm is calculated; this value seems to be significantly lower than the uncertainty component due to the intrinsic uncertainty of the measurand.

The realized measurement system was connected to different ultrasound equipments and was used by different human operators, and high sensitivity and noise rejection capability were observed. As for an example in Fig. 3 two zoom views of the results concerning the near and far wall analysis of a common carotid artery were reported. In fig. 3a) the near wall thickness trend diagram allows the thickening of the near wall to be detected; because of the its low magnitude, the phenomenon would be hardly recognized by a human eye analysis. Finally, in the carotid of Fig. 3. b) an evident plaque is detected on the far wall and its thickness is measured with an actual resolution of 0.01 mm.

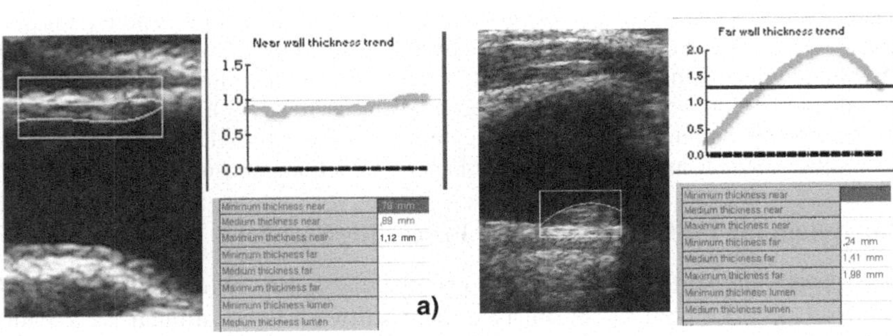

Fig. 3. Zoom of the output of the measurement system for different common carotid arteries a) near wall analysis, b) far wall analysis

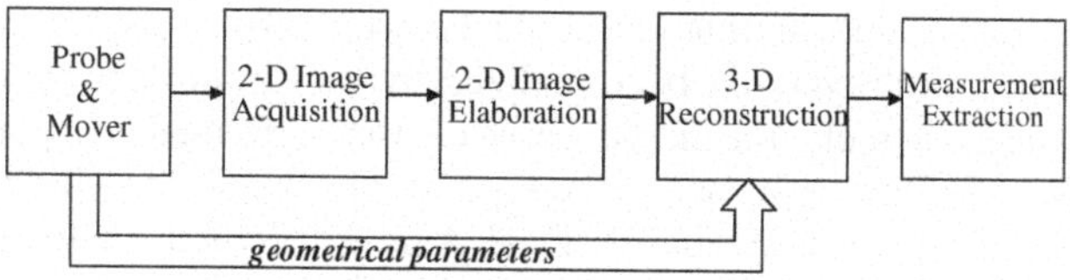

Fig. 4. Block diagram of a typical measurement system for 3-D ultrasound imaging

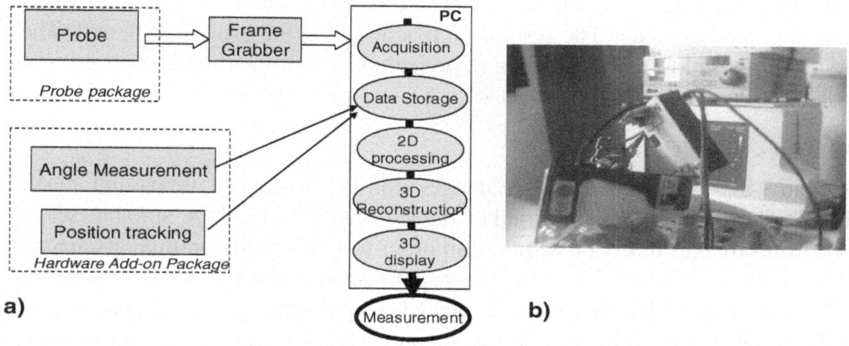

Fig. 5. (a) Scheme and (b) photo of the proposed prototype

2.5 The Instrument for 3-D Evaluation: Architecture

In Fig. 5 a scheme of a typical measurement system for 3-D ultrasound imaging is reported. The first block realizes the necessary hardware to capture the 2-D ultrasound images. The acquired two-dimensional images are processed in order to extract some information required for the 3-D reconstruction. Finally the 3-D image is processed to achieve the required measurements.

Different 3-D ultrasound instruments are present on the market; they can be classified into two classes: i) "add-on” devices, and ii) devices employing sensor arrays. i) "Add-on” devices exploit ordinary 2-D ultrasound probes, but have additional hardware in order to locate each image of the set of acquired images in the 3-D space. ii) Sensor array devices employ special probes, containing arrays of ultrasound emitter/receivers for which the image scanning plane can be changed electronically. Generally, add-on devices could be less expensive and smaller than sensor array devices, but the most of the system present on the market is able to make only qualitative measurements. The proposed measurement system (Fig. 5) is a low cost add-on instrument and is able to reconstruct the volumetric image of the carotid artery and to evaluate the plaque volume and the artery stenosis [13]-[14].

The probe and mover

A basic point in the realization of the whole prototype is the right choice of the ultrasound probe and the mover system. A good quality in the acquired images and accurate information on the probe position are necessary to obtain correct edge detection and accurate 3D reconstruction, respectively. For these reasons CPA imaging techniques have been selected; they highlight also slow blood fluxes to be evidenced whichever the ultrasound orientation could be with respect to the flux direction. A freehand scanning prototype has been realized. A very important parameter is the insonification angle, namely the angle between the start analysis plane and current scanning plane. In the proposed prototype it has been measured by a specific sub-system based on a AccuSwitch™ Dual Axis Tilt Switch inclinometer with 0.2° resolution and +/- 20° range. To allow a correct 3D image reconstruction also the information about the current position of the ultrasound probe is required. In order to reduce the cost of the overall system, instead of measuring an absolute position, a relative measurement is performed. This one is made using an *ad hoc* device, composed of a small encoder. The resolution reached is of the order of 0.1 mm. The only limitation imposed to the measurement procedure is that the doctor must not raise the ultrasonic probe from the patient neck.

2-D Images Acquisition

To acquire the B-mode and CPA images coming from the ultrasound machine a PAL/NTSC compatible frame grabber device has been used. A suitable set of 2-D CPA carotid images has been acquired in order to find out the optimal arrangement in terms of patient position, 2-D type images and acquiring set-up to allow the best 3-D reconstruction and reliability to be achieved. In particular, some choices have to be made: type of image and patient arrangement, scanning angles, number of images to acquire, timing in the image acquisition. Many experimental tests have been carried out and the consequently procedure for the 2-D image acquisitions is the following:

- The probe is suitably positioned on the patient neck;
- The operator handles the probe until both best longitudinal view and image quality are reached;
- The operator captures the image, and selects a box for the CPA modality;
- The probe is rotated around the pivot axis until another defined angular position is reached;
- The previous steps are repeated until the chosen number of acquisition is completed (the deviation is constrained in ±5 degrees).

2.6 The Instrument for 3-D Evaluation: Image Processing

Each 2-D image is processed in order to extract the four interfaces: media-intima, intima-lumen in near wall, lumen-intima and intima-media in far wall [13]-[14]. The 2-D processing algorithm outputs four arrays of the pixel coordinates of the points representing the contours of the four wanted interfaces, for each acquired image. Then, the 3-D reconstruction algorithm locates each point of each interface in the 3-D space with respect to a given reference system (O, x_i',y_i',z_i'), based on the knowledge of its pixel coordinates (x_i,y_i) and of the orientation and position values (θ_x, θ_y, d_x, d_y) of the probe, as measured during the acquisition of the considered image. By processing each point of the four interfaces of each image, the 3-D reconstruction of the carotid can be displayed and saved. The plaque volume evaluation is carried out considering cross sections of the reconstructed 3-D surface [13]. The 3-D object is sliced in the x direction, with a x-step, Δx, equal to a pixel. For each cross section in the z-y plane the area, A_p, delimited by the interpolated lumen-plaque and far wall lumen-intima

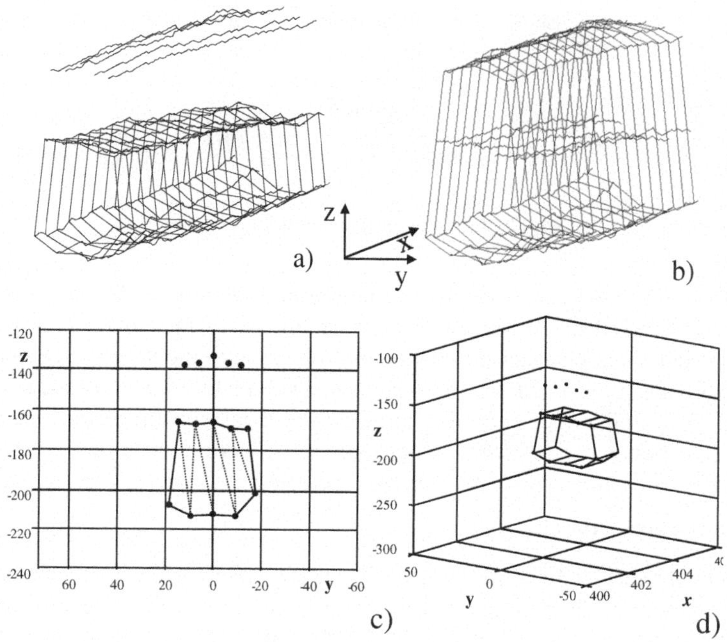

Fig. 6. An example of (a) plaque volume reconstruction, (b) carotid volume reconstruction, (c) plaque section measurement, (d) single slice plaque volume evaluation

interfaces Fig. 6a) is evaluated. The area of the so obtained polygon is computed by dividing it into adjacent triangles with consecutive points of the section contour (Fig. 6c). Finally the plaque volume is obtained summing the single slice volumes (Fig. 6d). The achieved measurement values are expressed in pixel ($pixel^2$, $pixel^3$); A suitable pixel to millimeter calibration has been performed [13], [14] finding a conversion factor, k, equal to 0.08mm/pixel. The volume of lumen, V_L is evaluated in the same way, but considering the intima-lumen interfaces (in the near and far wall), Fig. 6b).

A more helpful parameter, the occlusion percentage, $O_{\%}$, is determined; it is obtained as the ratio between the plaque and the lumen volumes.

2.7 The Instrument for 3-D Evaluation: Prototype Set-Up

A tuning phase has been performed to evaluate the optimal parameter choices and the system performance. In order to verify the performance of the measurement extraction phase, the software prototype has been tested on artificial images reproducing the human carotid, obtained in a 3D CAD environment. The artery is reproduced like a cylinder and a plaque with spherical shape is positioned inside it. Several plaques of different sizes were simulated and different positions of the carotid respect to the skin were considered (see Fig. 7). It has to be noted that in this way the lumen and the plaque volume are well-known, since they are built in a simulation environment. The 2-D acquisition process is emulated slicing the simulated artery: for each insonification angle a corresponding plane is built; the volume sections projected in the selected

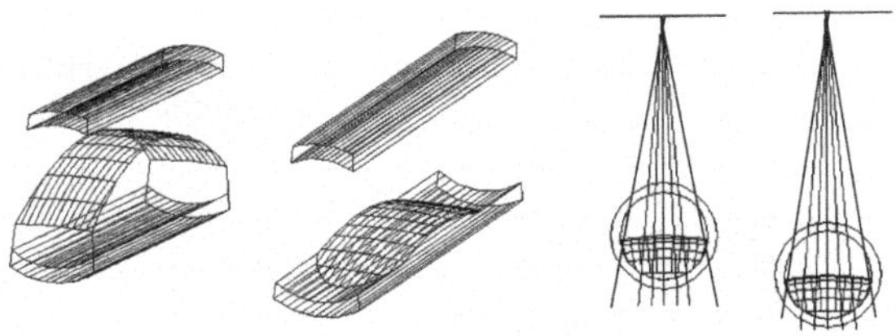

Fig. 7. The emulated artery with plaques

planes give the 2-D images. Five planes rotating in the interval [-5°,+5°] around pivot axis were considered. The prototype software processes the obtained artificial 2-D images in order to reconstruct the plaque and to evaluate the measurements. Following this approach the error of the measurement system on the carotid plaque volume is calculated. A systematic effect is highlighted: the measurement error is negative for each test (the system underestimates the volume). This effect is due, overall, to the poor number of acquisitions and can be corrected as: $V_p^c = c \cdot V_p^{mm} = c \cdot k^3 V_p$ where $c = 1 - \frac{e\%}{100}$ is the correction factor.

Some considerations have to be made about the value used for the correction factor. The obtained results highlight that the percentage error value depends on both carotid position and plaque dimension [13]-[14]. The same procedure has been also adopted to estimate the correction factor in the carotid volume measurement.

2.8 The Instrument for 3-D Evaluation: Experimental Results

It is important to evaluate the performances of the hardware and the software constituting the measurement system, and the uncertainty of the measurement obtained with the whole system, in order to let the physician to effectively compare measurement taken on the same patient in different phases of the medical treatment.

Metrological Characterization of the Hardware

The subsystem for the measurement of the angles has been characterized by a comparison with a reference computer-controlled pan-tilt unit having a 0.013° resolution. The obtained relationship between the averaged measurements X_c and the pan-tilt unit angles $\alpha_{\mu c}$ can be written as $\alpha_{\mu c} = m \cdot X_c + b = 0.0028 \cdot X_c - 71.097°$, with a m standard deviation of 0.0003 and a b standard deviation of 0.50°.

The position sensor has been characterized measuring the translation for the X axis and Y axis and comparing the results with those given by a micrometer measurement system. This one has a position resolution of 0.01mm. The experiment has been realized executing 20 consecutive tests in which the position sensor has been moved between two known positions fixed by the micrometer system. The obtained results show a position resolution of dx=30.00 mm/300 steps = 0.1 mm and a standard deviation of 0.05mm.

Metrological Characterization of the Software

Many experimental tests have been carried out. At first, the repeatability in the edge location was evaluated. In particular, for each acquired image, N=30 images are generated adding impulsive noise to it. On these new image sets, the segmentation algorithm is run and for each *x'* pixel coordinate, the variability of the extracted contour *z'* was evaluated. With reference to common quality images a repeatability of 100% was measured, while a 90% was observed for poor quality images. Then, the reliability of the procedure was evaluated comparing the localized edges with the ones achieved by a human technician. A very good agreement was observed. The correlation factor between the automatic and manual edge ordinates was never less than 0.98, and the distance among the two *z'* was always constrained in 5 pixels.

Errors and Uncertainty Analysis

A number of experiments in simulation environment were performed to estimate the systematic errors and the expected measurement uncertainties. To these aims, suitable numerical tests have been done on different emulated types of carotid, characterized by different values of stenosis (from 50 % down to 20 %) and by different skin-carotid distances (from 12 to 20 mm). For each one, the systematic errors and the overall measurement uncertainties were estimated.

As for the systematic errors, the maximum observed absolute errors were 0.02% in the plaque volume estimation, and 0.004% in the measurement of the stenosis, respectively. The worst performance was achieved for small plaque volumes, and small stenosis, whilst, mean error values of 0.01 % and 0.0002 %, respectively, were observed.

As for the uncertainty related to the final measurement results (plaque volume and stenosis) the uncertainty propagation law, as stated by the ISO EN 13005 [12], has been applied to the relationships involved in the determination of the measured quantities. It has included the contribution due to all components of the processing software (edge detection, 3D reconstruction, repeatability), as well as the contributions due to the hardware (frame grabber resolution, angle measurement system accuracy, probe shift measurement system accuracy). The maximum uncertainty values were 3 % for the volume measurements and 40 % for the stenosis. Also in this case, the worst performance were obtained for small plaque volumes and small stenosis, whilst, mean uncertainties of 2 % and 20 %, respectively, were observed.

An Application Example

A real clinical test is reported in Fig. 8, which is referred to a 32 years old patient with no evident plaque pathologies. The provided add-on package has been mounted on a

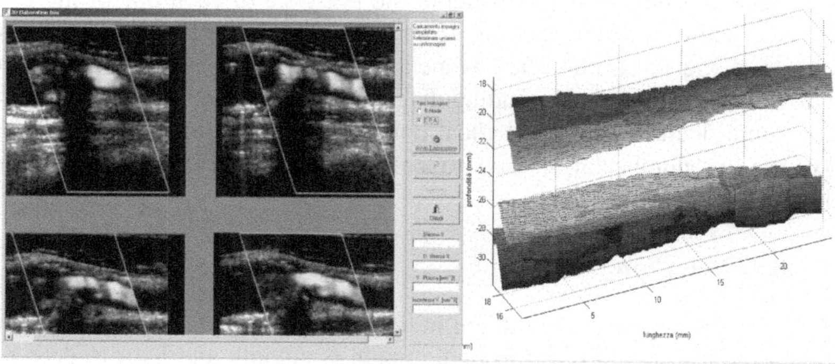

Fig. 8. a) The segmentation software; b) a rendered 3-D carotid reconstruction

carotid probe embedded with the ATL HDI 5000 ultrasonic machine. Fig. 8a) shows the output of the realized software during the 2-D edge segmentation procedure, and Fig. 8b) shows a particular of the rendered 3-D image. The measured volume value is 157 mm3 with a standard uncertainty of 16 mm3, and the stenosis is 0.75 with a relative uncertainty of 0.25.

3 ELM Image Processing for Melanocytic Skin Lesion Diagnosis Based on 7-Point Checklist

Malignant melanoma is one of the most frequent type of skin cancer and one of the most malignant tumors. A malignant melanoma diagnosed at an early stage can be cured without complications. Therefore, the early diagnosis is a crucial issue for dermatologists. As a consequence it is important to develop efficient schemes for the clinical diagnosis and to support the dermatologists with computer-aided diagnosis systems [15], [17]. Here an automatic measurement system for the diagnosis of melanoma based on 7-points check list applied on epiluminescence microscopy (ELM) skin lesion images is described. After a brief recall on the ELM technique and on the 7-points check list diagnosis method the processing procedures are presented. More in detail, the skin lesion border identification algorithm used for all the analysis, and the color lesion analysis for the identification of blue whitish veil and regression are described.

3.1 ELM Image

The images are obtained with the epiluminescence microscopy (ELM or dermoscopy). The technique uses a hand-held magnifying instrument, called a "dermatoscope" and uses oil immersion, which eliminates light refraction and makes subsurface structures of the skin more visible to the operator's eye [18]. The resulting images are much more detailed than clinical images, which show skin lesions seen through a magnifying lens, thus being considered the most suitable for digital processing.

3.2 The 7 Point Check List

The method classifies the lesion by means of a total score obtained reaching in the lesion image seven dermoscopic parameters. The presence of each one of the criterion determines the addition of the specific score to the total one. If a total score of 3 or more is given, the lesion is classified as melanoma. In Tab. 1 are summarized the parameters together with the corresponding diagnostic score. The ELM 7-point checklist provides a simplification (from a medical point of view) of standard pattern analysis because of the low number of features to identify and a very easy scoring diagnostic system [19]-[22].

3.3 Boundary Detection

Boundary detection is a critical problem in dermatoscopic images because the transition between the lesion and the surrounding skin is smooth and hard to detect accurately, even for a trained dermatologist. The realized segmentation algorithm for the

Table 1. The definitions of the seven ELM criteria

ELM criterion	Definition	Score
Major criteria		
1. Atypical pigment network	Black, brown, or gray network with irregular meshes and thick lines	2
2. Blue-whitish veil	Confluent, gray-blue to whitish-blue diffuse pigmentation associated with pigment network alterations, dots/globules and/or streaks	2
3. Atypical vascular pattern	Linear-irregular or dotted vessels not clearly combined with regression structures and associated with pigment network alterations, dots/globules and/or streaks	2
Minor criteria		
4. Irregular streaks	Irregular, more or less confluent, linear structures not clearly combined with pigment network lines	1
5. Irregular pigmentation	Black, brown, and/or gray pigmented areas with irregular shape and/or distribution	1
6. Irregular dots/globules	Black, brown, and/or gray round to oval, variously sized structures irregularly distributed within the lesion	1
7. Regression structures	White areas (white scarlike areas) and blue areas (gray-blue areas, peppering, multiple blue-gray dots) may be associated, thus featuring so-called blue-whitish areas virtually indistinguishable from blue-whitish veil	1

skin lesion border extraction is based on a image binarization with an adaptive threshold. From the source image (RGB standard color, 24 bits per pixel) the 3 different monochrome images (Fig.9 a), corresponding to the red, green and blue color components, are calculated. On each component the Otsu algorithm [23], is applied in order to isolate image background and foreground (corresponding to the histogram classes), then the thresholded image corresponding to the wider skin lesion area (the image foreground) is considered as binary mask for next processing. This choice is suggested by the experimental tests that showed a greater sensitivity than Otsu algorithm on the surrounding skin. Hence, a morphological closing operator fills isolated black points in the white regions (Fig. 9b). Finally, in order to extract the contour of the lesion, a simple blob-finding algorithm [24] is adopted for the binary image previously obtained: the tracking algorithm collects and sorts out the edges of the black-white

a) b) c)

Fig. 9. a) Monochrome image; b) binary mask; c) lesion border image into an ordered list. At this point, the border is superimposed on the color ELM image (Fig. 9.c) and displayed for visible inspection to the diagnostician. The binary image is also used in calculating the lesion dimension (number of white N_{tot}) that will be used in the lesion analysis.

The performance are evaluated on a large dataset constituted by 210 digital dermoscopy images obtained from the CD-ROM Interactive Atlas of Dermoscopy [25]-[26] (a collection of images acquired in three European university institutions). Three dermatologists well trained in 7-point checklist analysis have been asked for drawing the lesion contour for each digital image of the dataset; a satisfactory agreement with the visual inspection by the group of experts has been accomplished by the automatic procedure (percentage difference for the inner area of the pigmented lesion lower than 2%).

3.4 Algorithm for the Detection of: Blue-Whitish Veil and Regression

The Blue-whitish veil and regression are detected in the same step; both are characterized by the presence of gray areas, blue areas and a combination of both: regression structures can be seen like one previous phase to the blue-whitish veil. The distinction between them is based on the different distributions of the color areas. The proposed procedure [27] is composed of two main steps: the partition of the lesion into its different color components (lesion segmentation) and finally the recognition of the criterion of interest in the lesion map (region classification).

Lesion Segmentation

The lesion segmentation is carried out with the aim of splitting the internal area into multiple chromatically homogenous regions (the lesion map). The basic idea is the adoption of a suitable multithresholding of the color image.

The Principal Component Analysis (also known as the discrete Karhunen-Loeve Transform or Hotelling Transform [28]) is applied to the ELM image in order to reduce the problem dimensionality and complexity. By the Hotelling Transform equation a new 3D representation of the lesion pixels is obtained where the third component which contains most of the image noise can be neglected (Fig.10.c). Thus, just the first and second principal components (Fig. 10. a-b) which contribute most to the variance are used to compute a two-dimensional (2-D) histogram (Fig. 11.a). As enhancement against noisy due to the scarcity of data, the smoothing and down-sampling of the 2-D histogram are suggested.

The multithresholding is carried out by finding in 2-D histogram peaks with significant mass around them. It is expected that these peaks will correspond to the cluster

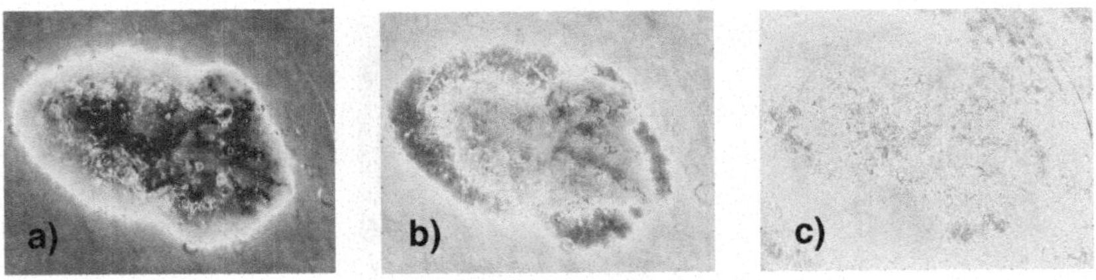

Fig. 10. Example of Hotelling Transformation: a) First-order Principal Component b) Second-order Principal Component, c) Third-order Principal Component

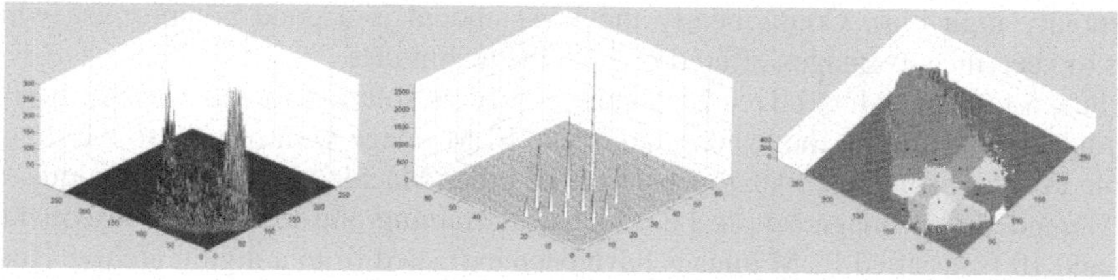

a) b) c)

Fig. 11. Construction and partitioning of the 2-D histogram: a) 2-D histogram; b) result of peak-picking method; c) partitioned 2-D histogram

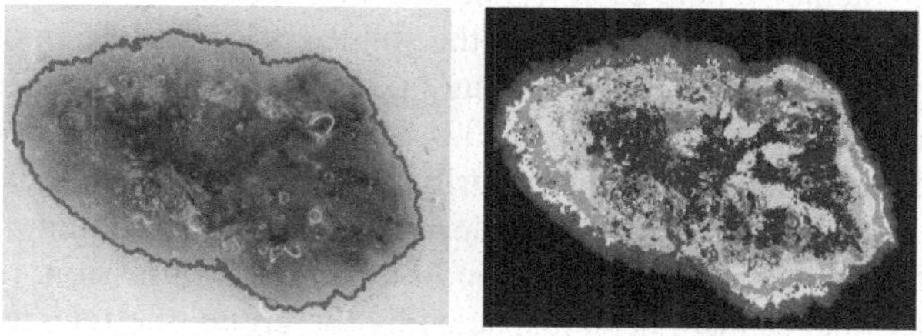

Fig. 12. Result of Lesion segmentation: a) ELM image; b) lesion map

centroids in 2-D space and consequently will be well-representative of corresponding color regions (or segments) in the starting image. The algorithm of Koonty [29] can be adopted as peak-picking method. The result achieved by applying the mentioned algorithm to the histogram shown in Fig. 11.a. is depicted in Fig. 11.b. when K equal to 14 is selected as maximum number of peaks.

Then, a 2-D histogram is partitioned using the peak bins and an assignment rule (gravity force) for each non-peak bin which takes into account the strength (height) of the peak and the distance from the pick to the histogram bin under consideration.

Once the partitioned 2-D is computed (Figure 11c) each pixel in the color image can be directly labeled by taking into account the corresponding values for the two principal components. An example is depicted in Fig.12 (false colors image) where different regions are easily identified.

Region Classification

The last stage of the procedure for the automatic detection of blue whitish veil and regression is the classification of the regions constituting the lesion map as areas characterized by the presence of the criterion of interest.

The classification can be viewed as a problem of data mining; among the different supervised Machine Learning techniques proposed in literature [30], a Logistic Model Tree (LMT) is used. The LMT is a combination of a tree structure and logistic regression models [31]. The Decision Trees represent a well-known class of solutions able to find non linear structures in the observed data and generally faster to train than Artificial Neural Networks and Support Vector Machines, though they can be concerned with stability and over-fitting problems. The linear logistic regression is a classification technique which fits a linear model to the data, resulting in a low variance but potentially high bias. Consequently the LMT model is a good compromise which aims to keep the advantages offered by both the techniques.

Thus, a Logistic Model Tree has been adopted as solution for the classification of the regions constituting the lesion map of the ELM image with respect to the dermoscopic criteria of interest. Analogously to the other Machine Learning technique, the LMT implies two image sets to be adopted as training and test sets. Consequently, 110 and 100 classified ELM images have been extracted from a digital archive (Interactive Atlas of Dermoscopy [25], [26]) and respectively adopted as training and test set for the classification of the blue veil and regression.

For each image the automatic lesion segmentation has been executed according to the proposed approach by considering *K* equal to 10 as maximum number of different color regions (segments). Then, each lesion map has been analyzed for identifying the main chromatic features corresponding to the single regions. More in details, for each region the components of the corresponding pixels in the *RGB*, *HSI* (Hue, Saturation and Intensity) and CIE *Luv* color spaces [32] have been considered to compute mean value and standard deviation as region attributes. In addition the area percentage of the region with respect to total area of the lesion is taken into account. In order to seek the attributes which are most related to the blue veil and regression, the lesion maps of the training set images have been inspected by the group of clinicians and a region-based classification has been achieved with respect to the dermoscopic criteria.

The Weka implementation [30] of the LMT algorithm has been adopted for obtaining the predictive models to be used in the automatic classification of the regions resulting from the lesion segmentation of the ELM images. About the classifier computed for the Blue Veil, three different logistic regression models have been computed according three ranges for the Hue mean value of the region to be analyzed (which can be interpreted as corresponding to blue, red or polychromatic "path"). The corresponding functions take into account the standard deviation for Hue component, the mean and standard deviation for Saturation and Intensity components to determine the probability that the color region belongs to an area characterized by the Blue Veil.

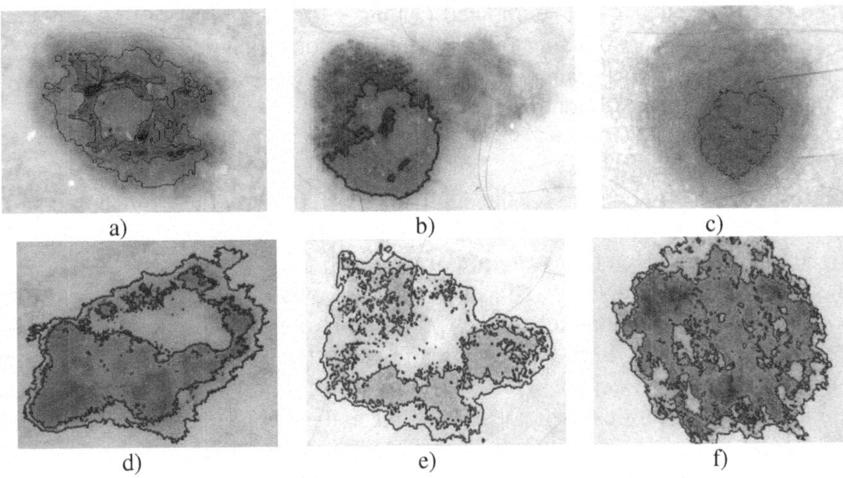

Fig. 13. Result of the automatic diagnosis: a), b), c) Detection of Blue whitish Veil; d), e), f) Detection of Regression area

Analogous considerations are hold for the classifier adopted for the detection of Regression: two different logistic regression models are computed according to the range for the mean value of the Saturation component.

About the lesion classification as whole with respect to the criterion of interest, the following strategy has been finally adopted: Blue Veil is detected within the lesion if the inner area obtained summing up the color segments classified as blue veil regions is wide enough (a threshold about the 7% of the lesion area has been determined by a ROC curve). Similar algorithm is also adopted for the detection of Regression (with a threshold equal to 5% of the lesion area). Very good results [27] have been obtained in classifying the test image set (65 and 40 cases characterized respectively by Blue Veil and Regression). Some results of the automatic detection are depicted in Fig. 13. More in detail, the Blue Veil has been correctly detected in 57 images against 5 lesions erroneously classified (corresponding to 0.87 and 0.85 respectively for *sensibility* and *specificity*). About Regression, only 7 missed and 9 faulty detections have been obtained, thus resulting in *sensibility* and *specificity* both equal to 0.85.

4 A Method for Detecting Cluster of Microcalcifications on Mammograms

Breast cancer is the most common cancer among women, excluding cancers of the skin. In 2007, breast cancer will account for nearly one out of every four cancer diagnoses in women. In the U.S., breast cancer is the second leading cause of cancer death among women as a whole (after lung cancer) [33].

Mammography is a radiological screening technique capable to detect lesions in the breast using low doses of radiation. It represents the only non-invasive diagnostic technique allowing the diagnosis of a breast cancer at a very early stage, when it is still possible to successfully attack the disease with a suitable therapy. For this reason, programs of wide mass screening via mammography for the female population at risk have been carried out in many countries.

A visual clue of breast cancer particularly meaningful is the presence of clusters of microcalcifications (see fig. 14). Microcalcifications are tiny granule-like deposits of calcium that appear on the mammogram as small bright spots. Their size ranges from about 0.1 mm to 0.7 mm, while their shape is sometimes irregular. Besides being arranged into clusters, microcalcifications can appear isolated and spread over the breast tissue, but in this case they are not indication of a possible cancer. However, even in the case of clustered microcalcifications their nature is not necessarily malignant, and thus the radiologist must carefully analyze the mammogram to decide if the appearance of the cluster suggests a malignant case. Such decision is taken on the basis of some properties (shape, size, distribution, etc.) related to both the single microcalcifications and the whole cluster.

Unfortunately, the low quality of mammographic images and the intrinsic difficulty in detecting likely cancer signs make the analysis particularly fatiguing, especially in a mass screening where a high number of mammograms must be examined

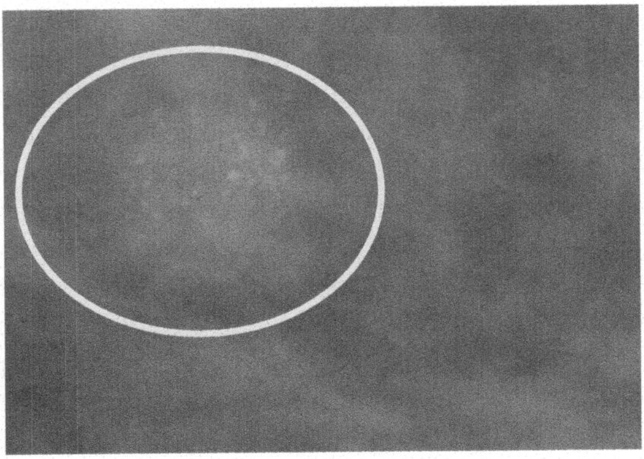

Fig. 14. Example of a cluster of microcalcifications

by a radiologist in a day. As a consequence, to avoid missed malignant cases, the behavior of the radiologist inclines to be quite cautionary in the case of doubt, prescribing to the patient a further diagnostic procedure much more invasive such as a biopsy. The drawback of such approach is a high number of non productive biopsy examinations, with very high economical and social costs.

In this case, a computer aided analysis could be very useful to the radiologist both for prompting suspect cases and for helping in the diagnostic decision as a "second reading". The goal is twofold: to improve both the *sensitivity* of the diagnosis, i.e. the accuracy in recognizing all the existing clusters and its *specificity*, i.e. the ability to avoid erroneous detections which can lead to unnecessary alarms.

4.1 State of the Art

In recent years there has been a significant development of *Computer Aided Detection* (CADe) systems for automated detection and classification of microcalcifications clusters in digitized mammograms. Two recent surveys by Cheng et al. [34] and Nishikawa [35] present a comparative analysis of various algorithms and techniques for the diagnosis of breast cancer on mammograms.

Two general approaches are commonly used [36]: the application of statistical techniques directly to the image data or the segmentation of the image through signal identification followed by a classification phase.

In the first approach several statistical classifiers have been applied such as artificial neural network [37], support vector machine [38] or relevance vector machine [39]. Furthermore, genetic algorithms [40] have been employed. The second approach usually entails an image transformation through several approaches, such as bank filters [41], wavelets [42], multiscale analysis [43] or higher order statistics [44]. In this case, the goal is to identify as many true signals as possible without an excessive number of false signals. An effective solution is given by the Markov Random Field model. The MRF model is a well known approach in image analysis [45] and in medical imaging [46] and in particular, on mammographic images for the detection of tumor mass [47]. In literature, there are also some papers applying MRF for the detection of microcalcifications clusters [48].

Once signals have been segmented, features of the extracted regions are evaluated through statistical classifiers to distinguish true from false regions. Many different types of classifiers have been employed such as neural networks [49], fuzzy logic [50], support vector machines [51] or multiple expert approaches [52]. Then, only in a following phase the selected regions are clustered with very simple rules based on proximity of the microcalcifications, to individuate those clusters that are important for the diagnosis.

4.2 CADe: An Example

The method here presented falls within the second approach described in the previous section; more details can be found in [53]. Since the size of the digitized mammographic image in input is greater than the region that contains the breast tissue, the first step consists in a pre-processing phase that recovers the region containing the breast tissue and deletes other signs or artifacts present in the image (see Fig. 15).

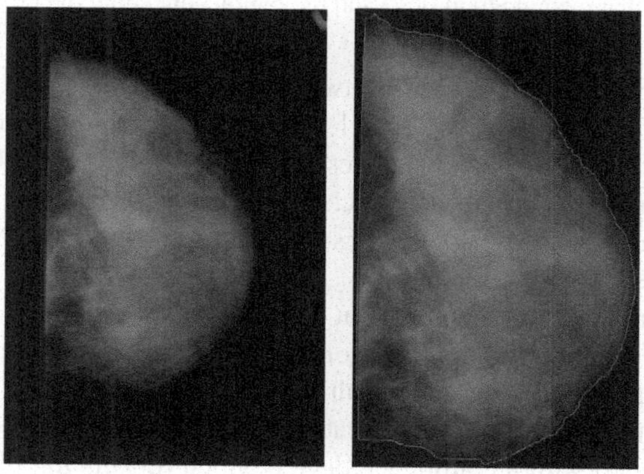

Fig. 15. The original mammogram (left) and the region containing the breast tissue (right)

Once the breast region has been found, a segmentation phase is performed to decompose the image in homogeneous regions. In particular, the proposed segmentation is based on the MRF model but tries to solve its major drawback, i.e. the high computational complexity, through the use of a tree structured model and therefore, a Tree-Structured Markov Random Field based segmentation (TS-MRF) [54]–[55] is employed so obtaining a fast and quite spatially adaptive segmentation process.

The segmentation step subdivides the mammogram in a huge number of homogeneous elementary regions (Regions of Interest - ROI) which are analyzed in the successive classification phase in order to identify the actual microcalcifications. To this aim, for each ROI, both geometrical and textural features are extracted according to the characteristics of the microcalcification.

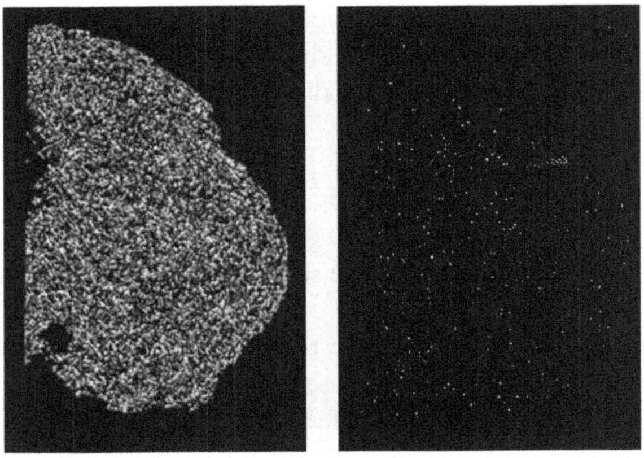

Fig. 16. The original mammogram (left) and the region containing the breast tissue (right)

The classification phase is made of two steps: in the first one a heuristic filter quickly discards the regions which are clearly non-microcalcifications. In particular, all the regions with size, shape or brightness strongly different from those typical for a microcalcification are classified as artifacts. The regions accepted in the first classification step have admissible characteristics for a microcalcification, but they still contain a large number of false positives.

The second classification step aims at locating the actual microcalcifications by means of a more refined (and computationally demanding) classification system. To this aim, a new approach based on a multiple classifier system has been adopted. In this way, at the end of the whole classification phase each region accepted by the heuristic filter is labelled with a confidence degree which reflects the likelihood of a microcalcification occurring in that region.

The last two steps of the CADe system are the grouping of ROI into clusters and their validation. The clustering is accomplished by analyzing the spatial coordinates and the confidence degree associated to each ROI and gives a possible partition in clusters. The clustering algorithm that has been employed is based on a sequential approach called Moving Leader Clustering with Merge [56]. Once the candidate clusters have been identified, the validation is made on each whole cluster according to the characteristics of ROIs that it groups. Some further features are extracted according to the spatial and textural characteristics of the clusters and using all these features a decision on each cluster is taken.

Figure 17 presents the flow chart of the whole method.

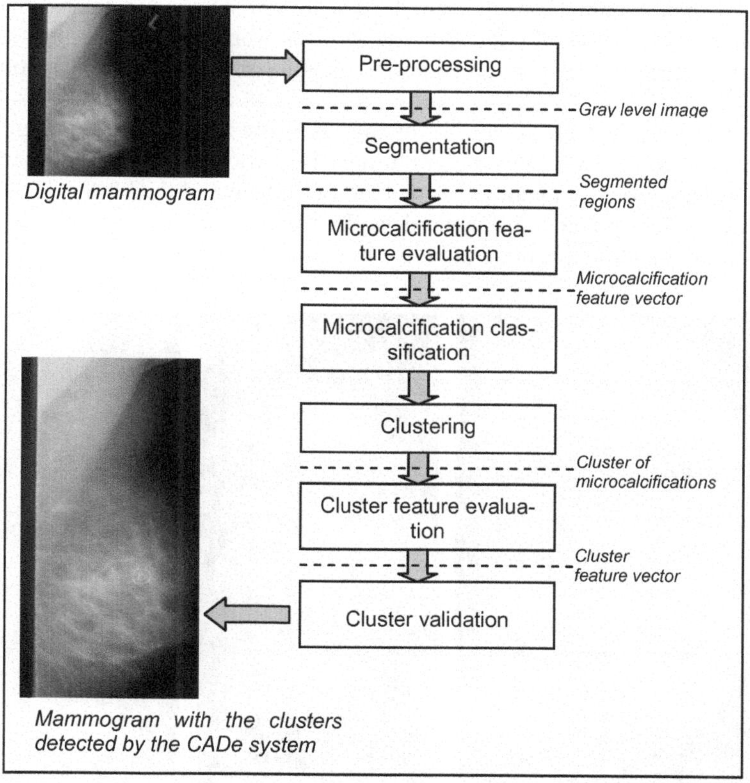

Fig. 17. The CADe system at a glance

4.3 Experimental Results

The system has been tested on a standard database, publicly available on the Internet, provided by courtesy of the National Expert and Training Centre for Breast Cancer Screening and the Department of Radiology at the University of Nijmegen, the Netherlands. It contains 40 digitized mammographic images composed of both oblique and craniocaudal views from 21 patients. All images have a size of 2048x2048 pixels and use 12 bits per pixel for the gray levels. Each mammogram has one or more clusters of microcalcifications marked by radiologists; each cluster is described by the centre and the radius of a circle totally containing it. The total number of clusters is 105, 76 of which are malignant and 29 benign. In order to have a training set adequately representative, in all the experiments we have adopted a "leave one image out( ) cross validation. According to this procedure we accomplish 40 different runs, in each of which 39 images are used for training and one for testing.

In order to assess the performance of the described method, we have considered the number of cluster correctly detected by the system (*True Positive*, *TP*) and the number of the erroneous detections (*False Positive, FP*): these parameters are the most interesting from the diagnostic point of view. In fig. 18 some outputs of the system are shown: the white blobs indicate the detected clusters, while the regular circles denote the clusters marked by the radiologists. It is worth noting that all the marked clusters are detected, while in the rightmost mammogram there are also some false positive.

The results obtained on all the set of images show a good behaviour of the described system in detecting the actual clusters. In particular, the system reached the 90% of clusters correctly identified with 1 FP per image and the 98% of TP with 2.2 FP per image. In summary the method works well in the detection of the clusters (good sensitivity) while it is quite effective in terms of specificity.

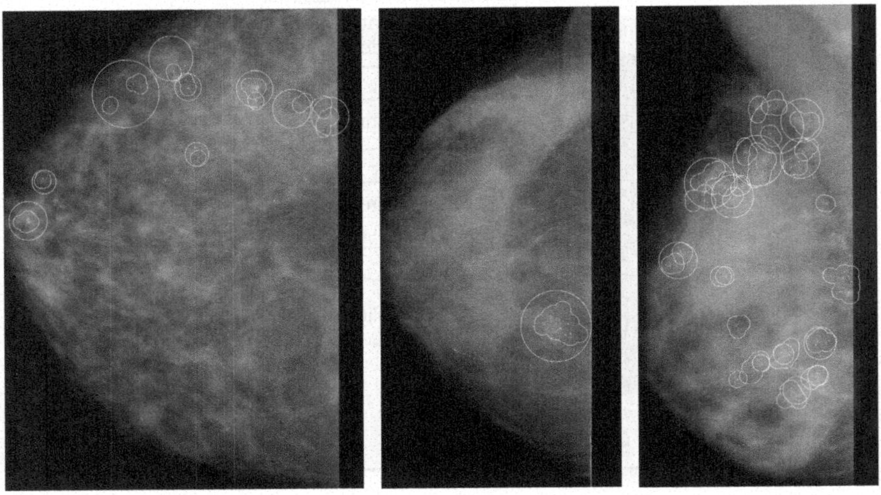

Fig. 18. The results obtained on some mammograms: the white blobs indicate the detected clusters, while the regular circles denote the clusters marked by the radiologists

References

1. Bond, M.G., Wilmoth, S.K., et al.: Detection and Monitoring of Asymptomatic Atherosclerosis in Clinical Trials. The Americal Journal of Medicine 86(suppl. 4A), 33–36 (1989)
2. Touboboul, P.J., Prati, P., et al.: Use of monitoring software to improve the measurement of carotid wall thickness by B-mode imaging. Journal of Hypertension 10(suppl. 5), S37–S41 (1992)
3. El-Barghouty, N.M., Levine, T., Ladva, S., Flanagan, A., Nicoladeis, A.: Histological Verification of Computerized Carotid Plaque Characterisation. European Journal Vascular Endovascular Surgey 11, 414–416 (1996)
4. Gill, J.D., Ladak, H.M., Steinman, D.A., Fenster, A.: Segmentation of ulcerated plaque: a semi-automatic method for tracking the progression of carotid atherosclerosis. In: Proc. of EMBS Conference, pp. 669–672 (2000)
5. Noritomi, T., et al.: Carotid Plaque Typing by Multiple-Parameter Ultrasonic Tissue Characterization. Ultrasound in Medicine & Biology 23(5), 643–650 (1997)

6. Fenster, A., Downey, N.M.: 3-D ultrasound imaging: A review. IEEE Engineering in Medicine and Biology 15(11), 41–51 (1996)
7. Hossack, J.A., Sumanaweera, T.S., Napel, S.: Quantitative 3D ultrasound imaging using an automated image tracking technique. In: Proc. of IEEE Ultrasonic symposium, pp. 1593–1596 (2000)
8. Rosenfield, K., et al.: Three-dimensional reconstruction of human carotid arteries from images obtained during non-invasive B-mode ultrasound examination. The American Journal of Cardiology 70(8), 379–384 (1992)
9. Wilhjelm, J.E., et al.: Estimation of plaque contents with multi-angle 3D compound imaging. In: Proc. of IEEE Ultrasonic Symposium, pp. 1077–1080 (1996)
10. Liguori, C., Paolillo, A., Pietrosanto, A.: An automatic measurement system for the evaluation of carotid intima-media thickness. IEEE Trans. on I&M 50(6), 1684–1691 (2001)
11. Betta, G., Liguori, C., Pietrosanto, A.: A structured approach to estimate the measurement uncertainty in digital signal processing algorithms. IEE Proc. Inst. Elect. Eng. Sci. Meas. Technol. 146(1), 21–26 (1999)
12. UNI CEI ENV 13005 (2000)Guide to the expression of uncertainty measurement
13. Capriglione, D., Ferrigno, L., Liguori, C., Paolillo, A.: Volumetric Carotid Plaque Measurements Based on Ultrasound Images: A Preliminary Approach. In: Proceedings of the 12th IMEKO TC-4 Part 1, pp. 255–260 (2002)
14. Ferrigno, L., Paciello, V., Paolillo, A.: A low cost measurement system for the 3-D evaluation of carotid plaque based on ultrasound images. In: Proceeding of 13th International Symposium on Measurements for Research and Industry Applications, IMEKO TC 4, pp. 468–474 (2004)
15. Ganster, H., et al.: Automated melanoma recognition. IEEE Transaction on Medical Imaging 20, 233–239 (2001)
16. Schmid-Saugeon, P., Guillod, J., Thiran, J.P.: Towards a computer-aided diagnosis system for pigmented skin lesions. Computerized Medical Imaging and Graphics 27, 65–78 (2003)
17. Hoffmann, K., Gambichler, T., et al.: Diagnostic and neural analysis of skin cancer (DANAOS). A multicentre study for collection and computer-aided analysis of data from pigmented skin lesions using digital dermoscopy. Br. J. Dermatology 149, 801–809 (2003)
18. Binder, M., Schwartz, M., et al.: Epiluminescence microscopy: a useful tool for the diagnosis of pigmented skin lesion for formally trained dermatologists. Archives of Dermatology 131, 286–291 (1995)
19. Fabbrocini, G., Argenziano, G., et al.: Epiluminescence microscopy for the diagnosis of doubtful melanocytic skin lesions: comparison of the ABCD rule of dermatoscopy and a new 7-point checklist based on pattern analysis. Archives of Dermatology 134, 1563–1570 (1998)
20. Di Leo, G., Fabbrocini, G., Liguori, C., Pietrosanto, A., Scalvenzi, M.: ELM image processing for melanocytic skin lesion based on 7-point checklist: a preliminary discussion. In: Proc. of the 13th IMEKO TC-4 Symposium, pp. 474–479 (2004)
21. Betta, G., Di Leo, G., Fabbrocini, G., Paolillo, A., Scalvenzi, M.: Automated Application of the 7-point checklist Diagnosis Method for Skin Lesions: Estimation of Chromatic and Shape Parameters. In: Proceeding of Instrumentation and Measurement Technology Conference, IMTC, pp. 1818–1822 (2005)
22. Di Leo, G., Liguori, C., Paolillo, A., Sommella, P.: An improved procedure for the automatic detection of dermoscopic structures in digital ELM images of skin lesions. In: IEEE International Conference on Virtual Environments, Human-Computer Interfaces, and Measurement Systems, VECIMS, pp. 190–195 (2008)

23. Otsu, N.: A threshold selection method from gray-level histogram. IEEE Transactions on System Man Cybernetics 9(1), 62–66 (1979)
24. Schmid-Saugeon, P., Guillod, J., Thiran, J.P.: Towards a computer-aided diagnosis system for pigmented skin lesions. Computerized Medical Imaging and Graphics 27, 65–78 (2003)
25. Stolz, W., Braun-Falco, O., et al.: Color Atlas of Dermatoscopy. Blackwell Science, Malden (1994)
26. Argenziano, G., Soyer, H.P., et al.: Interactive Atlas of Dermoscopy. EDRA Medical Publishing & New Media, Milan (2002)
27. Di Leo, G., Fabbrocini, G., Paolillo, A., Rescigno, O., Sommella, P.: Toward an automatic diagnosis system for skin lesions: estimation of blue-whitish veil and regression structures. In: International Multi-Conference on Systems, Signals & Devices, SSD 2009 (2009)
28. Gonzalez, R.C., Woods, R.E.: Digital Image Processing. Prentice Hall, New Jersey
29. Koonty, W., Narenda, P.M., Fukunya, F.: A graph theoretic approach to non-parametric cluster analysis. IEEE Transactions on Computer 25, 936–944 (1976)
30. Witten, H., Frank, E.: Data Mining: Practical Machine Learning Tools and Techniques. Morgan Kaufmann, San Francisco (2005)
31. Landwehr, N., Hall, M., Frank, E.: Logistic Model Trees. In: Lavrač, N., Gamberger, D., Todorovski, L., Blockeel, H. (eds.) ECML 2003. LNCS (LNAI), vol. 2837, pp. 241–252. Springer, Heidelberg (2003)
32. Levy, A., Lindenbaum, M.: Sequential Karhuen-Loeve basis extraction and its Application to images¸. IEEE Trans. on Image Processing 9(8), 1371–1374 (2000)
33. National Black Leadership Initiative on Cancer III: Community Networks Program, Breast Cancer Fact Sheet (Revised 10/21), `http://www.nblic.org`
34. Chan, H.P., Sahiner, B., et al.: Improvement of Radiologists' Characterization of Mammographic Masses by using Computer aided Diagnosis: An ROC study. Radiology (212), 817–827 (1999)
35. Cheng, H.D., Cai, X., Chen, X., Hu, L., Lou, X.: Computer-aided detection and classification of microcalcification in mammograms: a survey. Pattern Recognition 36, 2967–2991 (2003)
36. Nishikawa, R.: Current status and future directions of computer-aided diagnosis in mammography. Computerized Medical Imaging and Graphics 31, 1357–1376 (2007)
37. Wu, Y., Giger, M.L., et al.: Artificial neural networks in mammography: Application to decision making in the diagnosis of breast cancer. Radiology 187, 81–87 (1993)
38. El-Naqa, Yang, Y., Wernick, M.N., Galatsanos, N.P., Nishikawa, R.M.: A support vector machine approach for detection of microcalcifications. IEEE Transactions on Medical Imaging 21(12), 1552–1563 (2002)
39. Wei, L., Yang, Y., Nishikawa, R.E., Wernick, R.E., Edwards, A.: Relevance vector machine for automatic detection of clustered microcalcifications. IEEE Transactions on Medical Imaging 24(10), 1278–1285 (2005)
40. Jiang, J., Yao, B., Wason, A.M.: A genetic algorithm design for microcalcification detection and classification in digital mammograms. Computerized Medical Imaging and Graphics 31, 49–61 (2007)
41. Nakayama, R., Uchiyama, Y., Yamamoto, K., Watanabe, R., Namba, K.: Computer-aided diagnosis scheme using a filter bank for detection of microcalcification clusters in mammograms. IEEE Transactions on Biomedical Engineering 53(2), 273–283 (2006)
42. Strickland, R.E., Hahn, H.I.: Wavelet transform for detecting microcalcifications in mammograms. IEEE Transactions on Medical Imaging 15(2), 218–229 (1996)

43. Netsch, T., Peitgen, H.: Scale-space signatures for the detection of clustered microcalcifications in digital mammograms. IEEE Transaction on Medical Imaging 18(9), 774–786 (1999)
44. Gurcan, M.N., Yardimci, Y., Cetin, A.E., Ansari, R.R.: Detection of microcalcifications in mammograms using higher order statistics. IEEE Signal Processing Letters 4(8), 213–216 (1997)
45. Li, S.Z.: Markov random field modeling in image analysis. Kluwer Academic, Dordrecht (2001)
46. Medina, R., Garreau, M., et al.: Markov random field modeling for three-dimensional reconstruction of the left ventricle in cardiac angiography. IEEE Transactions on Medical Imaging 25(8), 1087–1100 (2006)
47. Suliga, M., Deklerck, R., Nyssen, E.: Markov random field-based clustering applied to the segmentation of masses in digital mammograms. Computerized Medical Imaging and Graphics 32, 502–512 (2008)
48. Yu, S.N., Li, K.Y., Huang, Y.K.: Detection of microcalcifications in digital mammograms using wavelet filter and markov random field model. Computerized Medical Imaging and Graphics 30, 163–173 (2006)
49. Tsujii, O., Freedman, M.T., Mun, S.K.: Classification of microcalcifications in digital mammograms using trend-oriented radial basis function neural network. Pattern Recognition 32, 891–903 (1999)
50. Cheng, H.D., Wang, J., Shi, X.: Microcalcification detection using fuzzy logic and scale space approach. Pattern Recognition 37, 363–375 (2004)
51. Papadopoulos, A., Fotiadis, D.I., Likas, A.: Characterization of clustered microcalcifications in digitized mammograms using neural networks and support vector machines. Artificial Intelligence in Medicine 34, 141–150 (2005)
52. De Santo, M., Molinara, M., Tortorella, F., Vento, M.: Automatic classification of clustered microcalcifications by a multiple expert system. Pattern Recognition (3), 1467–1477 (2003)
53. D'Elia, C., Marrocco, C., Molinara, M., Tortorella, F.: Detection of Clusters of Microcalcifications in Mammograms: A Multi Classifier Approach. In: Proc. 21st IEEE International Symposium on Computer-based Medical Systems, pp. 572–577. IEEE Computer Society Press, Los Alamitos (2008)
54. D'Elia, C., Poggi, G., Scarpa, G.: A tree-structured Markov random field model for bayesian image segmentation. IEEE Transactions on Image Processing 12(10), 1259–1273 (2003)
55. Marrocco, C., Molinara, M., Tortorella, F.: Algorithms for detecting clusters of microcalcifications in mammograms. In: Roli, F., Vitulano, S. (eds.) ICIAP 2005. LNCS, vol. 3617, pp. 884–891. Springer, Heidelberg (2005)
56. Peng, Y., Yao, B., Jiang, J.: Knowledge-discovery incorporated evolutionary search for microcalcification detection in breast cancer diagnosis. Artificial Intelligence in Medicine 37, 43–53 (2006)

Expanding the Metrological and Operating Characteristics of Cytofluorimeters

E. Balestrieri[1], D. Grimaldi[2], F. Lamonaca[2], and S. Rapuano[1]

[1] Dept. of Engineering, University of Sannio, Benevento, Italy
[2] Dept. of Electronics, Computer and System Sciences, University of Calabria, Rende (CS), Italy

Abstract. The chapter deals with an overview of the present stage of development of flow cytometers and with the description of a couple of proposals in order to overcome their limits. In particular, the possibility of deploying a flow cytometer almost everywhere without requiring a skilled operator on site has been studied and demonstrated feasible by realizing the prototype of a remotely controlled instrument based on satellite communications. Moreover, a method to process the cell images acquired by means of an image flow cytometer is presented in order to reduce the doubtful detection of micro nuclei in human lymphocytes in the flow cytometers.

Keywords: Flow cytometer, satellite communication, distributed measurement systems, image processing, micro nuclei, lymphocyte.

1 Introduction

The flow cytometry is a technology for measuring (i) the number of cells in a sample, (ii) the percentage of living cells in a sample, (iii) certain characteristics of cells such as size, shape, and (iv) the presence of markers on the cell surface. In particular, the analysis of biological material is executed by exploiting the light-absorbing or fluorescing properties of cells or sub-cellular fractions (i.e., chromosomes) passing in a narrow stream through a laser beam [1]-[5]. The flow cytometers (FC) count and classify cells through the energy scattering they produce when they are hit by the laser ray [6]. FCs can indicate relative cell size and density or complexity by measuring forward- and side-scattered laser light, respectively. In addition, they can measure relative fluorescence from fluorescent probes (fluorophores) which bind to specific cell-associated molecules [7]. Such measurement instruments are used in cell biology, immunology, hematology and oncology [8], [9].

Today there is a strong demand for a new advanced flow cytometry device able to: (i) reduce drastically the delay time of the classic cytometer data acquisition systems, in order to increase their time domain resolution, (ii) collect real time images of the flowing cells, in order to allow a microscope analysis of some cells, when necessary, and (iii) assure reliable and repeatable results. By fulfilling such requirements it would be possible to minimize the doubtful detections and the dependency of results from operators, as it happens today. A new FC architecture has been proposed in [10,11] introducing improvements to (i) the data acquisition system, (ii) the laser

S.C. Mukhopadhyay and A. Lay-Ekuakille (Eds.): Adv. in Biomedical Sensing, LNEE 55, pp. 210–239.
springerlink.com

beam automatic alignment system, (iii) the cell flow chamber, (iv) the digital signal processing procedure, and (iv) presenting a new methodology for pre-processing the cell images for a further classification [10,11,12].

The chapter starts with a brief introduction focused on the flow cytometry history and applications. Then, the FC operating principle is explained and its typical architecture described. An overview of commercial FCs currently available on the market as well as the proposals in this field coming from research in the later years are successively presented.

In the second part of the chapter two new contributes to the FC development are presented able to expand the metrological and operating characteristics of such instruments. The first proposal concerns the prototype of a mobile cytometry unit adding to the specifications of the compact FCs the remote control by satellite link in order to produce diagnoses on the field. Moreover, the proposed FC transmits the patient recordings and the analyzed data to a control centre for their storage [13]. Both the hardware and the software architecture of this new FC are described, focusing, in particular, on the mobility characteristics of the developed instrument. The second proposal concerns a new method for Image FCs (IFC) [14] giving to such instruments the capability of automatically recognizing and counting the Micro Nuclei (MNs) on the acquired images of human lymphocytes, in order to detect structural chromosome aberrations.

2 History and Applications of Flow Cytometry

Flow cytometry has developed over the last 60 years from single-parameter instruments measuring cell size only, to highly sophisticated machines capable of detecting 13 parameters simultaneously [15]. Various disciplines have been involved in the development of flow cytometry including biology, biotechnology, computer science, electrical engineering, laser technology, mathematics, medicine, molecular biology, organic chemistry and physics [16].

Flow cytometry developed from microscopy. Since the seventeenth century, microscopes have been used to examine cells and tissue sections. Successively (the end of the nineteenth century) stains have been developed making various cellular constituents visible. In 1934, Andrew Moldavan attempted photoelectric counting of cells flowing through a capillary tube, taking a first step from static microscopy toward a flowing system. In the 1940s and 1950s, fluorescence microscopy began to be used in conjunction with fluorescent stains for nucleic acids to detect malignant cells. Thanks to the development of fluorescent antibody technique, in fact, cell suspensions or tissue sections could be routinely stained with antibodies specific for antigenic markers of cell type or function [17]. In particular, Coons, Creech and Norman Jones labeled antipneumococcal antibodies with anthracene allowing to detect both the organism and the antibody in a tissue using UV excited blue fluorescence. In 1947, Gucker developed a FC for detection of bacteria in aerosols for the rapid identification of airborne bacteria and spores used in biological warfare. The basis of the first viable flow analyzer was developed in 1956 by Coulter, who carried out an electronic instrument for counting and sizing cells flowing in a conductive liquid with one cell at time passing a measuring point. In particular, the analysis was based on the amount by which the cells increased the electrical resistance of an isotonic saline solution while flowing through an orifice. Cells were thereby classified more or less on the basis of their volume since larger cells have greater electrical resistance [17]. The Coulter counters became the essential equipment in hospital hematology laboratories, allowing the rapid and automated counting of white and red blood cells [16]. In the mid-1960s, Kamentsky applied his knowledge in optical character recognition to the problem of automated cervical cytology screening. He described a two-parameter FC able to measure absorption and back-scattered illumination of unstained cells in order to determine cell nucleic acid content and size [15]. This instrument was the first multiparameter FC. The first cell sorter based on the sorting principle developed for the inkjet printer at Stanford in 1965, instead, was described by Fulwyler. In 1969, Van Dilla and other members of the Los Alamos Laboratory group developed the first fluorescence-detection cytometer, utilizing the principle of hydrodynamic focusing, 90° optical configuration and an argon ion laser excitation source [15, 16].

These sorting cytometers began to be used to look for distinguishing and separating white blood cells. By the end of the 1960s, they were able to sort lymphocytes and granulocytes into highly purified states [17].

An important improvement was reached by Herzenberg in 1972 with a cell sorter able to detect weak fluorescence of cell stained with fluorescence-labeled antibodies [16]. The remaining history of flow cytometry involves the elaboration of this technology, the exploitation of FCs for varied applications and the collaboration between scientists and industry for the commercial production of cytometers as user-friendly tools [17]. In particular the first clinical FCs were introduced in 1983, but flow cytometry was brought into routine use in 1990, when bench-top instruments with enclosed flow cells were developed [15]. In the mid-1990s FCs were able to measure a minimum of 5 parameters on 25000 cells in 1 s [15]. After that cytometry technology began to move simultaneously in two directions: toward increasingly sophisticated instruments able of measuring and analyzing more aspects of more varied types of particles with higher sensitively as well as able of sorting particles on the basis of these aspects at faster and faster rates, and toward streamlined instruments, that can be user-friendly and the essential equipment for many laboratory benches [17].

Nowadays, FCs are used in several different applications including molecular biology, pathology, immunology, plant biology and marine biology.

However, one of the largest applications of flow cytometry is in the clinical sciences, where the primary measurements are of fluorochrome-conjugated antibodies bound to cellular receptors. This is generally referred to as immunophenotyping since many of the cell types being studied are immune cells such as lymphocytes. In fact, almost every possible human cell has been evaluated by flow cytometry [18]. By conjugating fluorescent molecules to antibodies that recognize specific receptors, a population of cells binds that antibody and therefore that fluorescent molecule can be identified. With certain clinical syndromes, it is evident that a specific pattern will emerge when identifying which cells bind to certain antibodies [18]. One of the most significant findings in the early 1980s was that the identification of certain subsets of human T cells, the CD4 type, was important for the monitoring of the clinical status of AIDS patients [18]. Moreover, immunophenotypic analysis is critical to the initial diagnosis and classification of the acute leukemias, chronic lymphoproliferative diseases, and malignant lymphomas since treatment strategy often depends upon antigenic parameters. In addition, immunophenotypic analysis provides prognostic information not available by other techniques, provides a sensitive means to monitor the progress of patients after chemotherapy or bone marrow transplantation, and often permits the detection of minimal residual disease. Flow cytometric analysis of apoptosis, multidrug resistance, leukemia-specific chimeric proteins, cytokine receptors and other parameters may provide additional diagnostic or prognostic information in the near future [19].

3 Flow Cytometer Principle and Main Architecture

Flow cytometry simultaneously measures and then analyzes multiple physical characteristics of single particles, usually cells, as they flow in a fluid stream through a beam of light. The measured properties include a particle relative size, relative granularity or internal complexity, and relative fluorescence intensity [20]. These characteristics are determined using an optical-to-electronic coupling system that records how the cell or particle scatters incident laser light and emits fluorescence (Fig. 1).

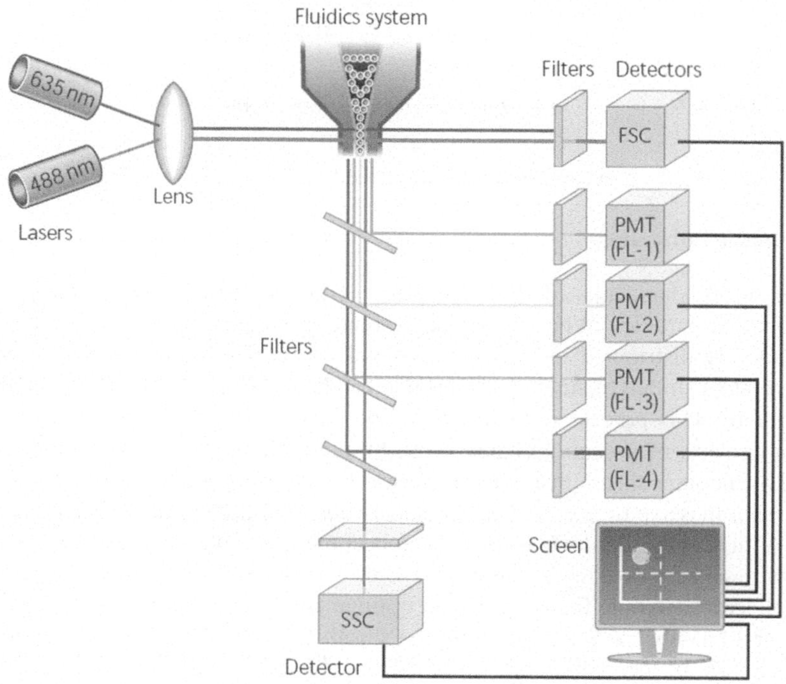

Fig. 1. Schematic overview of a typical FC setup [21]

A FC is made up of three main subsystems: fluidic, optic, and electronic (Fig.2).

- The fluidic subsystem transports the particles in a stream to the laser beam for interrogation.
- The optic subsystem consists of lasers to illuminate the particles in the sample stream and optical filters to direct the resulting light signals to the appropriate detectors.

- The electronic subsystem converts the detected light signals into electric signals that can be processed by the computer. For some instruments equipped with a sorting feature, the electronic system is also capable of initiating sorting decisions to charge and deflect particles [20].

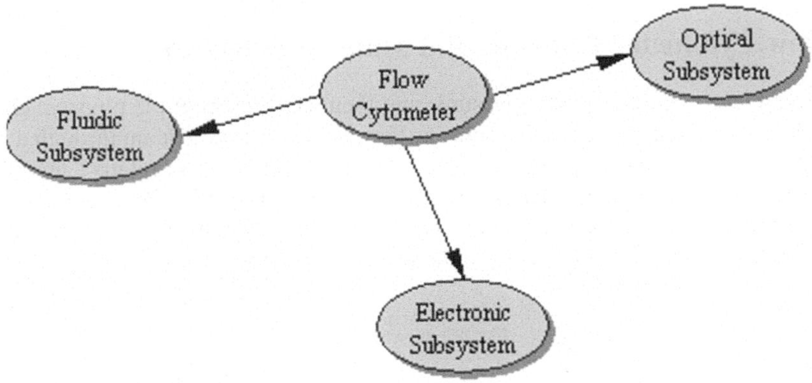

Fig. 2. Flow cytometer subsystems

In the following each system is briefly described.

3.1 Fluidic Subsystem

One of the fundamentals of flow cytometry is the ability to measure the properties of individual particles. When a sample in solution is injected into a FC, the particles are randomly distributed in three-dimensional space. The sample must therefore be ordered into a stream of single particles that can be interrogated by the machine detection system. This process is managed by the fluidic system. Essentially, the fluidic system consists of a central channel/core through which the sample is injected, enclosed by an outer sheath that contains faster flowing fluid. As the sheath fluid moves, it creates a massive drag effect on the narrowing central chamber. This alters the velocity of the central fluid whose flow front becomes parabolic with greatest velocity at its center and zero velocity at the wall. The effect creates a single line of particles and is called hydrodynamic focusing. Under optimal conditions (laminar flow) the fluid in the central chamber will not mix with the sheath fluid. Without hydrodynamic focusing the nozzle of the instrument would become blocked, and it would not be possible to analyze one cell at a time [21].

3.2 Optic Subsystem

After hydrodynamic focusing, each particle passes through one or more beams of light in the so called flow cell. Light scattering or fluorescence emission (if the particle is labeled with a fluorochrome) provides information about the particle properties. The laser and the arc lamp are the most commonly used light sources in modern flow cytometry. Light that is scattered in the forward direction, is collected by a lens known as the forward scatter channel (FSC in Fig.1). The FSC intensity roughly equates to the particle size and can also be used to distinguish between cellular debris and living cells. Light measured approximately at a 90° angle to the excitation line is called side scatter. The side scatter channel (SSC in Fig.1) provides information about the granular content within a particle. Both FSC and SSC are unique for every particle, and a combination of the two may be used to differentiate different cell types in a heterogeneous sample. Fluorescence measurements taken at different wavelengths can provide quantitative and qualitative data about fluorochrome-labeled cell surface receptors or intracellular molecules such as DNA and cytokines. Flow cytometers use separate fluorescence (FLn in Fig.1) channels to detect light emitted. The number of detectors will vary according to the machine and its manufacturer. Detectors are either silicon photodiodes or photomultiplier tubes (PMTs). Silicon photodiodes are usually used to measure forward scatter when the signal is strong. PMTs are more sensitive instruments and are ideal for scatter and fluorescence readings. The specificity of detection is controlled by optical filters, which block certain wavelengths while transmitting (passing) others. To detect multiple signals simultaneously, the precise choice and order of optical filters will be an important consideration issue [21].

3.3 Electronic Subsystem

When light hits a photodetector a small current (a few microamperes) is generated. Its associated voltage has an amplitude proportional to the total number of light photons received by the detector. This voltage is then amplified by a series of linear or logarithmic amplifiers, and digitized by means of analog to digital converters (ADCs). Log amplification is normally used for fluorescence studies because it expands weak signals and compresses strong signals, resulting in a distribution that is easy to display on a histogram. Linear scaling is preferable where there is not such a broad range of signals e.g. in DNA analysis. The measurement coming from each detector is referred to as a 'parameter' e.g. forward scatter, side scatter or fluorescence. The data acquired in each parameter are known as the 'events' and refer to the number of cells displaying the physical feature or marker of interest [21].

4 Commercial Flow Cytometers and Research Trends

There are two main types of FCs: the analysers and sorters. Sorters have the ability not only to collect data on cells (analyse cells) but also to sort cells with particular properties (defined by the FC operator) to extremely high purities (< 99%) [22].

Commercial analysers have been developed mainly for clinical analyses, being spurred on by the spread of AIDS and the need to assess the CD4 status (in particular) of patients infected with Human Immunodeficiency Virus (HIV). However they are also widely used for other analyses such as leukemia and lymphoma phenotyping, assessment of the immune status of patients, DNA analyses of tumour material, and a range of other clinical analyses. They are also increasingly used for research where there is no need to sort cells, since they are much cheaper than a sorter (approximately half or less in price). Most commercial analysers are built around an argon-ion laser as the light source [22].

There are two main types of sorters, droplet and fluidic. Droplet sorters operate by vibrating the flow cell at high frequencies (15–100 kHz) causing the sheath stream to break up into a regular series of droplets. When a particle is detected, its properties are compared to criteria entered into the computer by the FC operator. If the properties of the particle match the sort criteria, the last drop attached to the stream is electrically charged just as the particle to be sorted reaches it. The droplets pass between high voltage plates and any droplet that is charged is diverted out of the main stream into a catching vessel.

Fluidic sorters, or fluidic switching sorters, rely on a mechanism to divert the fluidic stream into the sort collection vessel at the time a particle that fulfils the sort criteria is detected. Fluidic sorters have a distinct advantage over droplet sorters for sorting biohazardous samples. In droplet sorters, in fact, the stream is broken up into a series of tiny droplets. This produces aerosols, which can potentially contaminate the operator. Fluidic sorters do not break up the stream into droplets, do not produce aerosols and thus are inherently safer for sorting biologically hazardous samples. However the speed of the stream switching mechanisms of fluidic sorters limits sorting rates, and the sorting purity is not as good as for droplet sorters [22]. A disadvantage of fluidic sorters is that they can only sort single populations whereas in droplet sorters the droplets can be charged either with a positive or negative charge allowing two populations to be sorted simultaneously [22].

Lasers used on commercial flow cytometry sorters can be divided into two types: air-cooled and water-cooled. In general the air-cooled lasers can be plugged into standard power outlets, are small, relatively cheap, have a single emission line, but have limited output powers. Water-cooled lasers generally can be tuned to different emission lines, require 3-phase power, are large, expensive (to buy and to run) and of course require water-cooling [22].

In the following some of the commercial FCs analyzers and sorters currently available in the market are presented, the FC software introduced and the main research proposals overviewed.

4.1 Commercial FCs

The main FC manufacturers are BD Biosciences, a division of Becton Dickinson, and Beckman Coulter comprising about 70% of the Research and Clinical areas of the cell-based flow cytometry market [23].

Examples of FC analyzers today available in the market are the Beckman Coulter Gallios and the BD LSR II.

Beckman Coulter Gallios FC, very recently introduced for the research market, houses up to three solid-state lasers in standard red and blue, with violet available as an option. Interchanged optical filters facilitate the detection of a variety of dyes and wavelengths. A selection of up to 62 parameters can be processed per analysis, at acquisition rates of 25,000 events-per-second.

BD Biosciences LSR II air-cooled four-laser (blue, red, violet, and UV) bench-top FC has the ability to acquire up to 18 colors, at an acquisition rate up to 20,000 events per second and can measure up to 8 fluorescent parameters.

Concerning FC sorters BD Biosciences introduced the BD FACSAria™ II flow cytometry system, which is designed to make advanced cell sorting easier for biologists across a range of applications and can sort 70,000 events per second.

Beckman Coulter also offers a powerful cell sorting systems, MoFlo XDP, which can analyze 100,000 events per second and sort 70,000 events per second.

A niche segment of the flow cytometry market that has recently gained prominence focuses on compact FCs [24]. Reducing dimensions, in fact, can spread the number of applications of an FC. By allowing the FCs to be deployed on the field it could be possible to carry out massive screening of the population health where no equipped hospitals are available. It could also be possible to carry out cell analysis on the sea or in the space.

Several companies participate in the compact FC market including Guava Technologies, Beckman Coulter, Partec and Accuri. Some of them sell small units that are more targeted to specific applications, such as measuring cell proliferation or apoptosis. In this way it is possible to spare cost along with dimensions and/or weight.

An example of compact FC today available in the market is Beckman Coulter's Cell Lab Quanta flow cytometry system that simultaneously measures electronic volume, side scatter and three fluorescent colors to provide good population resolution and accurate cell counting.

The Guava EasyCD4® System delivers the ability to monitor T-cells in patients with HIV easily by using a two-color, direct, absolute counting approach.

Partec also offers a range of dedicated small instruments for specific applications, such as the CyFlow® Ploidy Analyser for analysis of ploidy level in plants, CyFlow® CCA Cell Counter Analyser for DNA cell cycle analysis and cell counting, and the CyFlow® Counter for HIV monitoring and AIDS patient follow-up.

Finally, Accuri Cytometers offers a compact FC for general use. Their C6 FC also includes an intuitive software. The size and affordability of these smaller instruments, as well as the ease of use, have made it possible for them to be used in developing countries to monitor HIV infection in the population. Portability is a selling point for Cytobuoy's product line, which includes the CytoSub and CytoBuoy for in-situ ocean analysis [24].

4.2 Image FCs

In image cytometry a specimen or object of interest is placed on a glass slide or other rigid substrate and stained using fluorescence- or absorbance-based probes specific for the cellular substance or substances of interest. Analyses may be performed on tissue sections or individual cells. Image cytometry can be used to quantify probes (similar to flow cytometry) as well as to obtain morphometric and other information as permitted by its high optical resolution. Images may be acquired in two or three dimensions, permitting quantification and study of the distribution of substances throughout cells and tissues [25].

The basic image cytometer consists of a microscope, camera, computer, and monitor. Additional components such as a motorized stage, a motor for automated image focus, automated changing of objectives, and filter wheels for rapid selection of illumination and measurement wavelengths may be added depending upon the specific application(s). Instruments providing automated slide changing and even fully automated operation are also available [25].

Amnis Corporation pioneered multispectral image flow cytometry, a method that allows individual cells to be visualized as they zoom by the laser beam. Amnis' ImageStream 100 is similar to a FC with the important distinction that the traditional detectors have been replaced by a sensitive charge coupled device (CCD) camera, custom designed by the company. The camera takes as many as six independent images (brightfield, darkfield and up to four fluorescent images) of each cell at a rate of about 100 cells per second. Because it can image a large number of cells at once, the ImageStream is ideally suited to gather a statistically meaningful picture of a population of cells [26]. Even if much slower than conventional FC analysers, the IFC provide an image of each cell, allowing for a further direct examination looking for particular characteristics that cannot be measured by FCs, like the detection of MNs in human lymphocytes.

4.3 Software for Commercial FCs

Today most flow cytometry instruments use software for instrument control, data acquisition, analysis and display. Software control is provided for many or all instrument hardware functions including fluidics, lasers, electronics and sorting. Further, operations such as start-up and shutdown sequences, cleaning cycles, calibration cycles and self-test functions are monitored and regulated by software, so that instrument operation can be simplified and optimized for the user. Moreover, FCs can be used to provide population-based analysis on a large number of particles in a comparatively short period of time. The data generated from this analysis is usually presented in histogram form or on a dot or contour (bivariate) plot. The frequency histogram is a direct graphical representation of the number of events occurring for each detection channel (i.e. the number of particle events detected as a function of intensity of the received light). The dot plot is a two-dimensional extension of the frequency histogram. Each location on the dot plot corresponds to a measured signal at a first detector versus a second detector. Other techniques can be employed to visualize three or more parameters in highly multiplexed parameter analysis applications. Statistical analysis is performed graphically applying regions of interest to data sets. Additional gating schemes can be used to restrict only certain subpopulations of interest in order to investigate populations with minor variations in structure [27].

In the software packages that come with some imaging cytometers, such as Amnis' IDEAS data analysis software, the dot plots are linked to images of the cells. That means that a user can click on an individual dot to see the corresponding image or can draw a 'gate' around a population of dots to see what that cells that fall within the gate look like [26].

4.4 FC Data Format

Usually data from all flow cytometry systems are stored in so-called flow cytometry standard (FCS) format. The data stored in FCS format are usually "list mode"' data. This means that, in a four-parameter cytometer, four numbers are stored for each cell. A 10,000 cell data file will consist of a long list of 40,000 numbers, with each set of four numbers describing each cell in order it passed through the laser beam. By retrieving the stored data, each cell can be analyzed again. The intensity of each of the four signals for that cell will be known and can be correlated with each other or with the intensity of the four signals from any (or all) other cell(s). Another type of data storage is the so-called single-parameter data storage, that involves the storage of the intensity profiles for the population of cells in a sample for each parameter separately. The only information stored can be, for example, (i) the distribution of forward scatter signal intensities for the cells in the sample, (ii) the distribution of side scatter signal intensities for the cells in the sample, (iii) the distribution of red fluorescence signal intensities for the cells in the sample, and (iv) the distribution of green fluorescence signal intensities for the cells in the sample. In this case, however, no information has been stored about whether the bright green cells are the cells that are bright red or whether they are the cells that are not red. With this kind of storage, it is not possible to know whether the cells with a bright forward scatter signal are red or green or both red and green [17].

4.5 Research Trends

New fluorochromes, including UV-excited, complex of dyes ("tandem dyes"), and nanocrystals are under development, as well as a new generation of modular FCs using small, solid state lasers, robotics, and advanced, innovative bioinformatics software [19]. Over the last years research proposals for improving FC, have been related to software enhancements [28,29] and the reagents specificity versus a truly poor innovative hardware performance strategy [28,30]. However, customers and researchers request and expected true innovations concerning hardware and software aimed at performing tests in a variety of new applications with improved performance [12].

A FC measurement device, based on both hardware and software innovative architectures, optimized for diagnostic applications in the oncology field, has been presented in [10,11]. In particular, the proposed improvements concerned: (i) the flow chamber, (ii) the fluidic system, (iii) the ADC system, and (iv) the addition of an image acquisition system (IMAQ). A manual microscope scan will no longer be necessary when automatic cell sorting count is used, as in the case of lymphocyte MNs. Compared to the Imagestream approach the IFC proposed in [10,11] is able of taking photos of cells while analyzing them as in conventional cytometers, thus merging the advantages of both the approaches.

5 A Mobile FC Based on Satellite Communications: CYTOSAT

Some manufacturers of laboratory instruments are developing systems, called remote diagnostics, for enabling service engineers to troubleshoot laboratory equipment from afar and even allowing them to prevent downtime by monitoring the performance of the instrument via a modem or Internet connection. In particular, when it comes to either fixing existing problems or preventing future ones, remote diagnostics can be initiated in three ways: by the customer, by the manufacturer, or by the machine itself [31].

Most companies have developed remote diagnostics because of the obvious advantages, for both customer and manufacturer, of troubleshooting at a distance.

Future development of data transfer methods is focused on wireless connectivity, which may be more appropriate for point-of-care and self-care applications. Methods currently in use include infrared, spread-spectrum radio, and short-range cellular. Some communications methods already established are based upon the IRDA, OpenAir, 802.11, or Bluetooth wireless communications standards, all of which share the drawback of implementation cost. Infrared, although the least costly, has the additional drawback of a line-of-sight requirement [31].

An ideal remote diagnostics system features would include:

- A high-speed or network connection to the instrument
- The ability to look at a customer's screen without interfering with the operation
- A 128-bit or better encryption algorithm for security
- The ability to keep patient names or identifying information confidential
- The ability to track reagent usage, methods, and lot numbers, as well as replicated results
- The ability to do a cost-per-reportable result analysis
- The ability to access the instrument over the Internet
- Instrument logs that show diagnostic information or pinpoint problem areas
- The ability to notify the user of a possible problem
- The ability to upgrade software when it is released, thus reducing distribution costs
- Reliability and demonstrable value to users [31]

Remote diagnostic technology can provide many other useful services that flow cytometry applications can particularly benefit, for example installing software upgrades, training personnel, and automatic replenishment of reagents and other supplies [31]. Moreover, remote control capability can even overcome some actual FC limitations. Traditionally, in fact, exchange of information between flow cytometrists requires travelling to remote locations or mailing or faxing of plots and histograms. However, travel is expensive and time-consuming, and printed data presentations can be difficult to interpret accurately because they do not permit further analysis, for example, by changing of the marker settings and back-gating. In the past, the most effective communication between diagnostic centres has been accomplished by mailing computer diskettes containing data formatted to allow further analysis by the recipient (eg in list mode). However, physical transport inevitably introduces delays, which conflict with the requirements of modern clinical laboratory protocols [32].

A compact FC, self powered and light enough to be installed on a minivan or a car and with the possibility of being remotely controlled from a research centre with trained personnel, has been proposed in [13].

In particular, the architecture of the IFC proposed in [10,11] has been expanded into the new instrument called CYTOSAT satisfying the following new specifications: (i) reduced size, (ii) portability, (iii) satellite communication, and (iv) remote control. CYTOSAT can be used to realize a distributed measurement systems based on satellite communications for diagnosis and data collection campaigns on field. In particular, it can be useful (i) where medical units are far from patients, (ii) where the local personnel skills are inadequate, and (iii) for realizing a large database by taking information on site and analyzing it remotely, useful tool for researchers on blood-related illnesses, as AIDS.

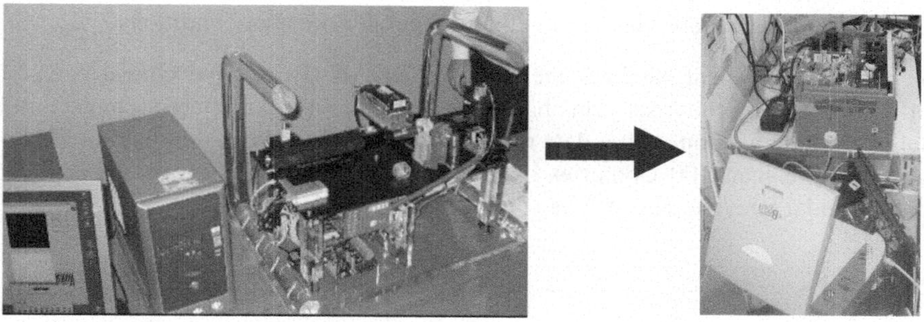

Fig. 3. The IFC proposed in [10,11] (right side) and an open top compact FC with the satellite modem (left side)[13]

Thanks to the reduced size and remote control functionality, an interconnected network of CYTOSATs can be used, in particular, for remote diagnosis and for continuously monitoring the illness progress in the third world Countries, which have limited coverage of communication infrastructures and a very limited number of operating FCs in the biggest hospitals and research centres. In the following subsections CYTOSAT communication hardware and software are presented.

5.1 Communication Hardware

In Fig.3 the IFC proposed in [10,11] and the new portable FC prototype described in [13] are shown. It can be seen the compactness of the new architecture, since all the digital components are integrated in a box having a laptop size width and length, and the satellite modem (the white component at the bottom of the figure) is placed on the top, making this device suitable to be carried by a vehicle.

CYTOSAT has been designed taking also into account the manufacturing cost reduction requirement in comparison with the classic FCs. This new FC is the result of considerable efforts devoted to the realization of a compact instrument with a scalable architecture and built-in satellite communication capabilities for reaching areas not covered by terrestrial digital communication networks.

A new data transmission and reception system has been added to the portable FC hardware architecture to provide it with satellite communications capability. By means of this relevant improvement the instrument is independent from the existing infrastructures of the particular place in which it has intended to work. Today, in fact, satellite communication systems can reach almost any spot of the earth surface.

CYTOSAT satellite communication system has been carried out by first selecting the appropriate hardware and then implementing the suitable communication software as described in [13].

Hardware selection has been conducted taking into account the following requirements:

- up-link bandwidth allowing the pre-processing data transfer at a short time
- size allowing the instrument portability
- data transmission and reception speed allowing the remote instrument functionality
- working capabilities even with unfavourable weather conditions

The chosen component has been the Regional BGAN (Broadband Global Area Network) produced by Inmarsat. This hardware system allows an internet connection by satellite communication with a data rate of 114 kbps (both in up-link and down-link) reaching more than 100 Countries over the world, including Europe, Africa, Latin America and a relevant part of Asia (Fig.4).

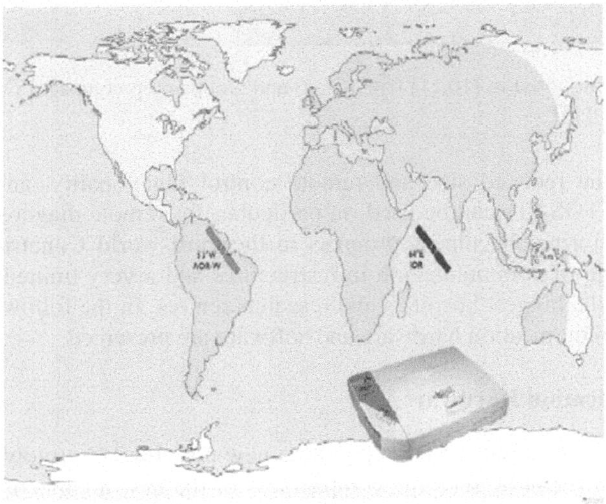

Fig. 4. Regional BGAN coverage map

Regional BGAN is light and compact, having reduced size (300mm x 240mm x 40mm), and a working temperature range (-10°C to 55°C) allowing its efficiency over almost all climatic conditions. Moreover this component is provided with the Bluetooth, Ethernet and USB connections.

The Regional BGAN has been therefore connected to the CYTOSAT processing and control unit through a USB port and integrated in the instrument prototype.

5.2 Software

CYTOSAT software has been designed to manage an interconnected network of more portable FCs by a client-server architecture. A control centre (CC) works as server keeping track of all the FCs connected to the network, that work as clients, managing the file transfer and storing all the data coming from the FCs. The remote control server allows the CC to check all the functions of the FCs, and/or to take charge of the data transmission. Data management as well as the automatic data record keeping are other features of the CC.

Concerning the CYTOSAT communication software the following requirements have been taken into account:

- to ensure a secure and reliable transmission of the processed data
- to store the received data in the CC
- to enable the interactive control of a CYTOSAT from the CC

Therefore, different integrated modules have been added to the classical FC analysis tools, as well as an encrypted data transmission capability and a remote control server.

5.2.1 FC Control Interface

The FC control interface has been designed to make almost all technical parameters, automatically set and manually changeable only as option for expert users. At the same time, biological parameters have to be directly accessible and easily set.

The FC control software, realized in LabVIEW environment, enables to set the reference parameters, acquire and process the diagnostic data, to display the results and to provide a final report (Fig. 5).

CYTOSAT interface (Fig. 5) can also provide specific graphs showing the cell samples by dot plot diagrams whose axis dimensions depend on the chosen cell characteristics. Moreover, single histograms are presented for each parameter.

The GUI has been designed to process 8 signals coming from the FC. The user can choose the parameters to be used as x and y axes for each of two plots. More functionalities are provided as those dedicated (i) to acquire the signals (start and stop acquisition), (ii) to reset the last information, eliminating the data obtained by the previous measurements, and (iii) to save the data acquired and processed to FCS files. Other important functions allow to set a cell detection threshold, and to choose the sample set to process.

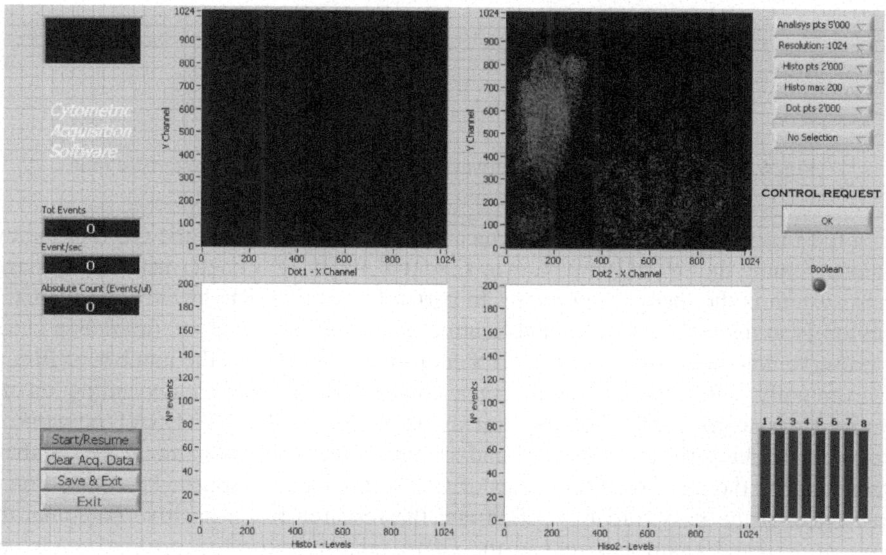

Fig. 5. CYTOSAT control GUI

5.2.2 Data Encryption and Transmission

Due to the type of information to be transferred, requiring the patient privacy safeguarding, the data encryption is provided before starting the transmission, that is successively speeded up by the data compression. In particular, the asymmetrical double-key cryptography, based on RSA (Rivest, Shamir, Adleman) algorithm, has been chosen to ensure the maximum information protection and implemented by a specific module developed in JAVA environment. To carry out the data compression, a lossless compression algorithm has been chosen, since a potential information loss could reduce the measurement reliability.

CYTOSAT communication software has been developed in LabVIEW. The communication starts by a handshake phase between the CC that is dedicated to manage the remote communications, and the mobile unit based on TCP/IP protocol. After that, the data transmission phase is carried out by the DataSocket Transfer Protocol (DSTP), that allows a real-time information sharing. After the data have been transmitted they are saved as FCS files in the CC database for storage or further analysis.

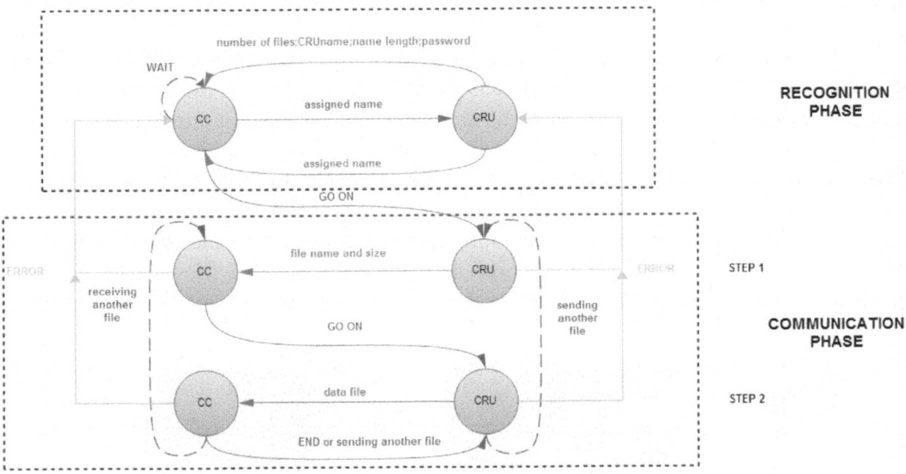

Fig. 6. Communication phases between the control centre and the remote unit

In the Fig. 6 the phases of communication procedure between the CC and the cytometer remote unit (CRU) are shown. Essential for the cytometer network management system is the recognition phase. In particular, every CRU has to send the CC a previously assigned password, identification name and name length, (used as a check to preserve this phase from errors), GPS coordinates, as well as the number of files to be eventually transmitted, to establish a connection. If the CC, depending on the received password, recognizes the unit as belonging to the CYTOSAT network, it proceeds checking for the password and searching the received name in the database containing all the identification names of CRUs that have established a connection at least once. If the password is verified and the research has a positive outcome, the name is assigned again and sent to the CRU, otherwise the CC assigns, sends and adds in the database a new CRU name. Obviously if the CC does not recognize the unit, the remote communication is cut off, and the CC waits for another CRU asking for communication. In this way, by the assigned name the CRU can be connected to the CC to start the communication phase, when one or more files need to be transmitted. The communication phase starts automatically asking the CC for communication when a CRU has registered in a file the data obtained by the measurements. The CC by the recognition phase decides to enable or not the connection, according to what reported above. The communication sequence is entirely controlled by the CC that, depending on the correct phase execution, allows the CRU to proceed with the next phase or not. Once successfully recognized, the CRU proceeds to send the CC, the name and size of the file to be transmitted. The file names are automatically assigned by the CRU, so that they include information about the hour, day, month and year in which the corresponding data measurement has been executed and can be easily extracted and sent according to a FIFO queue.

Once the CC has verified the received file name and length, checking that this last is different from zero, it allows the CRU to proceed the communication procedure by sending the file. The CC verifies the correctness of this operation by checking the dimension of the received file, after that this one is registered in the database. Then if there are no other files to be transmitted the CC ends the communication phase and waits for another CRU connection request, otherwise it allows the CRU to send the second file starting all over again the communication phase. Both the CC and CRU are designed to work together according to the sequence above reported, that is preserved also in case of errors. In particular, if at any time the CC would take notice that the CRU is out of line, it automatically cuts off the connection to wait for another CRU request. In the same manner if the CRU would take notice of a failed connection with the CC, it automatically comes back to the initial phase, that is asking CC for a connection. Moreover, to avoid the need of sending (from the CRU) or writing (from the CC) more times the same file because of a connection error that have not allowed to know the file has been already sent or written, three different directories are used during the communication procedure. In particular, the first directory is used by the CRU to record the file to be transmitted, after receiving notice from the CC that the file sending has been seen out, the file is moved on a second CRU directory storing all the file sent.

On the other side the CC allows the CRU to send the file, only if the file name received is not included in a directory storing all the file received and saved.

5.2.3 Communication and Control GUI

The communication and control GUI runs on the CC workstation. It has been designed to allow, (i) the geographic position identification transmitted by each mobile unit during the handshake phase, (ii) the file transfer, and (iii) the control of one or more FCs.

In particular the GUI shows (Fig.7) by a green box the geographical position of cytometry unit in transmission, by a yellow box the geographical coordinates of cytometry unit in connection and by a blue box the geographical coordinates of the CC. In Fig.7 a zoomed section of the GUI is shown to give an example of the coloured box indication. In particular, it can be seen that GUI is showing that the cytometry unit is connected and located in Africa (yellow box), for this example, while the CC is located in Italy (blue box).

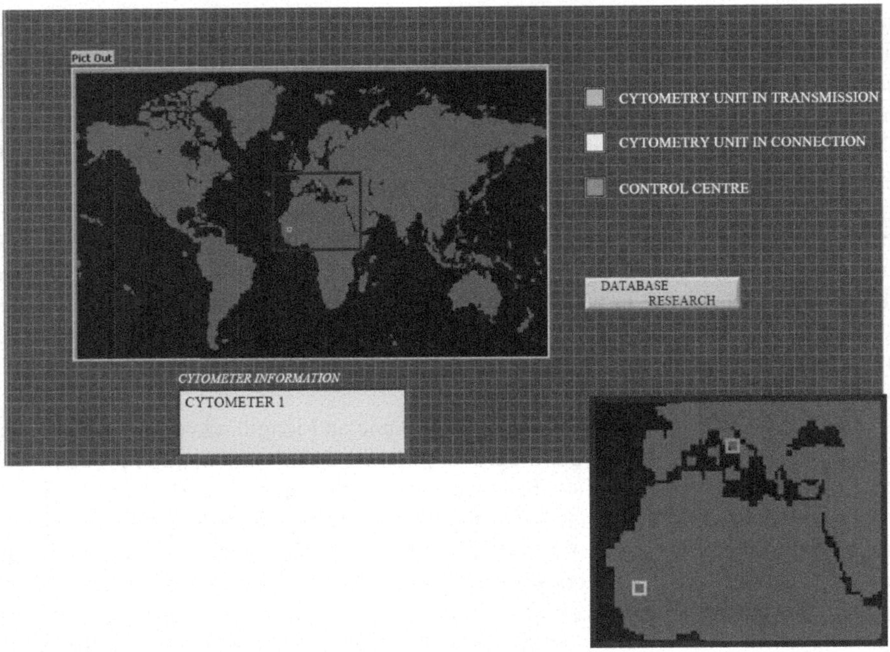

Fig. 7. CYTOSAT GUI for communication or controlling the mobile units (upper) and zoom of the section included in the blue rectangle (lower)

By means of such interface the system administrator can find the operating FCs within the covered area, observe their status, and gain the control of one of them. He/she can do that by simply double-clicking on the CYTOSAT icon. As a result the CC operator will access to the FC GUI shown in Fig.5 and observe how the FC is locally operated. Moreover, the CC operator can gain the control of the CRU and manage the FC as he/she was on the field.

The scalability of the whole system is granted by the possibility of adding or removing a new mobile FC without the necessity of any architecture change. By means of such a system the skills of the personnel operating on the field can be limited to the preparation of the blood samples and the input of patient data, while the doctor in charge for the diagnosis can even be in the control centre.

5.2.4 CYTOSAT Validation

In order to validate CYTOSAT communication system performance, several transmission tests have been carried out by varying the transferred file size and the transmission channel up-link and down-link bandwidth.

To analyze CYTOSAT transmission speed performance, the time elapsed to transmit the data files from the mobile unit to the control centre has been measured by placing the CC in the Executive Direction building of the National Cancer Research Center Foundation “G. Pascale” in Naples and the CRU in Roma square in Benevento, Italy.

Test results show that CYTOSAT system can ensure a throughput belonging to the range from 95 to 105 kbps.

In Tab.1 an example set of measured transmission times, by varying the transferred FCS file size from 100 kB to 1 MB, is reported together with the corresponding data rates. The obtained average speed is equal to 10.6 kB/s.

Table 1. Transmission throughput for a CYTOSAT validation

File Size (kB)	Elapsed Time (s)	Bit rate (in kB/s)
100	11	9.090
200	24	8.333
300	31	9.677
400	41	9.756
500	48	10.416
600	54	11.111
700	59	11.864
800	66	12.121
900	74	12.162
1000	85	11.764
	Average	**10.629**
	Standard Deviation	**1.373**

6 Automatic Count of the Micro Nuclei in Human Lymphocytes on Image Flow Cytometers

In [14] an improved method to recognize and automatic count the MNs on the IFC [10,11] acquired images of human lymphocytes has been pointed out. The method adopted in [10,11] to perform the MN detection is based on the pattern matching algorithm. Among the several pattern matching algorithms available in literature, that one implemented in the IMAQ Vision by National Instrument [33-35] has been selected to be used in the prototype of the new architecture of the FC [10,11] according to the following properties: (i) it is independent from the shape of the template to be detected, (ii) it is fast owing to the optimized learning phase, (iii) it offers the best results regarding to the correction of the alterations affecting the acquired image of human lymphocytes, and (iv) it is easy to integrate into the software environment pointed out to ménage the FC measurement device.

Biologists recommend specific criteria to identify in the cell one or more MNs [36]. In particular, the condition that the MNs are in the range [1/3, 1/16] of the surface of the associated nucleus is the fundamental constrain to perform the correct detection. The assessment of this condition is absolutely necessary and permits to distinguish the MN from the remaining objects in the image.

The disadvantage of the method based on the pattern matching algorithm is that it does not permit to translate into shape properties and geometrical features the detected object. Therefore, both the diameter and the surface of the detected MN cannot be accurately evaluated, the coarse evaluation of the geometrical features can be inferred, only. As a consequence, the detection of the MN can be affected by high level of uncertainty making the automatic count of the MNs not effective if referred to patients with immune diseases [12].

On the basis of these considerations, a new method based on the cascade of the pattern matching algorithm and another analysis tool able to evaluate the geometrical features of the MN and the nucleuses to improve the automated detection is proposed in [14].

The analysis tool proposed in the research is the Binary Large OBject (blob) analysis of the acquired image [37-41]. The blob is a group of connected pixels that have the same intensity. The image processing operates on these blobs to calculate the surface, perimeter, or to count the number of distinguishable blobs. Before to apply the blob analysis the image must be pre-processed by converting the grey scale image with 256 levels to the image with only two grey scales, zeros and ones. The objective is to separate the important objects, blobs, from the unimportant information contained in the background of the image. The technique based on the thresholding appropriately separates the blobs from the background. The result of the thresholding process is a binary image which is an image of pixel values of only ones and zeros. The blobs are represented by the connected pixels of ones, and the background is represented by the zeros.

Fundamental importance concerns with the appropriate estimate of the threshold value, because it influences the separation of the important blobs from the not important ones, and the location of the border. In literature this aspect is dealt with different approaches. In [37] the threshold is evaluated on the basis of a priori estimate meeting the recurring colours of the images. In [38] the evaluation is based on the estimate in advance from the tracking set of images. In [39] the threshold is stated on the basis of the difference between the expected surface of the target and that of image noise. In [40] a method is developed that combines the principles of thresholding with hysteresis in order to de-noise and to group the connected components. In [41] the histogram analysis is used to extract a list of significant local minima, and to choose the most indicated one for thresholding the image.

Each of these approaches is not effective and efficient for the MN image on the basis of the wide variation of the MN to be detected in the 256 levels of the grey scale, as a consequence of the characteristics of the human lymphocytes. Moreover, the acquired images by the FC can be affected by random alterations introduced by the movement of the lymphocytes and the acquisition system.

The method pointed out in the research is based on the adaptive thresholding process. In particular, the threshold is evaluated depending on the grey levels of the pixels constituting the MN detected by the pattern matching analysis. In this manner, it can be enhanced the confidence that in the image background represented by the zero grey level the MNs are not included.

In the following a brief overview of the method pointed out in the FC to detect the MNs is given. For sake of completeness, the aspects concerning the correct selection of the thresholding in the blob analysis is abstracted. Successively, the functional modality of the cascade of the pattern matching analysis and the blob analysis is presented. Finally, the results of the experimental tests are discussed.

6.1 Image Processing by Pattern Matching Analysis for MN Detection

The fundamental aspects concerning the method to detect the MN by the pattern matching algorithm [14] are discussed in the following: (i) the limited use for geometrical feature evaluation, and (ii) the detection and the correction of the contemporaneous alterations affecting the acquired image.

a. Geometrical feature evaluation by pattern matching

The pattern matching analysis is used to find and to check relevant structures [36] into a set. The detection of objects as MN and nucleus into the acquired images of human lymphocytes is performed by means of geometrical structure with circular shape.

In particular, nine different categories (Cat) of circular structures are defined, each one identified by the radius ρ pointed out in Tab. 2. These nine categories permit to detect the MN and the nucleus in very efficient way.

Two basic considerations can prove that the numerical procedure implementing the pattern matching algorithm permits to assess the coarse geometrical dimensions of the detected objects, only.

The first concerns the number of categories of circular structures. The increase of this number, and, consequently, the reduction of the step among the radii can contribute to improve the evaluation. Operating on this way the inconvenient is the enormous increase of the processing time.

The second one concerns the shape of the MN and the nucleus. Indeed, these shapes can be conveniently approximated by circular structure in the case of the pattern matching analysis. On the contrary, the real surface of these objects can include pixels leaved out from the circular structure.

More accurate evaluation of the geometrical features of the MN and nucleus can be achieved by means of the blob analysis.

b. Detection and correction of the contemporaneous alteration affecting the acquired image

The alterations affecting the acquisition system misstate the acquired image and reduce the Image Quality (IQ). It can occur that the MN cannot be detected by the pattern matching algorithm in consequence of the image alteration and not for its effective absence. Consequently, the number of the images both doubtful and rejected increases. These results clash with the important diagnosis requirements consisting of the minimization of the doubtful detections, and the confidence enhancement that the rejected images are not including the MNs. Both these requirements can be achieved by the image pre-processing before the MN detection by the pattern matching algorithm.

In [14] has been pointed out a method to estimate the IQ index and to correct the image alterations in order to increase the number of the correct detection of the MNs on the acquired images of human lymphocytes. In particular, the alterations taken into account are: (i) inadequate exposure (both the overexposure and the underexposure), (ii) the out of focus, and (iii) the Gaussian noise.

The method operates on the basis of the following steps: (i) evaluation of the IQ index estimating each image alteration, (ii) evaluation of the range of the IQ indexes in order to permit the image correction, and (iii) implementation of the convenient correction method for each image alteration in order to increase the IQ.

Table 2. Categories of circular structures and corresponding radius defined to find and check the MN

Cat	ρ [pixel]
A	17
B	15
C	13
D	12
E	11
F	10
G	9
H	8
I	7

Estimation of the IQ index: Three different IQ indexes are carried out on the basis of the three different image alterations taken into account. In particular: (i) to detect the Gaussian noise, the mean value of the minimum values of grey levels obtained by the high pass filter applied by windowing the original image is assumed as index, (ii) to detect the out of focus, the mean value of the higher values of grey levels obtained by the high pass filter applied by windowing the original image is assumed as index, and (iii) to detect the inadequate exposure, the mean value of grey levels is assumed as index.

Individuation of the variation range of the IQ index to correct the image: By referring to each IQ index previously evaluated, it must be established if it needs to correct the alteration and, successively, to process the image. The decision is carried out on the basis of the value of the IQ index compared with the admissible variation range.

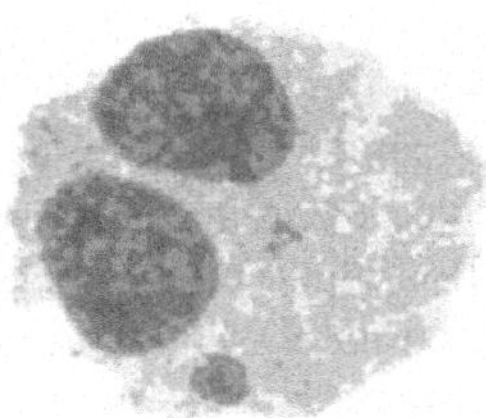

Fig. 8. Original uncorrupted image of the cell including two nuclei and only one MN

Image correction: On the basis of the three different IQ indexes, three different correction methods are pointed out. The correction depends on the disturbance type, and the value of the corresponding IQ index.

6.2 Thresholding Selection in the Blob Analysis

The blob analysis consists of a series of processing operations producing information about 2D shape in an image.

It can be used to detect a group of connected pixels that have the same intensity (blob) in the image and to make selected measurements of those pixels [35], [37] as the perimeter, the size, the surface and the number of blobs.

All pixels belonging to a blob are in the foreground state, all other pixels are in the background state. In the binary image, pixels in the background have values equal to zero, while every nonzero pixel is part of the blob.

The technique based on the thresholding enables to select ranges of pixel values in the 256 levels of the grey scale that separate the pixels in the foreground state from the background. Therefore, the result of the thresholding is a binary image that can be used to make particle measurements.

The thresholding is a subjective process, and the resulting binary image may contain unwanted information. The effect of the thresholding on the measurement of the blob surface is shown in Tab. 3.

Table 3. Effects of the different threshold values on the measurement of the blob surface

Binary image	Threshold value in the grey scale	% ratio of the surface of the blob and the surface of bigger one detected by blob analysis
	0-67	41,77 31,65 11,39 34,19 19,00 6,33 100,00 20,25 76,00 8,86 7,60
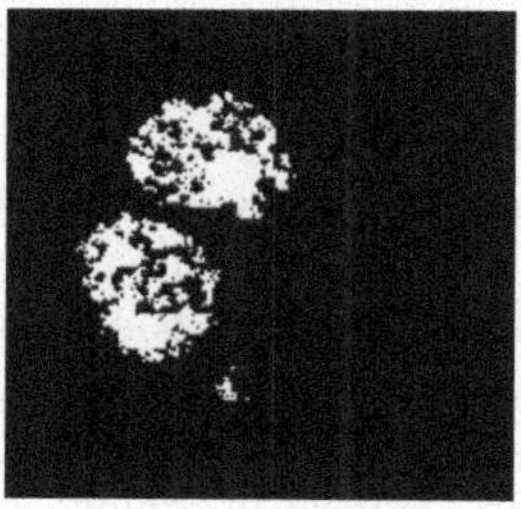	0-87	100,00 78,63 8,90 11,71

Table 3. (*continued*)

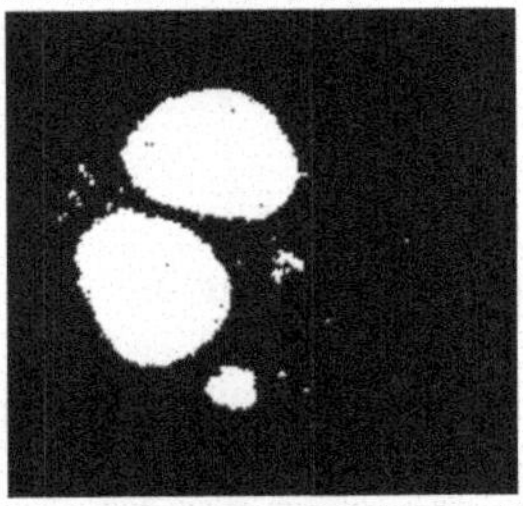	0-130	97,97 100,00 8,90
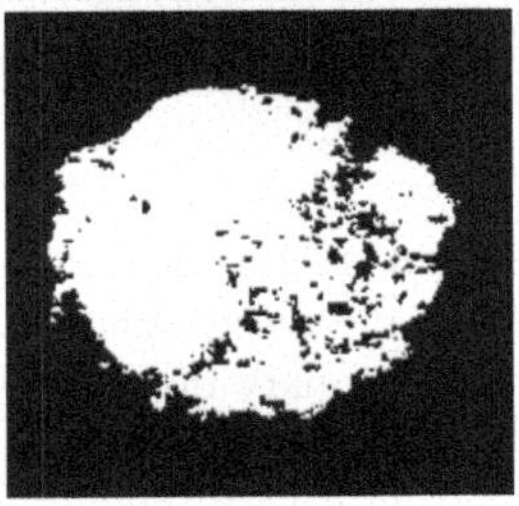	0-172	100,00

The original uncorrupted image in the grey levels of the cell including two nucleuses and only one MN is shown in Fig.8. On the basis of the different values of the threshold value, different number of blobs are detected. Tab. 3 points out the percent ratio between the surface of each detected blob and the bigger one.

The threshold value equal to 130 guaranties the correct evaluation of the ratio between the surface of the MN and that of the bigger nucleus.

By affecting the shape of the blobs, the correct evaluation of the threshold value can remove the unwanted information, thus improving the information in the binary image.

The difficulty in establishing a priori the threshold value depends on the fact that the characteristics of the image can change according to the alterations affecting the acquired image into the FC device.

In order to overcome this problem, the adaptive thresholding process is proposed. In particular, the threshold value is adapted to the mean grey level of the pixels in the neighbouring of the coordinate (x, y) corresponding to the MN detected by the pattern matching analysis.

In this manner, it can be enhanced the confidence that in the image background represented by the zero grey level the MNs are not included.

6.3 Image Processing for Surface Measurement of MN

Before the pattern matching analysis, the pre-processing phase is performed in order to detect and to correct the alterations of Gaussian noise, out of focus, and bad exposure affecting the images acquired into the flow cytometer device. This phase is performed according to the method presented in [14]. Successively, the corrected image is processed in order to detect the MNs. In the case the MN is detected, the coordinate (x, y) of the centre of the circular structure and the parameter Cat are furnished.

Fig. 9 shows the set of information furnished by the pattern matching algorithm and the convention used to establish the coordinates (x, y). This set of information is sent to the successive block for the threshold process.

Image	Pattern matching analysis
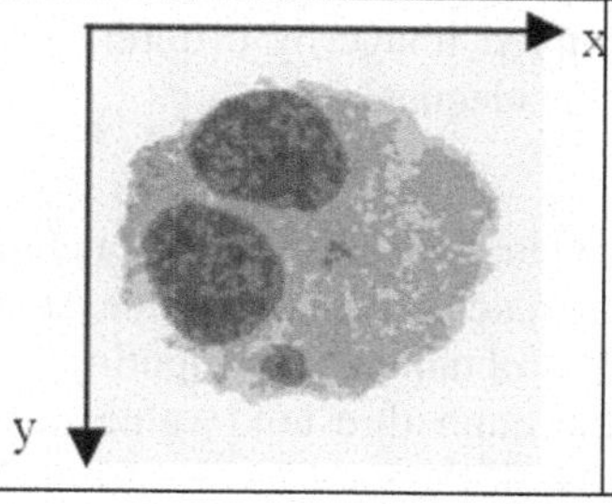	MN Cat F coordinates: x = 88 pixel y = 146 pixel

Fig. 9. Coordinates (x, y) of the centre of the circular structure and the parameter Cat furnished by the pattern matching analysis

a. Threshold process

The threshold process starts once MNs are detected. In general, more MNs can be detected in the same image. For each detected MN, information about coordinates and category Cat is stored.

The steps performed are the following:

Step 1: for each detected MN the Region of Interest (RoI) is computed on the basis of the coordinates (x, y) and the category Cat previously stored. The borders of the i-th RoI (with i=1,…, N), are:

$$x_{right}=x_i+\rho;\ x_{left}=x_i-\rho;\ y_{top}=y_i+\rho;\ y_{bottom}=y_i-\rho; \quad (1)$$

Step 2: for each RoI the mean grey level value mRoIi is computed;

Step 3: the mean value MRoI of mRoIi, i=1,…, N, is computed. MRoI is assumed as the upper bound of the threshold range, the lower bound is constant and it is set equal to zero;

Step 4: the resulting binary image is created by assuming the minimum threshold value equal to 0, and the maximum threshold value equal equal to MRoI.

At the end of the threshold process the binary image is available and the blob analysis can be performed.

The blob analysis permits to evaluate the surfaces of both the MN and the nucleuses with high accuracy.

This result is used to assess the condition that the MN is in the range [1/3, 1/16] of the surface of the associated nucleus.

6.4 Experimental Results

The experimental tests were performed to assess the fitness of the method pointed out to evaluate the threshold values and to reduce the grey levels from the range [0, 255] to the range [0, 1].

The tests show that the threshold values computed by considering the mean grey value of the pixels constituting the MN are valid to remove the unwanted information and to improve the information in the binary image.

Three set of tests were performed according to the different alterations affecting the acquired image of human lymphocytes into the FC device. The tests were executed by affecting the original uncorrupted image by different levels of the alterations: Gaussian noise, out of focus, and inadequate exposure.

a. Image affected by Gaussian noise

Fig.10a shows the image affected by Gaussian noise with standard deviation s=30. Without image correction the pattern matching analysis does not detect the MN. Therefore, the binary image of Fig.10b is obtained by computing the threshold values on the full image of Fig.10a. The minimum threshold value is equal to zero, the maximum corresponds to the mean value of the grey levels of Fig. 10a and it is equal to 225. These threshold values do not permit to remove the unwanted information. As a consequence the MN is always hidden and the blob analysis cannot be correctly performed. In the contrary, after the image correction [14] the pattern matching analysis is able to detect the MN. The corrected image is shown in Fig.10c. The binary image of Fig.10d is obtained by computing the threshold values on the basis of the coordinates and the category Cat furnished by the pattern matching algorithm. This information corresponds to that shown in Fig.9. The minimum threshold value is equal to zero. The maximum corresponds to MRoI equal to 122. These threshold values permit to remove the unwanted information as shown in Fig.10d. As a consequence, the MN is always distinguished and the blob analysis can be correctly performed. The percent ratio between the surface of the MN detected and the surface of the bigger nucleus is equal to 8.17%.

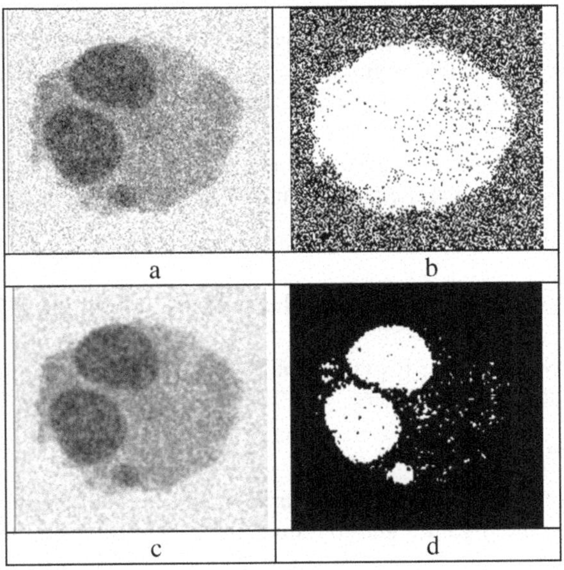

Fig. 10. a) image affected by Gaussian noise, b) binary image obtained by the threshold values computed on the original image a), d) binary image obtained by the threshold values computed on the pixels of the MN detected by the pattern matching analysis on the corrected image c)

b. Image affected by out of focus

Fig.11a shows the image affected by out of focus. Without image correction the pattern matching analysis does not detect the MN. Therefore, the binary image of Fig.11b is obtained by computing the threshold values on the full image of Fig.11a. The minimum threshold value is equal to zero, the maximum corresponds to the mean value of the grey levels of Fig.11a and it is equal to 196. These threshold values do not permit to remove the unwanted information. As a consequence, the MN is always hidden and the blob analysis cannot be correctly performed. In the contrary, after the image correction [14] the pattern matching analysis is able to detect the MN. The corrected image is shown in Fig.11c. The binary image of Fig.11d is obtained by computing the threshold values on the basis of the coordinates and the category Cat furnished by the pattern matching algorithm. This information corresponds to that shown in Fig.9. The minimum threshold value is equal to zero. The maximum corresponds to MRoI equal to 172. These threshold values permit to remove the unwanted information as shown in Fig.11d. As a consequence, the MN is always distinguished and the blob analysis can be correctly performed. The percent ratio between the surface of the MN detected and the surface of the bigger nucleus is equal to 8.25%.

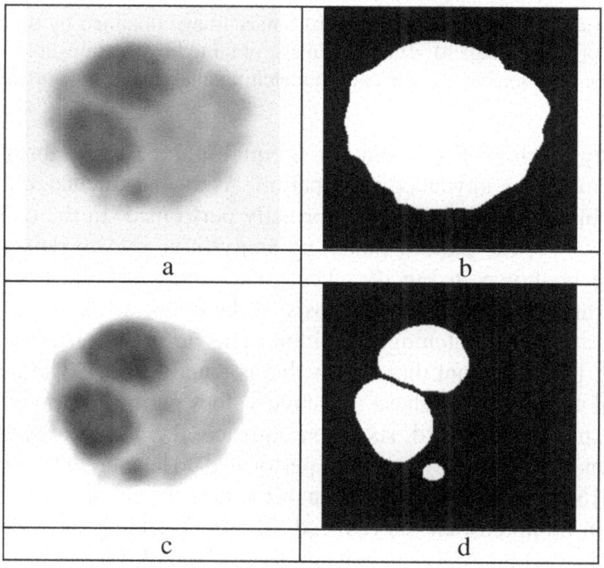

Fig. 11. a) image affected by out of focus, b) binary image obtained by the threshold values computed on the original image a), d) binary image obtained by the threshold values computed on the pixels of the MN detected by the pattern matching analysis on the corrected image c)

c. Image affected by bad exposition

Fig.12a shows the image affected by over exposure. Without image correction the pattern matching analysis does not detect the MN. Therefore, the binary image of Fig.12b is obtained by computing the threshold values on the full image of Fig.12a. The minimum threshold value is equal to zero, the maximum corresponds to the mean

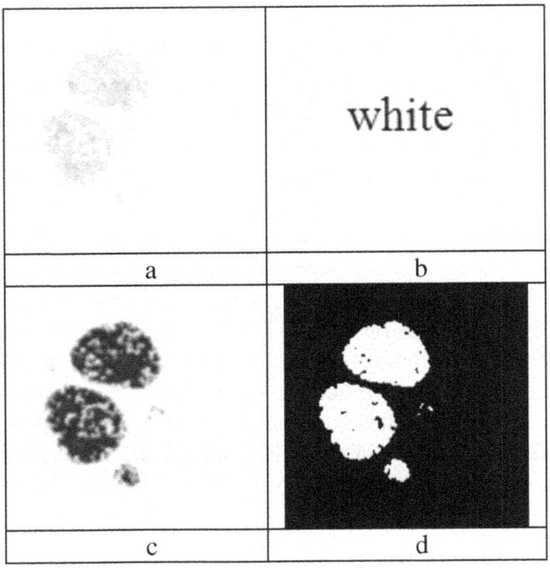

Fig. 12. a) image affected by over exposure, b) binary image obtained by the threshold values computed on the original image a), d) binary image obtained by the threshold values computed on the pixels of the MN detected by the pattern matching analysis on the corrected image c)

value of the grey levels of Fig.12a and it is equal to 255. These threshold values do not permit to remove the unwanted information. As a consequence, the MN is always hidden and the blob analysis cannot be correctly performed. In the contrary, after the image correction [14] the pattern matching analysis is able to detect the MN. The corrected image is shown in Fig.12c. The binary image of Fig.12d is obtained by computing the threshold values on the basis of the coordinates and the category Cat furnished by the pattern matching algorithm. This information corresponds to that shown in Fig.9. The minimum threshold value is equal to zero. The maximum corresponds to MRoI equal to 228. These threshold values permit to remove the unwanted information as shown in Fig.12d. As a consequence, the MN is always distinguished and the blob analysis can be correctly performed. The percent ratio between the surface of the MN detected and the surface of the bigger nucleus is equal to 10.07%.

Finally, it can be noted that the reduced difference among the values of the ratio between the surface of the MN detected and the surface of the bigger nucleus depends on the fact that (i) the correction method [14] reduces the influence of the alteration, only, and (ii) both the pattern matching analysis and the blob analysis are sensitive to the presence of the Gaussian noise, out of focus and bad exposition alterations.

7 Conclusion

Flow cytometry is a technology that has impacted both basic cell biology and clinical medicine in a very significant manner. However, today commercial and research FCs still presents some limitations, as their dimensions, cost and skill requirements as well as the troublesome exchange of information. To provide significant results, a FC should be managed by a skilled personnel. Most of conventional FCs are expensive, bulky and heavy and require to be connected power mains in order to work. For such characteristics, most FCs, in particular those not targeted to specific applications, can be used in hospitals and research centres being not transportable where they could be really useful, as for example in third world countries. In this chapter, in order to expand the metrological and operating characteristics of FCs overcoming the above quoted limitations, the prototype of a mobile cytometry unit, presenting the same specifications of the traditional FCs, and remotely controlled to produce diagnosis on the field is presented.

A method to enable IFCs to recognize and automatically count the micro nuclei (MN) on the acquired images of human lymphocytes has also been presented. The proposed method allows to overcome the disadvantages of conventional cytometers as i) the manual microscope scanning of all the bi-nucleated cells in order to estimate the MN frequency, ii) the operator dependency of test and iii) the length and the high cost of the measurement method.

References

[1] Cram, L.S., Martin, J.C., Steinkamp, J.A., Yoshida, T.M., Buican, T.N., Marosiorone, B.L., Jett, J.H., Salzman, G., Sklar, L.: New flow cytometric capabilities at the National Flow Cytometry Resource. Proc. of IEEE 80(6), 912–917 (1992)

[2] Liu, Y., Fisher, A.C.: Human erythrocyte sizing and deformability study by laser flow cytometer. In: Proc. of Ann. Int. Conf. of the IEEE Eng. in Medicine and Biology Society, vol. 1, pp. 324–325 (1992)

[3] Maguire, D., King, G.B., Kelley, S., Durack, G., Robinson, J.P.: Computer-assisted diagnosis of hematological malignancies using a pattern representation of flow cytometry data. In: Proc. of 12th South. Biom. Eng. Conf., vol. 1, pp. 153–155 (1993)

[4] Ormerod, M.G.: Flow Cytometry: a practical approach, 2nd edn. IRL Press at Oxford University Press, Oxford (1994)

[5] Cossarizza, A.: Advanced methodologies in flow cytometry. University of Modena, Graphical Press of University of Modena (1997)

[6] Altendorf, E., Zebert, D., Holl, M., Yager, P.: Differential blood cell counts obtained using a microchannel based flow cytometer. In: Proc. of TRANSDUCERS 1997, Chicago, vol. 1, pp. 531–534 (1997)

[7] Melamed, M.R., Lindmo, T., Mendelsohn, M.L.: Flow Cytometry and Sorting, 2nd edn. Wiley, Chichester (1990)

[8] Godavarti, M., Rodriguez, J.J., Yopp, T.A., Lambert, G.M., Galbraith, D.W.: Automated particle classification based on digital acquisition and analysis of flow cytometric pulse waveforms. Cytometry 24, 330–339 (1996)

[9] Abate, G.F., Bavaro, F., Castello, G., Daponte, P., Grimaldi, D., Guglielmelli, G., Martinelli, U., Mauro, F., Moisa, S., Napolitano, M., Rapuano, S., Scerbo, P.: Tomography System to Acquire 3D Images of Cells in Laminar Flow: Hardware Architecture. In: Proc. of International Workshop on Medical Measurement and Applications, Benevento, Italy, April 20-21, pp. 68–73 (2006)

[10] Abate, G.F., Bavaro, F., Castello, G., Daponte, P., Grimaldi, D., Guglielmelli, G., Martinelli, U., Mauro, F., Moisa, S., Napolitano, M., Rapuano, S., Scerbo, P.: Tomography System to Acquire 3D Images of Cells in Laminar Flow: Software Architecture. In: Proc. of International Workshop on Medical Measurement and Applications, Benevento, Italy, April 20-21, pp. 74–79 (2006)

[11] Grimaldi, D., Palumbo, A., Rapuano, S.: Hardware and software improvements in flow-cytometry measurements. Measurement 36, 111–119 (2004)

[12] Rapuano, S., Zoino, F.: A mobile cytometry unit based on satellite communications: cytosat. In: Proc. of International Workshop on Medical Measurement and Applications, Warsaw, Poland, May 4-5, pp. 1–4 (2007)

[13] Carnì, D.L., Grimaldi, D., Lamonaca, F.: Pre-processing correction for micro nucleus image detection affected by contemporaneous alterations. In: Proc. of IEEE Instrum. and Meas. Tech. Conf., Sorrento, Italy, April 19-23 (2006)

[14] Macey, M.G.: Flow Cytometry: Principles and Applications. Humana Press

[15] Radcliff, G., Jaroszeski, M.J.: Flow Cytometry Protocols: Basics of Flow Cytometry. Methods in Molecular Biology 91, 1–24 (1997)

[16] Longobardi Givan, A.: Flow Cytometry: First Principles. John Wiley & Sons, Chichester (2001)

[17] Paul Robinson, J.: Flow Cytometry: Theory, `http://www.cyto.purdue.edu`

[18] Riley, R.S., Idowu, M.: Principles and Applications of Flow Cytometry, Hematology/oncology lymph node seminar for medical ii students, Department of Pathology-Medical College of Virginia/VCU Health SystemsVirginia Commonwealth UniversityRichmond, VA

[19] BD Biosciences, Introduction to Flow Cytometry: A Learning Guide, Manual Part Number: 11-11032-01 (April 2000)

[20] Rahman, M.: Introduction to Flow Cytometry, Serotec (2005)

[21] Chapman, G.V.: Instrumentation for flow cytometry. Journal of Immunological Methods 243(1), 3–12 (2000)

[22] Biocompare, 2008 Flow Cytometry Report: Market Overview and Industry Survey (August 2008)

[23] Instrument Business Outlook, The flow cytometer market diversifies (June 2004)

[24] Robinson, J.P., Watkins, S.: Image Cytometry Instrumentation in Current Protocols in Cytometry. Wiley Interscience, Hoboken (2008)

[25] Bonetta, L.: Flow Cytometry smaller and better. Nature Methods 2(10) (October 2005)

[26] Dako Educational Guide. Flow Cytometry (2006)

[27] Huh, D., Tkaczyk, A.H., Wei, H.H., Grotberg, J.B., Takayama, S.: Towards portable flow cytometry: study on the use of air-sheath-based volume-efficient two-phase microfluidic systems. In: Proc. of 2nd IEEE Joint EMBS/BMES Conference, Houston, TX, USA, October 23-26, pp. 1642–1643 (2002)

[28] Kothari, R., Cualing, H., Balachander, T.: Neural network analysis of flow cytometry immunophenotype data. IEEE Trans. on Biomedical Engineering 43(8), 803–810 (1996)

[29] Miyake, R., Ohki, H., Yamazaki, I., Takagi, T.: Investigation of sheath flow chambers for flow cytometers (Micro machined flow chamber with low pressure loss). JSME Int. J. B. – Fluids Ther. Eng. 40, 106–113 (1997)

[30] Reder, A.: Remote diagnostics for laboratory instruments, IVD Technology Magazine (January 2001)

[31] Lorenzana, R., Coustan-Smith, E., Antillon, F., Ribeiro, R.C., Campana, D.: Simple methods for the rapid exchange of flow cytometric data between remote centers. Leukemia 14(2), 336–337 (2000)

[32] Zilmer, N.A., Godavarti, M., Rodriguez, J.J., Yopp, T.A., Lambert, G.M., Galbraith, D.W.: Flow cytometric analysis using digital signal processing. Cytometry 20, 102–117 (1995)

[33] Russ, J.C.: The Image Processing Handbook, 4th edn. CRC Press, Boca Raton (2002)

[34] `http://zone.ni.com/devzone/devzoneweb.nsf/`

[35] Marquez, A., Buenfil, J.A., Gonsebatt, A.E., Lozano, M.E.: Pattern analysis of cell micronuclei images to evaluate their use as indicators of cell damage. In: Proc. of the IEEE 25th Annual Int. Conf. Engineering in Medicine and Biology Society, September 2003, vol. 1, pp. 731–734 (2003)

[36] Lu, S., Tsechpenakis, G., Metaxas, D.N., Jensen, M.L., Kruse, J.: Blob analysis of the head and hands: a method for deception detection. In: Proc. of the 38th Annual Intern. Conf. on Syst. Scien., Hawaii, January 3-6, pp. 20c–29c (2005)

[37] Aggarwal, J.K.: Understanding of human motion, actions and interactions. In: Proc. of IEEE Conference on Advanced Video and Signal Based Surveillance, AVSS 2005, September 15-16, pp. 299–305 (2005)

[38] Yonemoto, S., Nakano, H., Taniguchi, R.: Real-time human figure control using tracked blobs. In: Proc. of 12th Int. Conf. on Image Analysis and Processing, September 17-19, pp. 127–132 (2003)

[39] Castro, A., Arcay, B., Dafonte, C., Santos Suarez, A.: Development of an analysis system of the X-rays of bones for prosthesis placement. In: Proc.of the IEEE 22nd Annual Int. Conf. of Engineering in Medicine and Biology Society, July 23-28, vol. 3, pp. 1795–1798 (2000)

[40] Tobias, O.J., Seara, R.: Image segmentation by histogram thresholding using fuzzy sets. IEEE Trans. on Image Proc. 11, 1457–1465 (2002)

Biomedical Sensors for Ambient Assisted Living

Eric T. McAdams[1], Claudine Gehin[1], Norbert Noury[1], Carolina Ramon[1], Ronald Nocua[1], Bertrand Massot[1], Aurélien Oliveira[1], André Dittmar[1], Chris D. Nugent[2], and Jim McLaughlin[2]

[1] Biomedical Sensors Group of the Nanotechnologies Institute of Lyon, INSA Lyon, France
[2] NIBEC, University of Ulster, Northern Ireland

1 Context

The percentage of the population classified as being elderly has been predicted to increase dramatically in size over the next 30-40 years. Figures produced by the World Health Organisation (WHO) anticipate an increase from around 600 million in the year 2000 to close to 2 billion by the year 2050[1]. By 2050, 22% of the world's population will be over 60[2] in Europe it will be over 30%[3]. In addition, according to the WHO, approximately 10% of the population experience some form of disability. Already 21% of people above the age of 50 have severe vision, hearing and/or mobility problems.

There is a recognised need to radically change Healthcare provision, presently based on an acute care model, to make it more suited to the ongoing management of chronic disease and hence move towards a preventative care model. This has to involve changing the existing Health Care systems and their associated technologies.

The technologies required for such a Health Care revolution, are related to the wearable and point-of-care sensors required to enable ambient monitoring of patients, helping patients manage their conditions more effectively and keeping them out of hospital.

It is widely agreed (e.g. by the WHO and the EC) that it is imperative that healthcare leadership implement a more sustainable form of care, improving healthcare quality while reducing unnecessary costs. This is to be achieved by shifting away from today's reactive model of care to an integrated approach which enables, encourages and supports individuals and their families to continuously monitor and manage their health from the comfort of their homes, cars, work place, etc., avoiding, to a great extent, costly acute intervention. The emphasis now is therefore on "self-management", "personalised health", "pervasive healthcare" and "preventative healthcare"; terms reflecting key aspects of the new approach.

It has been shown that home-based or Ambient healthcare delivery, where appropriate, is much less costly coupled and the patient's perceived quality of

[1] Ageing and life course, World Health Organization, http://www.who.int/ageing/en/
[2] United Nations "Population Aging 2002"
[3] http://www.continuaalliance.org/

S.C. Mukhopadhyay and A. Lay-Ekuakille (Eds.): Adv. in Biomedical Sensing, LNEE 55, pp. 240–262.
springerlink.com

life is generally much higher. It is interesting to note that, even in cases where the elderly know that they are more at risk, when given the choice, they prefer to continue to live at home, on their own, rather than opting for some form of institutionalised care [5].

Home-based care is less expensive on a 'per day' basis and more appreciated by the patient. More importantly, it is generally more effective, therapeutically and financially, to encourage and support the patient to manage their chronic condition at home; to detect and act on symptoms as early as possible.

1.1 Potential Solutions

The development of sensor-based technologies which enable the monitoring of a patient's health and lifestyle at home and "while on the move" promise major improvements to the quality of healthcare while driving down costs. There is therefore a need of revolutionary AAL systems which involve novel sensor technologies, mobile technology, home-based medical equipment, embedded systems, wearable systems, ambient intelligence, etc. which are capable of conveniently, discreetly and robustly monitoring patients in their homes and while performing their daily activities without interfering significantly with their comfort or lifestyle.

In the modern era, where patients see themselves as customers, patient compliance is a key aspect of the success of any long term, patient-centred approach. In this scenario, Healthcare can be made to address the individual patient's needs via innovative healthcare delivery programs that link the patient to their care, no matter where they are. Early studies indicate that these technology-enabled, personalised programs can, with further developments, form a complete solution for sustained, efficient, quality health care.

It has been estimated that the prevention and effective management of chronic conditions is 20% medicine 80% 'other'[4]. By concentrating on the 80% non-medicinal aspects, better healthcare can be provided without the escalating cost. Solutions to the monitoring and treatment of many chronic diseases can therefore be offered from a number of technological perspectives, all of which aim to provide a level of independence within the home environment (and beyond) and the effective early detection and treatment of conditions before they necessitate costly emergency intervention. According to the EC definition, AAL is the consolidation of the necessary technologies, systems and services required to delivery this new healthcare provision.

One of the key goals of AAL is the delivery of preventative support rather than (simply) reactive intervention. This involves the intelligent processing of information gathered concerning the patient's environment, lifestyle and vital signs. The patient (and/or family) is/are effectively encouraged to take some responsibility for their healthcare, and medical conditions can be more effectively managed to the benefit of all concerned. The services which AAL can offer within the home environment should also be available once the person leaves their home,

[4] Nakita Vodjdani, 1st Global Village Workgroup, February 7th, 2008 Paris, France.

goes for a walk, gets into their car, goes to their GP, etc. Future health care must be "Personalised" and "Connected".

1.2 Basic "Telemonitoring" Systems

To date, many states have concentrated on the development of "Telemedicine", essentially voice-dataimage networks involving videoconferencing and the transmission of images and medical records/files[5]. Although such technologies have given rise to significant advances in patient care, they form only a part of the potential of technology lead innovation in Ambient Assisted Living.

Many examples of simple home-based, telemonitoring technologies exist: blood pressure cuffs, glucose meters, medication reminders, weight scales and heart monitors are already on the market or are available within a range of recent or ongoing studies. People with heart disease or diabetes have been "transmitting" verbally or digitally their "vital signs" (blood pressure, heart rate, glucose levels, etc), from their homes to their health care professional, and getting (real-time) feedback on their condition.

Using such "telemonitoring" devices, the Connected Cardiac Care program in the US has helped hundreds of heart failure patients avoid re-hospitalisation and improve their health and wellbeing. "Our goal is to help patients manage their disease, once they are discharged from the telemonitoring program. Through daily monitoring, teaching and raising awareness of their condition, we are helping our patients recognize signs and symptoms of a problem, and to seek early intervention, to help keep them out of the emergency room." [6] In addition to helping avoid hospitalisations, patients in such programs report high levels of satisfaction with the care they have received. Similar successes have been reported around the world and for a range of chronic illnesses.

Basic home-based "Telemonitoring" systems are therefore enabling providers at some leading medical centers around the world to more effectively keep patients healthy and out of the hospital, empowering patients to make healthy lifestyle choices and enjoy a more patient-oriented healthcare delivery.

1.3 Some Early Ambient Monitoring Systems

The sensor systems used in SHL Telemedicine's portable CardioBeeper™ and CardioPocket™ was successfully designed and developed at NIBEC[14]. These systems have been marketed around the world by PHTS Telemedizin, Philips and Raytel Cardiac Services. The CardioBeeper12/12™ is a compact, handheld trans-telephonic ECG electrode and transmitter (Figure 1). It is designed to be used by an individual to transmit a 12-lead ECG from any location, in real time, to a physician's office, hospital or monitoring centre, over a standard telephone line, for the purpose of remote real time diagnosis of arrhythmia, ischemia, and MI. The patient simply puts the device on his/her chest when symptoms are felt

[5] Lareng. http://www.telemed.ru/rfrs/sem2000/art/lareng.html

[6] Partners Centre for Connected Health. http://www.connected-health.org/media/158623/2007%20progress%20report.pdf

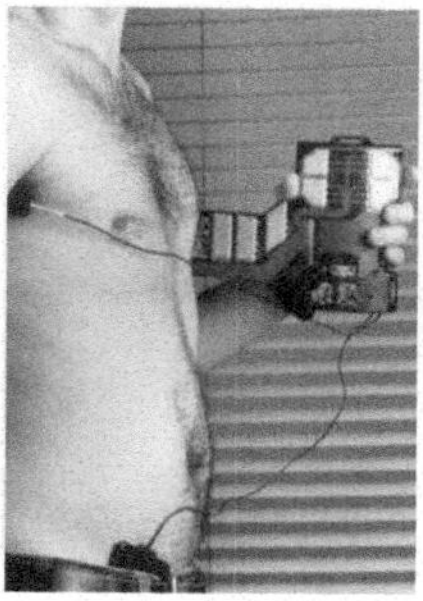

Fig. 1. Early version of CardioBeeper™

Fig. 2. CardioPocket™

or during routine evaluations and transmits the ECG by pushing a button[7]. The clinician can then advise the patient to use an auto-injector to self-administer a "clot-busting" drug, send emergency services or take whatever other action is appropriate.

In order to improve patient compliance and convenience, the CardioPocket™ was later developed (Figure 2). This is an attractive leather wallet which also serves as a one-lead ECG transmitter for diagnosing heart rhythm disturbances. By simply placing the wallet against the chest and using any telephone or cellular phone, the user can transmit a real-time ECG strip within seconds to the remote monitoring center for immediate consultation.

Roth et al [19] recently studied the outcome and cost-effectiveness of the cardiac programs carried out by Shahal (SHL) over the previous 19 years. In chronic heart failure a 66% reduction of hospitalisation days was observed, together with an improvement in quality of life. The authors concluded that management of coronary artery disease with appropriate telemedicine is not only effective, but has a huge potential for cost savings, improvements in quality of life and in prognosis of heart disease. They reported that Taunus BKK, a large German Healthcare Insurance Company, calculated that the Shahal cardiac program could lead to a saving of at least 5 million Euro per year in Germany alone.

1.4 Current Wearable Systems

Many patients would greatly benefit from continuous ambulatory monitoring of a range of vital signs over a prolonged period in order to optimally observe and treat a chronic condition or to supervise recovery from an acute event or surgical procedure (Figure 3). The multi-biosignal systems presently envisaged in numerous EC and industrial projects unfortunately often involve unwieldy wires between the monitoring system and the various standard electrodes/sensors distributed over the patient's body. Elderly patients, especially those with physical disability, have great difficulty in correctly positioning and attaching the traditional sensor designs used. The multiple wires limit the patient's activity and

[7] http://www.shl-telemedicine.com/

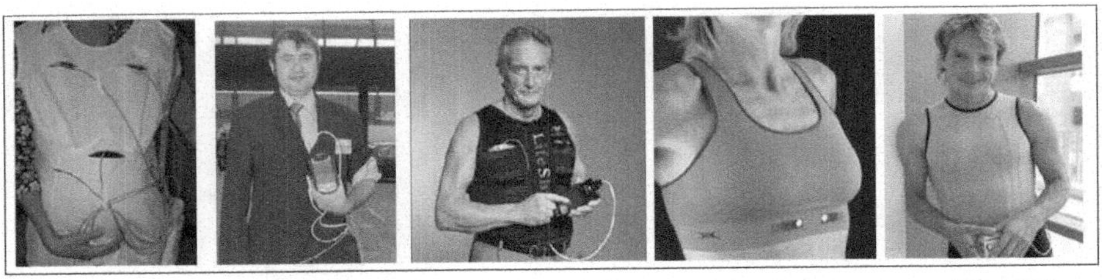

Lifebelt *Mobihealth* *Lifeshirt* *Numetrex* *WEALTHY*

Fig. 3. Current wearable systems from EC and industrial projects

level of comfort and thus negatively affect patient compliance. The wires, coupled with inappropriate sensor design, give rise to large amounts of motion-induced biosignal artefacts, essentially rendering many of the "ambulatory" monitoring systems less than suitable for this demanding application. Recent technology advances in wireless networking, including Personal Area Networks (PAN) or Body Area Networks (BAN), promise the possibility of removing the connecting wires and thus minimising many of their associated problems. However, the challenges of sensor positioning and motion-induced artefacts still remain and the latter may even be increased.

Some sensors have been successfully built into tight fitting garments such as bras (e.g. numetrex-bra) which ensure a quality contact, apparently without problems arising from skin friction. However these tend to be simple, one-channel ECG systems capable only of monitoring heart rate. More promising multi-parameter designs, involving the development of textilebased systems for Wearable health, include "Vêtement de Té*l*é*a*ssistance Mé*d*icale Nomade" VTAMN [17], and, most recently WEALTHY[8].

WEALTHY is a wearable, fully integrated system, able to acquire physiological parameters such as ECG, respiration, posture, temperature and a movement index. Sensors and connections are integrated into the fabric structure, and these include six fabric ECG electrodes, four fabric Impedance electrodes and two embedded temperature sensors. However, although WEALTHY is probably the best wearable system to date, the electrodes still suffer from artefact problems. Work is ongoing in this area with the manufactures of the "WEALTHY" garment (Smartex[9]) in the European project ProeTEX[10] 6th Framework IST Integrated Project in order to develop higher performance electrode systems.

Several companies have recently introduced wireless monitoring platforms (Figure 4). For example, SHIMMER[11] (Sensing Health with Intelligence, Modularity, Mobility, and Experimental Reusability) is a small wireless platform

---

[8] http://cordis.europa.eu/data/PROJ_FP5/ACTIONeqDndSESSIONeq112422005919ndDOCeq2475ndTBLeqEN_PROJ.htm

[9] http://www.smartex.it/

[10] http://www.proetex.org/

[11] http://www.intel.com/healthcare/research/portfolio.htm?iid=health+lhn_hriportfolio

designed by Intel to support wearable monitoring applications. It features a large storage capacity and low-power wireless communication technologies which should facilitate wearable or wireless sensing.

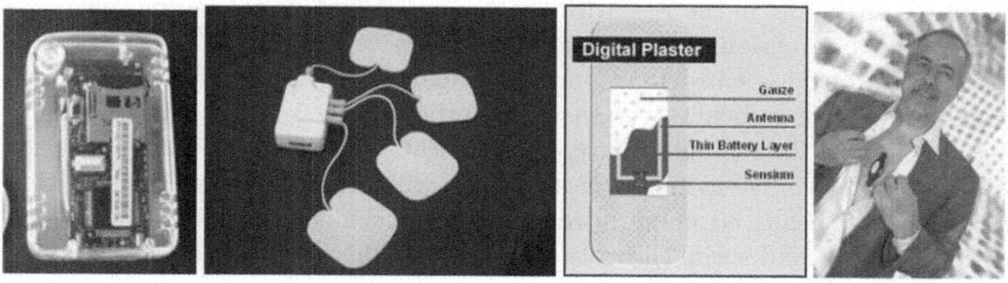

SHIMMER *SHIMMER & ECG Electrodes* *Sensium* *Welch Allyn*

Fig. 4. Wireless monitoring platforms

Toumaz has also introduced a wireless infrastructure for intelligent, non-intrusive continuous monitoring of patients at home or on the move called "Sensium"[12]. Incorporated within the Sensium platform is the capability of intelligently processing data locally and of providing real time feedback to the patient, if required.

Although these monitoring platforms are very exciting and pave the way forward for widespread research in the area of mobile monitoring of patients, the electrodes and sensors are still largely the same, and thus the bottle-neck to acceptable performances remains essentially unchanged.

2 Barriers to the Success of Ambient Assisted Living

Although early trials, pilot studies and a few major programs have had promising results, the widespread adoption of such "telemonitoring" and AAL has to date been very slow, much too slow to radically improve health and quality of life and dramatically reduce healthcare costs. Several key barriers exist to the more widespread use of such exciting technology.

These include

1. Connected HealthCare Systems
2. Device Interoperability
3. Suitable Sensor Technology.

2.1 Connected HealthCare Systems

Any single organization or group does not have the resources needed to address the complex public health issues that need to be resolved to facilitate the introduction and implementation of effective and efficient management of chronic

[12] http://www.toumaz.com/

disease and the ageing. New "Connected Health" coalitions will need to be established between the various sectors involved to ensure that the advancements in prevention, control, and treatment of chronic diseases benefit all. Fortunately, the critical mass of interest and political will is now coming together in many countries, including in the EU with the goal of connecting the various elements of Healthcare provision required for this "disruptive" approach[13].

The "European Centre for Connected Health" was recently established in Northern Ireland by local Government, with the encouragement of the EC, "linking up" of all of the relevant "players" in HealthCare. "The European Centre for Connected Health, based in Northern Ireland, will focus on developing the region as a connected health economy, introducing new technologies and working closely with the health and social care system to deliver this"[14]. It is anticipated that, once successfully implemented in Northern Ireland, which is effectively a good-sized, real-world, test-bed for the introduction of new systems and technologies, the "Connected Health" model will be eventually be adopted throughout Europe and possibly elsewhere following adaptation of aspects of the model to make it compatible with the varying social preferences and the regulatory issues which exist on the national or regional level across Europe.

2.2 Device Interoperability

If the more sophisticated monitoring devices and associated systems are to be successfully introduced into the general Healthcare, the devices must be fully interoperable with each other and with other information sources. "Interoperability" is the ability of a system or a product to work with other systems or products without special effort on the part of the patient/customer. For medical devices to be able to function efficiently with other products they all have to adhere to some form of widely accepted standards. No standards presently exist that fully define interoperability among medical devices and systems, thus the market is unable to fully exploit the technical solutions presently available. International interoperability standards will require the collaboration of all the key participants involved in the Healthcare provision and will involve the development of public policy, regulations, re-imbursement, etc. without which such standards will not gain widespread acceptance. Policy makers, regulators and industry leaders must collaborate to remove policy barriers to standards development worldwide and create new policies and incentives to advance harmonisation.

The Continua Health Alliance was formed in 2006 by technology, medical device and health and fitness industry leaders with the aim of "establishing an ecosystem/market of connected personal health and fitness products and services making it possible for patients, caregivers and healthcare providers to

[13] A disruptive technology or innovation is one that, when introduced, either radically transforms markets, creates wholly new markets or destroys existing markets for other technologies/strategies.

[14] http://www.eu-cch.org/index.htm

more proactively address ongoing health care needs" [15] and to thus accelerate the healthcare technology transformation. The alliance now includes over 100 of the top companies, including IBM, Philips, Dell, Welch Allyn, Partners HealthCare, GE, Medtronic, Intel, Boston Scientific, Motorola and Samsung. By bringing together all the different levels of healthcare technology, from the devices up to the healthcare providers, the Alliance is creating a system of standards using currently available technologies. The Continua Health Alliance recently announced the production of its first set of guidelines which will enable most available devices to seamlessly function together. Products are now being certified and marked with the Continua Health Alliance symbol to indicate that a given device is compatible with all other Alliance certified products.

2.3 Suitable Sensor Technology

The major remaining stumbling block to the successful design and development of monitoring systems capable of being applied to and functioning reliably on real patients, under real conditions for an extended period of time with sufficient robustness is the design of the sensors [10] [18].

This rather surprising observation is a consequence of the standard approach taken by engineers while designing wearable/portable monitoring devices. Invariably the engineers involved, who are generally specialists in IT, Electronics etc., start with the monitoring device, hardware, software, telemetry, etc and leave the "simple" sensors to be added on at the end. Unfortunately, electrodes are not just 'pieces of conductive metal' as is often evidently assumed (for example, silver chloride, the best electrode material for most non-invasive biosignal monitoring applications, is in fact very resistive), nor are all sensing problems miraculously solved by making sensors smaller.

Key problems to be overcome include the convenient attachment of sensors in their correct locations, the comfortable and discrete wearing of the sensors for prolonged periods and the avoidance of motion-induced artefacts. Microfabrication obviously does have its place and advantages, however only when the "macro" problems are understood and solved will the benefits and additional challenges of microfabrication be advantageously explored.

3 Sensors

New methods of recording biosignals from the body are required for demanding AAL applications. For many "wearable" applications, the biosignals must be recorded from non-traditional sites which lend themselves to sensors incorporated into smart clothes, body worn patches, wrist watches, etc. For this to be possible, novel transduction mechanisms are often required. For example, blood oxygen saturation is traditionally measured using Pulse Oximetry (SpO2). A light source and photodetector are enclosed within a probe and attached to the

[15] http://www.continuaalliance.org/

finger or ear lobe. Such a sensor is fairly bulky and not ideally suited for ambulatory use. The necessary connecting wires from such a sensor to a body worn device or patch are unacceptable for widespread clinical use and patient compliance. An alternative method of detection must therefore be developed which enables the monitoring of blood oxygen saturation levels through the skin under the integrated monitoring patch, located for example on the chest. This requirement also applies to the other "vital signs" sensors (e.g. ECG, respiration, temperature) as well as those used to monitor emotional (e.g. stress), behavioural (e.g. motion) and environmental parameters. For a given patient's condition, one must be able to monitor all key biosignals and parameters of interest from within a small patch "footprint" with sufficient accuracy and with a minimum of artefact. This will involve a major change in the way vital signs are traditionally monitored. Below we present some of the work of our research group.

3.1 Bioelectric Sensors

Novel electrode systems have to be developed for non-traditional measurements of electrical biosignals such as ECG, EMG and Galvanic Skin Response. These applications require a good understanding of the electrical properties of electrodes and human skin. Motion artefact, a major problem in ambient monitoring, has to be investigated in order to avoid, minimise or remove it using appropriate electrode design and location; and suitable signal processing. The electrode-electrolyte interface has be studied and modelled in order to optimise the surface topography and thus minimise electrode interface impedance. This is a key concern, although generally ignored, when miniaturising sensors, for example for point-of-care or lab-on-chip sensing. As the electrode area is decreased, the interface impedance increases causing measurement problems such as signal distortion, often misinterpreted in the literature as intriguing reactions taking place on the sensor surface. By modelling the surfaces and, based on this theoretic work, creating optimal nano- micro-surface features, the interface impedance can be dramatically reduced and/or properly understood and the experimental results correctly interpreted. Tissue impedance has also to be studied and modelled to help in the development of high-performance impedance-based sensors for non-invasive monitoring of, for example, respiration, diabetes and wound healing. Particularly in the case of diabetic monitoring using impedance sensors, although the use of such a non-invasive technique is highly attractive, the underpinning science has not yet been clearly demonstrated.

Electrode/Sensor Materials

The electrodes used in smart clothing, for example, would greatly benefit from having a lower-profile and being flexible. One solution to this problem is the use of suitable textile or profiled electrodes and an interface layer. Unfortunately, at present the textile electrodes used in smart garments are made from such "inert" materials as stainless steel fibers/threads for their "washability", but give rise to poor electrical performances. Research must therefore focus on the development of more appropriate materials for electrode interface performance, rather than simple conductivity. As standards for "dry" textile electrodes presently do not exist, the AAMI standard for disposable ECG electrodes[16] (which involves testing electrodes gel-to-gel and hence are inappropriate) has to be modified. One promising material is carbon which was used successfully in the past as a standard ECG electrode[21] and was used in early versions of SHL's CardioBeeper (Figure 1). The performances of these solid carbon electrodes have been found to compare reasonably well with those of Silver/Silver Chloride electrodes, the biomonitoring "gold standard" electrode material[13]. Carbon has so far been largely ignored as a textile electrode material or inappropriate versions have been used. However, various forms of carbon fibers and the weaving thereof are currently being investigated to generate the most appropriate interfacial electrochemistry and electrode surface topography.

Woven electrode materials can be made to produce a porous, 3-dimensional structure, greatly increasing the effective electrode surface area and decreasing the electrode-sweat interface impedance. For example, in the EC's ProeTEX project, Smartex Ltd, the leading manufacturer of "Smart Garments", and NIBEC (University of Ulster) are developing and testing silver fabric electrodes and initial results are encouraging (Figure 5a).

Fig. 5. (a) ProeTEX silver fabric Electrodes (b) Silver/silver chloride functionalised CNTs

As Silver/Silver Chloride is recognised as the best material for surface bio-electrical monitoring, research is also concentrating on the coating/functionalisation of fibers and surfaces with Silver/Silver Chloride[22]. Promising electrode structures using carbon nanofibre (CNF) arrays functionalized with silver/silver chloride (AgAgCl) coreshell nanoparticles (Figure 5b) are currently being developed. The coreshell AgAgCl nanoparticles on the CNF surfaces have been shown to enhance transduction in ionic media as demonstrated by electrical impedance spectroscopy. With further work, these functionalized CNF arrays could be utilised as part of miniaturised electrochemical biosensors and possibly as dry electrophysiological sensors for bio-potential monitoring applications (subject to biocompatibility studies).

[16] ANSI/AAMI EC 12:2000 "Disposable ECG electrodes".

Sensor Inter-layers

Although the properties of the electrode-sweat interface are important, the electrical properties of the skin are a much more significant problem in electrical biosignal recording. Dry electrodes, when in contact with the skin, give rise to unacceptable motion artefact due to the deformation of the dry skin and its resulting piezoelectric effect.

There is therefore the need of a textile or similar interlayer which effectively holds moisture and acts as an electrolyte reservoir, accumulating sweat and hydrating the skin. Such a textile interlayer, if correctly profiled, could provide optimal contact between the electrode and the skin while ensuring patient comfort by minimising friction and avoiding the use of adhesives. The use of occlusive backings which help trap the moisture under and around the electrode should also be investigated. Key to the success of these approaches are the material properties of the textile and backing layers, the weave of the fabric and the overall "wicking effect" achieved. These properties have to be tailored to the individual monitoring application. For ECG and impedance pneumographic measurements, the electrode-skin impedance must be reduced, hence the skin must be hydrated as much as possible. For thermo-neurovascular measurements one wishes to measure actual properties of/at the skin and thus it is important not to "drown" these out by the active hydration of the skin.

3.2 Thermal Sensors

Thermal parameters of interest in AAL are mainly body temperature, skin temperature, and body heat flow. From these parameters, one can assess many physiological phenomena, such as metabolism, thermal comfort, skin blood flow, skin hydration, skin thermal conductance, skin infection, core temperature, fever, muscle activity, autonomic nervous system (ANS) activity and respiration rate.

Skin Temperature

In ambulatory conditions, the easiest, and hence the most frequently recorded parameter used to assess the thermal state of the subject is skin temperature. Generally a high precision thermistor is attached onto the skin, most often it is integrated directly into the given ambulatory monitoring device. Although this measurement is quite easy to perform, interpretation of the data is not trivial and the skin temperature alone can be insufficient to evaluate the thermal state of the subject. Two points have to be taken into account:

1 - The skin temperature varies with numerous parameters such as the initial thermal state of the subject, his/her physical activity, climatic conditions, etc. Typically, skin temperature can range from 15°C in a cold environment of 1°C to 38°C in a hot location during intense exercise. These large variations can be considered as normal variations depending on the environmental and physical conditions of the subject but can also indicate a dangerous health state (hypothermia or hyperthermia) requiring urgent intervention.

2 - Skin temperature is not homogenous over the body and the location of the thermistor has to be correctly chosen. As the human body is thermo-regulated,

in order to maintain its core temperature (around 37°C), peripheral members (arms and legs) act as regulators. As a consequence, the temperature of the latter tends to vary. A thermal sensor located on the wrist is therefore not necessarily relevant in assessing the thermal state of a subject.

To adress these problems, multiparametric monitoring is necessary. For example, the measurement of skin temperature can be completed by that of core temperature to obtain thermal state information such as vasodilation and vasoconstriction. To obtain a reliable measurement of core temperature, it is possible to insulate the surface thermistor with a low thermal conductivity material and thus minimize the thermal gradients. In this case, the choice of measurement location is crucial, the distance to the core has to be minimized (e.g. axillary position).

Brain and Core Temperature

As the thermoregulation centres are deep in the brain, cerebral temperature is one of the most important markers of fever, circadian rythms as well as physical and mental activities. However, due to the lack of accessibility, brain temperature is not often measured [1]. Unfortunetly, axillary, buccal, tympanic and rectal temperatures do not accurately reflect cerebral temperature, rectal temperature is generally considered the most reliable indicator of core body temperature. Brain temperature can be measured using NMR spectroscopy, microwave radiometry, near infrared spectroscopy, ultrasound thermometry, etc. However, none of those methods are amenable to long term ambulatory use outside a laboratory or hospital during normal daily activities.

The brain core thermometer "BCT" sensor, developed by the Biomedical Sensors Group of the Nanotechnology Institute of Lyon at INSA Lyon is a flexible active sensor using the "Zero heat flow" principle. The sensor is attached to the

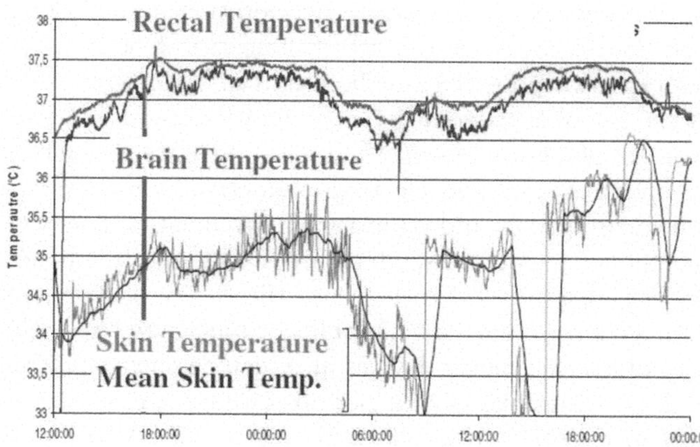

Fig. 6. Rectal, Brain and Skin temperatures over a 48 hours period recorded at the Center for Study and Treatment of Circadian Rhythms at Douglas Hospital in Montreal (Canada)

temple and its accuracy is of the order of $\pm 1/10^oC$. The sensor has been tested using several physical models and phantoms simulating the thermal parameters of the layers under the sensor such as skin, bone and brain [7]. The threshold of sensitivity is so small that $1/1000^o$C changes can be detected.

The sensor has been used in a range of experiments, recording brain temperature during mental activity and in hospitals for the study of circadian rythms [3]. The results are in good agreement with measurements made using a rectal probe (Figure 6). Two prototype versions of this sensor are available: a non-ambulatory one for use in hospitals and an ambulatory version using teletransmission. The autonomy of the ambulatory version is about 24 hours. This wearable biomedical sensor (WBS) can be used for circadian assessment, for chronobiology studies and in medical therapies such as chronotherapy.

3.3 Actimetry

Importance of Movement

The ability of an elderly or handicapped person to live independently in their own home greatly influences their health, well being and happiness. It is important to be able to assess a person's ability to perform the key basic activities involved in independant daily living. Movement of the human body is a complex phenomenon involving many factors of physiology, mechanics, psychology etc. Information on the quality or quantity of movement has the potential to provide an invaluable source of knowledge to more accurately diagnose, treat and manage a range of medical conditions. One means to successfully monitor and record human movements in real environments is through the direct use of accelerometers.

Direct Measurement of Body Movements by Accelerometry

Accelerometers are devices that measure acceleration acting along a sensitive axis and can be used to measure the rate (intensity) of body movement in up to three planes (anterior-posterior, mediolateral and vertical) [20]. Morris [15] suggested that the use of accelerometry as a quantitative measure to completely define the movement of a body in space had many advantages over the then commonly used kinephotography and electrogoniometry. Moreover, accelerometers respond to both the frequency and intensity of movement and, as a result, are superior to actometers or pedometers, which are attenuated by impact or tilt [12]. Accelerometers can also be used to measure tilt (body posture) making them superior to motion sensors that have no ability to measure static characteristics [11]. Advantages of accelerometer devices also include their small size and their ability to record data continuously for periods of days, weeks and even months. Recent advances in microelectromechanical systems (MEMS) have enabled the size and cost of accelerometer devices to be greatly reduced while ensuring that these devices are of a high quality and reliability.

Manufacturers can now provide uni-axial accelerometers (to record acceleration in a single direction) and bi-axial or triaxial devices (that can act along two/three orthogonal axes and so can thus provide a representation of acceleration in two to three dimensions).

Accelerometers are normally attached to the part of the body whose movement is being studied [12]. For example, accelerometers attached to the ankle are used to study leg movement during walking and accelerometers attached to the wrist have been used in the study of Parkinsonian tremor. In many circumstances it is necessary to study "whole body" movements. This is best achieved by placement of a sensor as close as possible to the center of mass of the patient's body [4].

The acceleration signal carries two components: the first is static and the second one is dynamic. These two components can be separated using frequency analysis techniques. A low pass filter extracts the low frequency components of the static acceleration due to gravity; higher frequencies hold the information of the movement. If the subject is at rest, or moving slowly, the output of the accelerometer reflects its orientation relative to the gravitational vector. If the orientation of the accelerometer relative to the person is known, then the resulting accelerometer recordings can be used to determine the orientation of the subject relative to the vertical or gravitational direction.

Current Accelerometer-Based Movement/Mobility Monitoring Systems

A number of studies have been carried out involving the application of accelerometers in the area of movement and mobility analysis. The primary aim of the work is to develop an appropriate means of movement-and-mobility-detection from body-mounted accelerometer-based sensors by using a multiple or single locations. Mathie and al [12] discussed a number of key considerations when considering sensor configuration, including the design tradeoffs between the number of sensors used, the cost, the usability and the transferability of the monitoring system. The design choices will be determined by the purpose and duration of the monitoring. In long-term unsupervised monitoring environments, subject compliance is essential if the system is to be used, with the monitor being as comfortable and unobtrusive as possible. Multiple sensors increase the complexity of the monitoring system. The use of one sensor attached at a single location on the body is a more straightforward approach. This significantly simplifies the design and use of the monitor but also reduces the quantity of information on the movements.

Since the first adoption in the early nineties in human movement analysis, large heavy equipment with multiple body-mounted sensor configurations have been replaced by smaller and lighter sensors and single body location placement. To compensate for this reduction in sensor attachment more complicated signal-processing approaches have also been introduced. There are now many commercially available PA monitors on the market for academic research or for individual health care monitoring that incorporate MEMS accelerometers (Table 1).

Table 1. Some portable movement monitors that incorporate MEMS accelerometers

Name	Placement	Functionalities	Dimensions (mm)	Manufacturers
IDEEA	Multiple: upper and lower leg, wrist, sternum, foot	Gait analysis (15 parameters), functional capacity identification, duration, frequency		MiniSun, California
PAM	Lower leg (amputees)	Daily activity, walking patterns	85x38x32 (50g)	Dynastream innovations, Canada
AMP331	Ankle	Distance, velocity (runners, walkers), energy expenditure, time spent	71x24x37.5 (50g)	Dynastream innovations, Canada
StepWatch3	Lower leg above ankle	Number of steps (2 months)	75x50x20 (38g)	Cyma corp., USA
GT1M	Waist (activity) or wrist (sleep)	Steps, activity level, Sleep cycles and latency	38x37x18 (27g)	Actigraph LLC, Florida
activPAL	Upper tight	Number of steps, time spent in postures	35x53x7 (20g)	PAL technologies, Glasgow
RT3	Waist	Metabolic activity (30 days)	71x55x27 (65g)	Stayhealthy Inc., California
ACTIM3D	Trunk under armpit	Fall detector, postures, walk (45 days)		BSE, France

Treating Accelerometry Data for Clinical Purposes

The basic treatment of accelerometry data include the use of simple mathematical operators such as the mean and standard deviation. These methods have been used in conjunction with thresholds to define basic motoric activities such as sitting, lying (supine, prone, left and right), standing and walking where the orientation of the accelerometer is rotated within the range of 0g to -1g [11]. More modern and complex accelerometer data analysis has involved the use of frequency spectrum analysis and the examination of dominant frequencies using fast Fourier transforms. These traditional spectral analysis methods provide information on the frequency components contained in a signal. Fourier analysis is a global tool providing a description of the overall regularity of a signal and copes well with naturally occurring sinusoidal behaviour. However, the Fourier transform does not provide the time at which these frequency components occurred. Thus, a more comprehensive tool is required that can analyse the accelerometer signal in more detail (in both time and frequency domains). Time-frequency analysis is important in analysing non-stationary signals, such as the acceleration pattern during human movement, where varied accelerations with sharp high-frequency transients are present at certain time instances and frequencies. Multi-resolution analysis (MRA) can carry this out as it analyses the signal at different frequencies with different resolutions. As human movement is generally associated with low frequencies (approximately 2.5 Hz for walking), MRA can be considered to be ideally suited for translating accelerometer data into comprehensive clinical information (quantitative and qualitative data). Wavelet analysis with the continuous wavelet transform (CWT) and the discrete wavelet transform (DWT) are two such MRA techniques.

Problems Associated with Direct Measurement of Body Movements with Accelerometry

The output of an accelerometer worn on the body is dependent on 4 factors: 1) the position at which it is placed, 2) its orientation relative to the subject, 3) the posture of the subject and 4) the activity being performed by the subject.

As stated earlier, for "whole body" movements, the best placement of the accelerometer is close to the center of mass of the body, which is sometimes inconvenient for long term wearing. Placement on the trunk is sometimes preferred [16], although it leads to increased complexity in the algorithms.

The orientation of the accelerometer relative to the body segment under study must be taken into account. It is preferably established at the start of the recording session or later deduced/reassessed during an easily recognized activity (e.g. the trunk can be assumed vertical during walking).

The accelerometer, if not perfectly attached to the body segment, is subjected to its own movement relative to the body segment which may give rise to false detections. Special signal processing can help identify/minimize/overcome this problem.

The compliance of the wearer is of major importance, especially in long term monitoring applications, and therefore the sensor should be as unobtrusive as possible. The integration of accelerometers into garments is generally well accepted by the subject, but they must be removed at certain stages (washing, toileting). A different approach is the direct placement of the accelerometers onto the skin surface by means of an adhesive patch. This latter approach can generally only be tolerated for several days without causing skin irritation problems.

4 Multisensor Devices

To meet the challenges of AAL, a range of new sensing devices must be developed that enable the robust, reliable, comfortable, user-friendly monitoring of the key vital signs and additional parameters required for a given application. These sensing systems should preferably (i) group (new) sensors together in one convenient location on the patient (e.g. on the wrist or in a body-worn adhesive patch) or (ii) combine dispersed sensors and their connections in garments such as bras, vests or T-Shirts. The following multisensor devices are some of those presently being developed by the authors in their respective laboratories and companies.

4.1 Smart Wrist

The Biomedical Sensors Group of the Nanotechnologies Institute of Lyon at INSA Lyon, France, has specialised for many years in the area of thermo-neurovascular sensing and they have been involved in several academic and industrial-led research projects e.g. for the study of drivers'reactions under real driving conditions; the study of smell, taste, touch and thermal comfort; the study of athletic performances and for the study of mental imaging [8]. The Group has developed a wrist worn device for neuro-physiological investigations (Figure 7). The "Emosense" wrist device is a prototype ambulatory monitoring and recording system [6] comprising sensors as well as circuitry for amplification and wireless data transmission. The device includes a range of integrated sensors for measurement of skin blood flow, skin temperature, skin conductance and skin potential. These measurements have enabled the monitoring and study of autonomic nervous system activity; providing information on emotional and sensorial reactivity, vigilance and mental state. The Emosense device can be used in conjunction with a "Smart T-shirt" such as the WEALTHY shirt and a PC which records the data. The "Smart T-shirt" can include sensors for ECG, Rib cage and abdominal respiration, core temperature and body heat flow, thus expanding the range of possible medical applications.

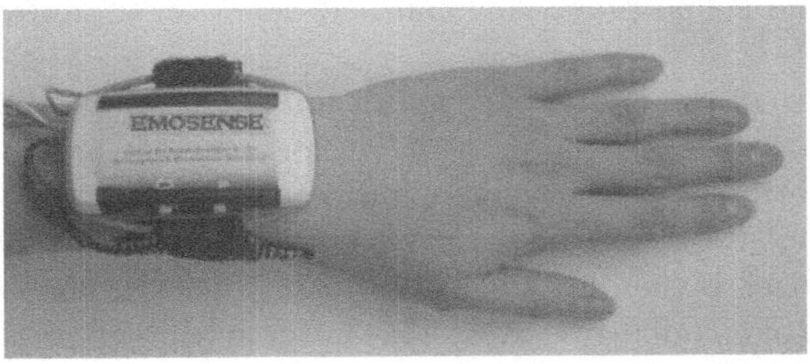

Fig. 7. The wrist ambulatory monitoring device Emosense for the monitoring and the study of autonomic nervous system activity, provides information on the emotional and sensorial reactivity, vigilance and mental state

4.2 "Body Worn Patch"

A University of Ulster spinout company called Sensor Technology and Devices (STnD)[17] has been working on the development of appropriate body-worn sensor patches for a range of companies and is presently developing its own high performance sensor patch capable of discreetly monitoring a range of vital signs (Figure 8).

[17] http://www.stnd.com/

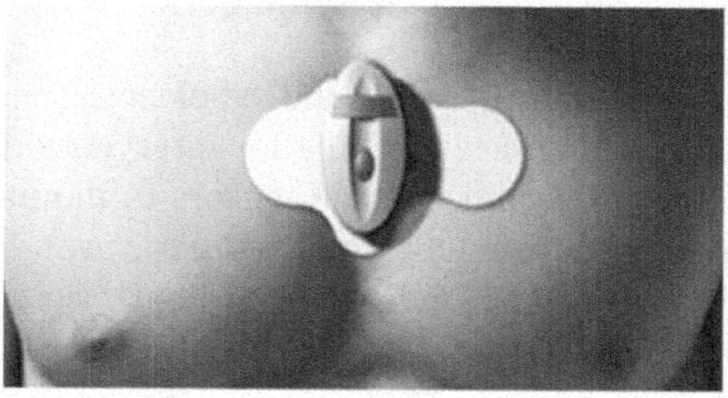

Fig. 8. Sensor patch for the monitoring of a range of vital signs STnD

This is a versatile wireless monitoring system which is highly miniaturized, light weight and has extremely long battery life. The system includes:

1. a miniaturised short range body-worn wireless monitor with on board intelligence to monitor for and trigger on medical events e.g. cardiac arrhythmias.
2. a matching belt-worn device using cellular links to send data immediately to the clinician.
3. a non-irritant, easy-to-apply, disposable sensor patch for high quality collection of the vital signs.

As stated previously, in order to develop a small, discrete monitoring patch with totally integrated sensors which would be acceptable to a typical patient, one must move away from the standard sensor types and locations. This requirement applies not only to 'vital signs' sensors (e.g. ECG, respiration, temperature) but also to those used to monitor emotional, (e.g. stress) behavioural (e.g. motion) and environmental parameters. For a given patient's condition, one must be able to monitor all key biosignals and parameters of interest from within the small patch "footprint" with sufficient accuracy and with a minimum of artefact.

An outstanding challenge to patch-worn sensing is biocompatibility and skin irritation - a limitation which is generally ignored in the enthusiasm for the promising approach. The sensor patches are adhered to the patient using medical-grade adhesive foam backing, or similar. The same materials are used in the construction of ECG electrodes and hence are well tested. However, adhesive patches, including ECG electrodes, can only remain attached to a patient for several days without falling off or causing skin irritation. Patch worn devices are therefore most appropriate for short term ambulatory monitoring applications. For longer term applications, wearable "Smart garments" should be required. An alternative solution, suitable for mid-term monitoring applications i.e. for a couple of weeks is the periodic repositioning of the patch on a fresh skin site to enable the previous site to recover. The challenge in this approach is to develop sensor systems (and associated hardware/software) that are capable of monitoring the same parameters from a range of locations on the patient.

4.3 Smart Clothing – EC Project ProeTEX

To improve rescuer safety, coordination and efficiency, the European program ProeTEX[18] aims at developing new clothing for intervention staff. This clothing (Figure 9) involves integrated smart textile sensors to monitor physiological parameters and the subject's environment (gas, temperature). It also includes GPS, an acquisition module, a transmission module and batteries [2]. Physiological parameters monitored are ECG, respiratory rhythm, posture, internal temperature, external temperature, heat flux, etc. The contribution of the Biomedical Sensors Group, INL, INSA Lyon to the ProeTEX project is the monitoring of thermal parameters for the prevention of heat stroke. Considering the harsh environment meet by fire fighters, monitoring thermal parameters is of prime interest. Internal temperature, external temperature and heat flux are relevant parameters to assess when seeking to prevent heat stroke in fire fighters exposed to intense fires.

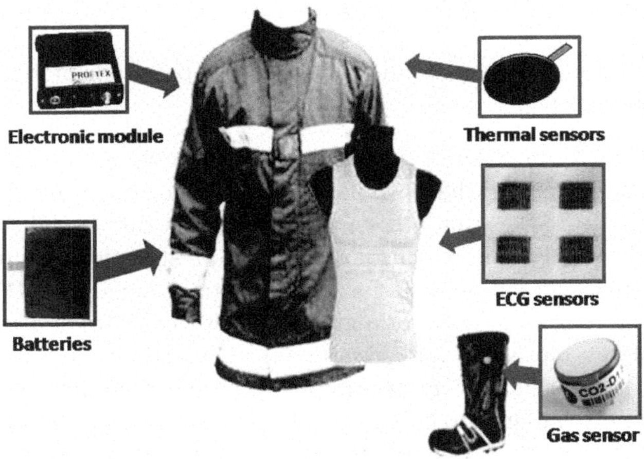

Fig. 9. The newly developed garment integrates sensors, an acquisition module, batteries etc. to monitor not only vital signs such as ECG, respiratorion and internal temperature but also the subject's environment (gas, temperature). All parameters are wirelessly sent to a laptop for real time monitoring.

Thermal Parameter Measurement on Fire Fighters

Due to his/her protective clothing, the firefighter cannot lose all of the metabolic heat produced during exertion (basal metabolic heat and physical exercise heat). The special jacket (Figure 9) acts a thermal barrier in two ways [9], (1) from outside to inside, insulating the rescuer from the external environment; (2) from inside to outside, preventing the fire fighter dissipating his/her metabolic heat.

[18] http://www.proetex.org

The latter is obviously not desirable. The external heat sources add to the problem: fire radiation, convection and in the case of outdoor fires, solar radiation. These heat sources can induce a significant and dangerous increase in internal body temperature which can lead to fire fighters having heat strokes. The monitoring of thermal parameters in this project seeks to minimize this danger.

In the above case, the most important parameter is internal temperature. When core temperature increases by 1.5°C in relation to the temperature at the beginning of the intervention, the firefighter is ordered to come back. To minimize risks linked to core temperature increase, external and skin heat flow, yarn temperature (from the jacket), and skin temperature are also monitored (Figure 10).

Skin temperature gives information on the state of vasodilation/vasoconstriction of the subject. Heat flux provides information on the heat exchanges between the subject and his/her environment. When heat flux is positive, heat goes out of the body. That is the usual case. The subject removes the metabolic heat produced to avoid internal temperature increase. When heat flux is negative, heat enters the body. These parameters have been recorded during fire fighting, in field trial conditions, and highlight, on one hand, that the outer garment of fire fighters effectively insulates the fire fighter from the external environment, and on the other hand, that the thermal monitoring is relevant.

In future, development of algorithms for thermally-at-risk situations will be considered. These algorithms will enable the activation of alarms to automatically warn of risk situations.

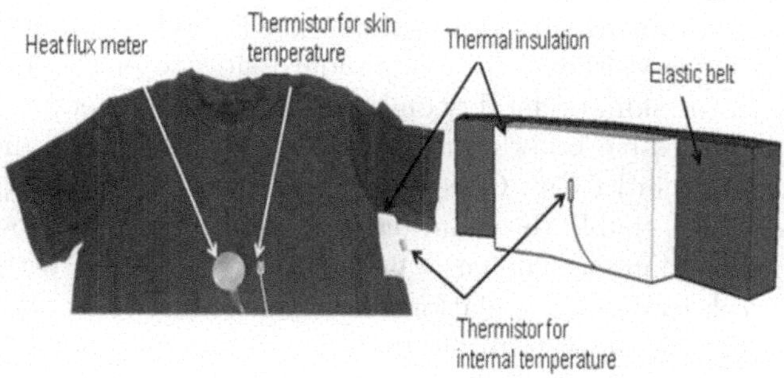

Fig. 10. Sensors location in a T-shirt. Skin heat flux and skin temperature are monitored on the chest while core temperature is measured in the axilla. The right hand sketch shows the elastic belt over the sensors to ensure firm contact.

5 Conclusion

It is suggested that one of the major stumbling blocks to the successful design and development of robust monitoring systems is the design of the sensors.

The veracity of the above comments is sadly evidenced by the disconcerting lack of commercialised "wearable" monitoring systems, in spite of the vast sums of money spent in this area by funding agencies such as the EC. Although around the world there have been a large number of trials and several thousand papers have been published in scientific journals, the picture remains one that is dominated by uncommercialised devices and "proof of concept" trials. The devices which have been built and reported in the literature, although performing well when carefully used in laboratory conditions, do not work acceptably well on patients in real-life situations. There has therefore been little clinical uptake of or patient compliance with these unwieldy assortments of hardware, connecting wires and standard vital sign sensors.

The technology must accommodate the needs of the patient, not the other way round. To fully harness the potential of Ambient Assisted Living, it is important to develop a new generation of sensor-driven technologies.

In order to develop a small, discrete monitoring patch with totally integrated sensors which would be acceptable to a typical patient, one must move away from (literally in this case) the standard anatomical locations and sensor types. For a given patient's condition, one must be able to monitor all key biosignals and parameters of interest from within a small patch "footprint" with sufficient accuracy and with a minimum of artefact. This will involve a major change in the way vital signs are traditionally monitored. Similar, if not more demanding, requirements exist for sensors embedded into clothing. Although, in theory, it is possible to locate electrodes anywhere on the clothed body - even over body sites traditionally used in standard monitoring (e.g. sites for standard 12-lead ECG monitoring) - relative movement between loose fitting clothing and the skin give rise to problems associated with quality of contact, motion artefacts and patient comfort. It is often not possible or desirable to make firm sensor contact with traditional monitoring sites (for example due to body contours). Generally, sensors have to be repositioned in non-standard sites to enable firm, comfortable contact with the skin, preferably on sites that do not generate artefacts due to, for example, excessive body hair, muscle noise (i.e. EMG), and body flab (i.e. motion artefacts in ECGs). Once again, novel sensing technologies are generally required which enable the monitoring of vital signs from novel locations using novel transduction mechanisms. One must therefore start with the desired biosignals and seek to develop a platform of novel sensing technologies enabling the monitoring of vital signs from alternate locations using novel transduction mechanisms. "It makes sense to start with the Sensors"[19]. This approach should not only lead to systems that actually work under the required constraints, but should also lead to novel, patentable innovations. This latter point is imperative if major industries and SMEs are to take up, further develop and commercialise these innovations and if the inventions are thereby to have widespread clinically use. In this way these novel sensing technology platforms will help solve some of the outstanding challenges to Europe's aspiration to improve the quality of

[19] Eric McAdams, "Robust, Artefact-free Monitoring. It makes Sense to start with the Sensors" Workshop on Smart Wearable Health and Healthcare Systems and Applications. 27th Annual International Conference of the IEEE Engineering in Medicine and Biology (EMBS) Shanghai September 1-4, 2005.

health care while decreasing costs through its introduction of Ambient Assisted Living programmes.

References

[1] Benzinger, T.H., Taylor, G.W.: Cranial measurement of internal temperature in man. In: Temperature: Its Measurement and Control in Science and Industry, pp. 111–120. Reinhold, New York (1972)
[2] Bonfiglio, A., Carbonaro, N., Chuzel, C., Curone, D., Dudnik, G., Germagnoli, F., Hatherall, D., Koller, J.M., Lanier, T., Loriga, G., Luprano, J., Magenes, G., Paradiso, R., Tognetti, A., Voirin, G., Waite, R.: Managing catastrophic events by wearable mobile systems. In: Löffler, J., Klann, M. (eds.) Mobile Response 2007. LNCS, vol. 4458, pp. 95–105. Springer, Heidelberg (2007)
[3] Boudreau, P., Shechter, A., Dittmar, A., Gehin, C., Delhomme, G., Nocua, R., Dumont, G., Boivin, D.: Cerebral temperature varies across circadian phases in humans. In: Proc. 30th Annual International Conference of the IEEE Engineering in Medicine and Biology Society, EMBS 2008, August 20-25, pp. 4856–4858 (2008)
[4] Bouten, C., Koekkoek, K., Verduin, M., Kodde, R., Janssen, J.: A triaxial accelerometer and portable data processing unit for the assessment of daily physical activity. IEEE Transactions on Biomedical Engineering 44(3), 136–147 (1997)
[5] Cook, D., Das, S.: How smart are our environments? an updated look at the state of the art. Pervasive Mob. Comput. 3(2), 53–73 (2007)
[6] Dittmar, A., Axisa, F., Delhomme, G., Gehin, C.: New Concepts and Technologies in Home Care and Ambulatory Monitoring. In: Wearable eHealth Systems for Personalised Health Management: State of the Art and Future Challenges, pp. 9–35. IOS Press, Amsterdam (2004)
[7] Dittmar, A., Gehin, C., Delhomme, G., Boivin, D., Dumont, G., Mott, C.: A non invasive wearable sensor for the measurement of brain temperature. In: Proc. 28th Annual International Conference of the IEEE Engineering in Medicine and Biology Society, EMBS 2006, August 30–September 3, pp. 900–902 (2006)
[8] Guillot, A., Collet, C., Dittmar, A., Delhomme, G., Delemer, C., Vernet-Maury, E.: Psychophysiological study of sport concentration: mental imagery evaluated through neurovegetative indices. Journal of Human Movement Studies 48, 417–435 (2005)
[9] Havenith, G.: Heat balance when wearing protective clothing. Ann. Occup. Hyg. 43(5), 289–296 (1999)
[10] Lymberis, A., De Rossi, D. (eds.): Wearable eHealth Systems for Personalised Health Management: State of the Art and Future Challenges. IOS Press, Amsterdam (2004)
[11] Lyons, G.M., Culhane, K.M., Hilton, D., Grace, P.A., Lyons, D.: A description of an accelerometer-based mobility monitoring technique. Medical Engineering & Physics 27(6), 497–504 (2005)
[12] Mathie, M.J., Coster, A., Lovell, N., Celler, B.G.: Accelerometry: providing an integrated, practical method for long-term, ambulatory monitoring of human movement. Physiol. Meas. 25(2), R1–R20 (2004)
[13] McAdams, E.T.: Bioelectrodes. In: Encyclopedia of Medical Devices and Instrumentation, pp. 120–166. Wiley InterScience, Hoboken (2006)
[14] McLaughlin, J., Anderson, J., Mcadams, E.T.: Profiled biosignal electrode device (1998)

[15] Morris, J.R.: Accelerometry–a technique for the measurement of human body movements. J. Biomech. 6(6), 729–736 (1973)
[16] Noury, N., Barralon, P., Flammarion, D., Vuillerme, N., Rumeau, P.: 18.19 an embedded microsystem for early detection of the fall – methods and results. Gait & Posture 21(suppl.1), S117–S118 (2005)
[17] Noury, N., Dittmar, A., Corroy, C., Baghai, R., Weber, J.L., Blanc, D., Klefstat, F., Blinovska, A., Vaysse, S., Comet, B.: Vtamn - a smart clothe for ambulatory remote monitoring of physiological parameters and activity. In: Proc. 26th Annual International Conference of the IEEE Engineering in Medicine and Biology Society, IEMBS 2004, vol. 2, pp. 3266–3269 (2004)
[18] Nugent, C.D., McCullagh, P.J., McAdams, E.T., Lymberis, A.: Personalised Health Management Systems: The Integration of Innovative Sensing, Textile, Information and Communication Technologies. IOS Press, Amsterdam (2005)
[19] Roth, A., Korb, H., Gadot, R., Kalter, E.: Telecardiology for patients with acute or chronic cardiac complaints: The [']shl' experience in israel and germany. International Journal of Medical Informatics 75(9), 643–645 (2006); International Council on Medical and Care Compunetics (ICMCC)
[20] Schutz, Y., Weinsier, R.L., Hunter, G.R.: Assessment of free-living physical activity in humans: an overview of currently available and proposed new measures. Obes. Res. 9(6), 368–379 (2001)
[21] Spekhorst, H., Sippensgroenewegen, A., David, G.K., van Rijn, C.M., Broekhuijsen, P.: Radiotransparent carbon electrode for ecg recordings in the catheterization laboratory. IEEE Transactions on Biomedical Engineering 35(5), 402–406 (1988)
[22] Watts, P.C.P., Mendoza, E., Henley, S.J., Silva, S.R.P., Irvine, J.K., McAdams, E.T.: Coreshell silver/silver chloride nanoparticles on carbon nanofibre arrays for bio-potential monitoring. Nanotechnology 18 (2007)

Biosignal Processing to Meet the Emerging Needs of Telehealth Monitoring Environments

Nigel H. Lovell[1] and Stephen J. Redmond[2,1]

[1] Graduate School of Biomedical Engineering,
University of New South Wales, Sydney, Australia, 2052
[2] School of Electrical Engineering and Telecommunications,
University of New South Wales, Sydney, Australia, 2052

Abstract. The distributed health care model of telehealth promises to meet the emerging medical needs of the world's aging population, whilst also improving quality of care for sparsely distributed societies. However, the unsupervised nature of the acquisition technology creates a number of challenging problems, with regards to biosignal interpretation and biosignal quality validation. This chapter serves to provide an overview of existing unaddressed issues related to remote biosignal recording and telehealth management. We highlight the difficulties associated with unsupervised recording, particularly in an unobtrusive manner, and suggest some possible avenues of investigation which may prove fruitful in the solution of these problems. Finally, we hypothesize on how the future direction of telehealth will likely be driven by the success of decision support technologies.

Keywords: Telehealth, biosignal processing, signal artifact, feature extraction, remote monitoring, clinical measurement, unsupervised monitoring.

1 Introduction

For almost half a century, biosignal processing techniques have been utilized to extract parameters from acquired physiological measurements, with an initial major focus on automated electrocardiogram (ECG) interpretation [1]. Since that time a vast literature base has grown around automated feature extraction measures [2-9] across the spectrum of clinical measurements, as well as around algorithms for artifact detection and removal in controlled recording environments [10, 11].

Should excessive noise or artifact be present in the observed signal, in a controlled recording environment, a trained operator, health care worker or clinician monitors the recording process and typically adjusts a setting or sensor position, and the measurement is repeated. This high-level quality control is conducted before any automated signal analysis and parameter extraction (such as deriving heart rate from the ECG). In a similar manner, a measurement is repeated should the recording technique be compromised, for example, if a forced expiratory breath maneuver is prematurely terminated. Thus the operator performs a critical role in the quality control of the acquired signal, both in terms of the level and occurrence of noise, as well as correcting erroneous or poor measurement technique.

S.C. Mukhopadhyay and A. Lay-Ekuakille (Eds.): Adv. in Biomedical Sensing, LNEE 55, pp. 263–280.
springerlink.com

Traditional health-care models behave in a reactive manner, with typical patient presentations being episodic in nature, relating to the onset of morbidity crises. It is becoming well-recognized that these models are often unsuitable for managing patients with chronic and complex disease living in the community [12, 13]. One health service delivery approach that is gaining favor is home telecare – often referred to more generally as telehealth [14].

In simple terms, telehealth is the provision of health services at a distance [15, 16]. Typically this occurs in unsupervised environments such as patients' homes and community care settings, whereby a number of commonly acquired biosignals, indicative of health status, are routinely recorded by the patient, as scheduled by the health care professional. These data are fed back to a central database, which the health-care professional can access in order to monitor the well-being of the patient, on a time-scale closer to real-time than existing health-care models. Patients observed to have a worsening condition may then be more actively managed.

One of the central tenants of telehealth is that the subject is involved in the active management of their health and wellness. In contrast with hospital and clinic-based recordings, the majority of clinical measurement recordings are performed without supervision, i.e. without operator or clinician presence or intervention. This results in significant issues relating to the quality of the recorded waveform in terms of noise levels, movement artifact and poor measurement technique.

While a vast range of biosignal processing approaches have been suggested and documented for clinical measurements, the key assumption tends to be that the recorded waveform is of sufficient quality to allow extraction of appropriate measurements and diagnostic features from the signal. Therefore existing algorithms reported in the literature tend to deal with minimizing the influence of noise and movement but they do not provide an over-arching assessment on whether the waveform is suitable for further processing.

Another fledgling area of development in the telehealth arena – different from those outlined above, which require the subject to engage with the measurement device – revolves around the unobtrusive monitoring of a patient's condition, by means of a ubiquitous array of sensors embedded throughout the residential environment, commonly referred to as a sensor network; although, with the advent of wireless solutions, the goal of a *wireless* sensor network (WSN) is often seen as a more desirable endpoint. Similar, or perhaps more challenging, biosignal processing problems are emerging, regarding the interpretation of signals recorded in such an unobtrusive manner; the difficulty arising from the fact that patient action generating the signal is not only unsupervised but also undirected. Decoding such information, in the context of patient wellbeing, will require the intelligent fusion of features extracted from multiple sensor signals, with each sensor in turn perhaps being stimulated by multiple sources (patients) simultaneously.

This chapter will focus on a discussion of the some of the unaddressed challenges facing the interpretation of telehealth recordings which are recorded in the unsupervised environment, in either a directed or undirected manner. The importance of signal quality indicators for unsupervised directed measurements at the feature extraction phase is highlighted; particularly in signals arising from clinical measurements, including ECG, pulse oximetry, blood pressure (BP) and signals recorded from a triaxial accelerometer as a determinant of ambulation.

The other main measurement recorded in the telehealth environment is that of a forced expiratory maneuver used in the assessment of lung function. This waveform must conform to strict criteria in order to be acceptable for feature extraction and these criteria have been coded into standards which are relatively easy to encode in a computer algorithm and have been for over a decade [17]. There is very little difference therefore in performing this measurement in the supervised versus the unsupervised setting in terms of being able to detect incorrect measurement technique and artifacts.

This chapter will also focus on preliminary results from the automated detection of poor signal quality in the ECG in the unsupervised setting. These signals constitute the main time-varying waveforms that are recorded in the telehealth setting. Other single point measurements such as weight, temperature and blood chemistry are not amenable to the treatments described above. However, from a telehealth knowledge management viewpoint, they, along with the features extracted from the previously mentioned waveforms, do comprise a longitudinal measurement record of the physiological state of the subject.

The issues of signal quality are superseded, in the unsupervised-undirected model of the WSN, by the sheer complexity of the challenge of mapping multiple signals to an indicator of subject health; we shall provide a brief discourse on some examples highlighting such obstacles.

The final sections of this book chapter will briefly deal with the perceived future direction of the telehealth care model, with a discussion of: how signal processing algorithms can be applied to detect trends embedded within the longitudinal records of extracted features; how the evidence base required for the automatic interpretation of these data trends is not yet available; and how a decision support system might be developed, given a substantial evidence base upon which to train such a system. It is this final point which touches on one of the major challenges facing telehealth systems – that of providing effective decision support facilities to enhance the interpretation of features extracted from remotely acquired clinical measurement records.

2 Patient Managed Unsupervised Recording

As stated in the introduction, the core model for telehealth care involves the self-management of patients. The necessity for the patient to engage with the measurement equipment in a directed manner leads to a host of signal quality problems, which in the supervised clinical environment would have been quickly recognized and amended by a trained technician or clinician. While education is a key factor in the acquisition of high quality telehealth data, the unavoidable truth is that, since the majority of patients are ignorant to the technical principles at work during biosignal measurement and interpretation, records are frequently inadvertently spoiled and methods to identify these poor quality records are imperative. The following sections discuss some of the quality issues associated with unsupervised-directed measurement, as alluded to above.

2.1 Electrocardiography

One chronic disease condition which exhibits a pattern of repeated hospital admission, under existing care models, is congestive heart failure (CHF). One particular biosignal which finds application in the monitoring of this condition is the ECG.

In the clinical environment, after performing a short ECG recording, the heart rate is estimated, typically under the premise that the recorded ECG signal is of a suitable quality and contains no noise or artifact interference components; due to movement during recording, or other external influences. Conversely, movement artifact, line noise, and muscle tremor are extremely common during the unsupervised acquisition of ECG [18-20] in the standard residential telehealth environment.

For the ECG in particular, a number of proprietary techniques have been developed and are instantiated in the form of online embedded analysis software systems in clinical ECG monitoring devices, or offline PC-based packages. Being proprietary these techniques have not stood the test of peer review to appear in the scientific literature; as such, no critical analysis of these methodologies has been possible, beyond a cursory higher level comparison of pseudo-indicators of performance (such as false alarm rate), between common ECG monitoring systems. The fact that there is a definite paucity of scientific literature relating to such algorithms is surprising, given that over 30 years of research has been invested in ECG signal processing and analysis.

To help highlight the extent of the issue of motion artifact in telehealth recordings, some of our recent work has focused on the development of a rudimentary algorithm to identify motion artifact in a single lead ECG recording, acquired using a TMC-Home (TeleMedCare Pty. Ltd., Sydney, Australia) device. Fig. 1 provides an example of a typical artifact event generated by patient hand and arm movement in a remotely acquired unsupervised ECG recording. During ECG acquisition the patient rests both hands lightly on an ECG plate comprising three Ag/AgCl dry electrodes. As no conductive gel is used the recording is typically noisier than in a supervised environment. Fig. 1 also illustrates the ECG artifact detection as determined using our algorithm, described in [21].

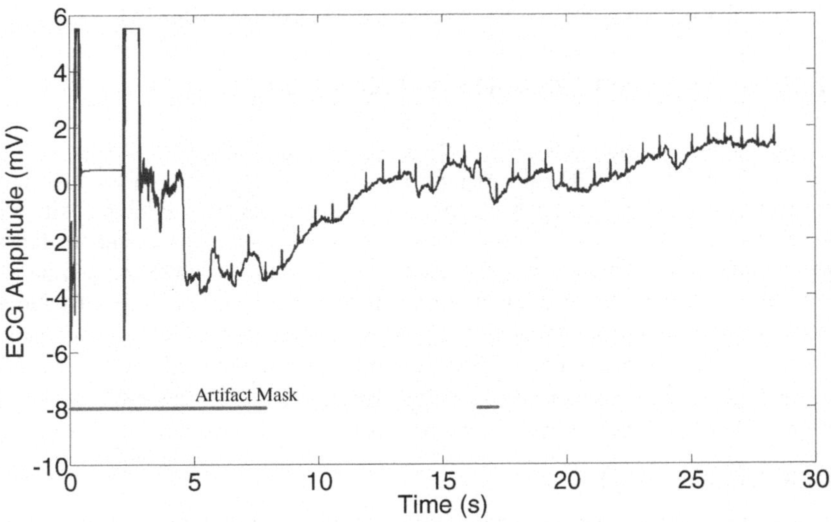

Fig. 1. Example of movement artifact in a sample ECG recording. Saturation of the front-end ECG amplifiers is evident in the first five seconds of recording. Also shown as horizontal line segments is the artifact detection determined by the algorithm, described in [21].

Examining 4751 single Lead-I ECG recordings, made by 24 chronically ill patients in the home environment, in an unsupervised manner, demonstrates the extent of the artifact events generated in such measurement signals. Of all 4751 records which were analyzed, 3506 (74%) contained 10 seconds or more of continuous ECG, uninterrupted by motion artifact; 1344 (28%) of all signals contained no artifact; a significantly large number of 875 (18%) were considered completely unusable, with the remaining records distributed quite uniformly over the remaining percentages. This is an important result, that 18% of all records should be considered unsuitable for any further interpretation. In standard telehealth systems this would result in at least 18% of all heart rates being erroneous. The results indicate that, in the majority of cases, the capture of ECG in an unsupervised home environment is achievable once movement artifact is correctly identified.

2.2 Pulse Oximetry Measurement

Some pulse oximetry manufacturers, such as Nellcor (Boulder, CO, USA) and Masimo (Irvine, CA, USA) have recognized the importance of identifying motion artifact and have implemented proprietary algorithms in their equipment to do so. However, being proprietary, their techniques are not open to critique; although the medical literature often quotes comparisons of performance between various 'off-the-shelf' pulse oximeters [22], the underlying signal processing is, again as with the ECG, hidden from scrutiny. Some published pulse oximetry quality assessment techniques employ the use of the ECG as a reference signal and require that valid sections of the pulse oximetry signal contain beats which have a corresponding beat in the ECG; or some simply require that the mean heart rate estimated from both signals over a short duration is approximately equal. However, the hypothesis here is slightly circular, as when a discrepancy occurs between the signals it is difficult to designate blame to one or other signal without some measure of the inherent quality of each.

Pulse oximetry is notoriously prone to several sources of error; for instance, relative motion between finger tip and the probe alters the optical path between the LEDs and the photodetector causing significant variations in the recorded light intensity at the detector; since all variations in the light intensity are assumed to be due to the pulsatile arterial blood in the finger tip, the resulting estimate of the arterial oxygen saturation (SpO2) value is often rendered invalid by such motion artifact. Much research has been directed at recovering the underlying signal by suppressing any artifact components through the use of accelerometers or additional light sources, which serve as a measure of probe movement. However, still very little attention has been paid to the scenario whereby the artifact is too extreme for successful suppression. In such an event, the failure to recover the signal must be recognized.

2.3 Blood Pressure Measurement

Similarly, automated remote BP measurement is made possible through the use of an audio microphone implanted in a pressure cuff, which is placed on the subject's arm. The cuff is automatically inflated and then deflated slowly. During the deflation phase, the microphone records the auscultatory (Korotkoff) sounds, normally observed by a clinician with a stethoscope. The changing intensity of these sounds as the cuff deflates is a function of the systolic and diastolic blood pressures. In the unsupervised recording environment undesired additions to the audio recording are common; relative movement between the arm and the cuff creates a rustling sound; ambient background noise is common; if the initial cuff pressure is not high enough the blood vessels will not be completely occluded, making it impossible to estimate the systolic pressure; if recording stops before the Korotkoff sounds have ceased, the determination of the diastolic pressure is not possible. All of these artifactual effects are easily discernable to a trained ear. In addition, from our experience with clinical trials of home telecare devices, we have observed trained patients frequently place the pressure cuff on their arm upside-down or in a rotated position, causing the incorrect placement of the microphone away from the brachial artery. When automatic analysis proceeds to estimate the blood pressure, fallacious results are often generated as a consequence of the unrecognised poor signal quality.

2.4 Triaxial Accelerometry (TA)

While the biosignals described above – ECG, BP and pulse oximetry – form the basis for a subset of the standard battery of tests utilized in the remote management of chronic cardiovascular diseases, like Chronic Obstructive Pulmonary Disease (COPD) and CHF, unsupervised monitoring is breaking ground in areas traditionally reserved for the clinical setting. Ambulatory monitoring, in the form of wireless body-worn devices, is emerging as a natural extension of the standard telecare repertoire, which previously was typically realized as a console-style device [23]. TA-based ambulatory monitoring has been demonstrated to offer a means to anticipate an increased risk of falling through an observed deterioration in functional ability, as measured by a TA-based assessment or directed-routine (DR) [24]. The DR is simply a small set of physical tasks the subject would perform, such as 1) standing and sitting down five times in quick succession, 2) rising from a chair walking three meters and returning to the chair in a seated position, or 3) standing in front of small step and placing the left foot on the step and then back to the floor, then putting the right foot on the step and back to the floor, and repeating this alternating step pattern several times – a so-called alternate step test (AST).

Early results indicate that when the subject performs the DR correctly, as confirmed by a supervising researcher, timing and waveform features from the resulting acceleration signals can be used to automatically estimate the risk of the subject falling in the near future [24]. However, when these DR functional ability tests are migrated to the unsupervised environment, analysis of these signals will only succeed if the subject has adhered to the correct execution of the test. Through an analysis of the raw acceleration signals it may well be possible to determine whether the subject has correctly performed test, before the signals are use to determine a subject's falls risk. It is a considerable challenge to recognize the correct performance of a particular test in the DR given the wide range of movements possible and only a single TA to capture the event. However, the repetitive nature of the chosen tests lends itself to discrimination of a valid test from undirected movement. A successful algorithm which does so would permit the capture of these signals in a fully unsupervised manor.

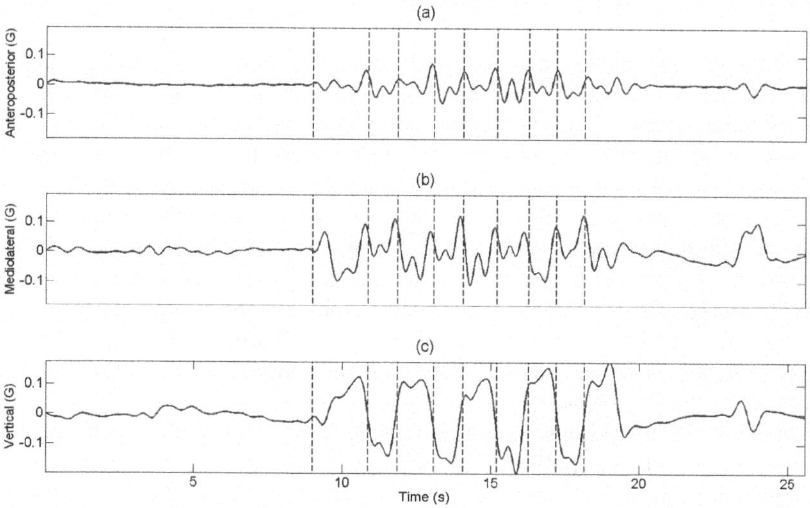

Fig. 2. Shown are the three accelerometry signals for the AST. G is acceleration due to gravity: 9.81 m/s/s. The signals have been annotated by an observer marking the data stream at the start/end of each stepping motion. We also note that this particular subject incorrectly performed an additional, ninth, step at approximately 18 s.

Fig. 2 provides an illustrative example of the acceleration pattern seen during the performance of the AST. Note that this subject was instructed to perform four repetitions of the left-right step pattern, but placed the left foot upon the step one extra time. This is a minor deviation from the normal performance of the DR; however, since these signals are destined for automatic segmentation and analysis, this and more serious abnormalities in the DR execution must be recognized before a fallacious falls risk estimate is calculated.

2.5 Quality Measures by Data Fusion

In the previous sections, quality issues associated with unsupervised directed measurement have been highlighted. The methodology thus far proposed involves the analysis of each signal independently. However, it is apparent that improved estimates of signal quality may be attained through concurrent recording and fusion of multiple signals. Given an independent quality measure for the ECG, good quality sections of an ECG signal may be employed to validate the quality of a simultaneously recorded pulse oximetry, or blood pressure signal.

Certainly in the supervised setting (aside from telehealth), this is a very active research area; for example, a recent paper by [25] attempts to derive quality measures for heart rate estimation from concurrent ECG and arterial blood pressure recordings, for use in ICUs – but as of yet, no parallels exist for unsupervised monitoring.

3 Unobtrusive Patient Assessment

When aiming to holistically manage the health and wellness of a patient, it is well-recognized that physiological measurements alone are not entirely satisfactory. More comprehensive forms of activity monitoring systems, particularly in the form of unobtrusive assessment are needed [26-28]. The reasons for this are multi-factorial but include the fact that vital signs monitoring, even in the unsupervised mode, is still an interventionist technique and thus the person may become unduly worried or perturbed by the measurement process – thus influencing the actual measurement. Moreover, changes in the person's daily routine that are indicative of a decline in functional health capacity may precede significant changes in physiological measurement parameters.

In this section, the specific research area of wellness monitoring as a tool for functional health assessment in home and residential aged care settings using unobtrusive technologies based around wireless sensor networks will be examined. The objectives of ubiquitous monitoring using WSNs for wellness management is three-fold; to provide the tools and training to promote self care; to identify the onset of certain conditions before obvious symptoms develop [29]; and to ensure immediate treatment or hospital admission of chronically ill or frail elderly patients when the condition requires urgent intervention of treatment (for example in the case of an incapacitating fall). The first two objectives address the growing importance of prevention or early detection of a medical condition in order to reduce morbidity crises, while the latter focuses on social alarms and emergency responses systems in the case of urgently needed intervention.

In certain cases, particularly in people with motor and neuro-cognitive disorders, including early-onset dementia, it is often infeasible for the subject to self-perform routine clinical measurements. In these cases, the use of pervasive and ubiquitous networks of sensors for physiological parameter measurements may be appropriate. Sensors can be categorized into three groupings; ambient environmental sensors, body area network sensors, and location/usage sensors. Ambient environmental sensors are used to monitor a given geographical region. They include: passive infrared (PIR) sensors used commonly to detect the presence of moving heat sources (typically to identify human presence) [26]; ambient light sensors to assess day/night sleeping patterns; and ambient temperature sensors. The second category of body area network sensors includes various types of monitoring transducers with wireless communication capability. The main differentiating point of these sensors is that they must be worn by the subject. They include: TA-based devices used to determine the ambulation and activity-level (a surrogate of energy expenditure) [30, 31] of a subject, as well as falls propensity and falls risk [23, 32, 33]. Other sensor devices are designed to integrate into a wearable garment for long-term and continuous monitoring [34] of a diverse array of body parameters.

The third category of sensors relate to resource utilization and location detection, typically used to monitor the location and specific movements of a subject, including their interaction with other objects, people or resources (including time spent toileting, receiving care and medication, etc.). As these latter sensors represent the interaction between various people, resources and objects, they may be fixed to a particular object in the facility, located in a specific position in a room or worn by a person. Radio frequency identification (RFID) tag technology is often used for these tasks. Examples include load sensors installed on bed frames for weight measurement and ECG systems installed in bathroom facilities [35]. In both these cases, the sensor interaction involves subject localization and identification, and thence unobtrusive sensing of a physiological variable.

In general, vital signs monitoring using this ubiquitous sensing approach is restricted to development in the laboratory. Large-scale clinical deployments are yet to occur as there exist technical challenges such as data transfer security, sensor link reliability, system cost and sensor battery life which need to be addressed. Moreover from the viewpoint of biosignal processing, all the difficulties outlined in Section 2 apply, with the added complexity that there is no opportunity to train the subject to perform better or more correct measurement techniques – hence the likelihood of noise and signal artifact is even higher in these systems. The advantage one has though is that the measurements are oft repeated many times per day, so assuming certain quality measures can be attached to any particular signal, it should be possible to be highly discriminatory as to which data to choose to perform feature extraction to derive physiological parameters.

The literature often refers to activity assessment as activities of daily living (ADL) monitoring. The end aim of ADL monitoring systems is to build a user-specific profile of the daily activity pattern of a subject and use this as a baseline for comparison against irregular patterns. It is hypothesized that such systems will be useful in early detection of the onset of clinical problems. For instance, a gradual decline in usage of kitchen facilities in an otherwise healthy patient may suggest a digestive disease, while increased episodes of wandering can be signs of the onset of dementia. In the latter case, the potential benefit of ubiquitous monitoring becomes apparent as it is unlikely that an early onset dementia patient would be strictly compliant to unsupervised measurement protocols, whereas the pervasive and unobtrusive nature of an appropriately designed WSN would mean that the subject is totally unaware of the fact that they were being monitored and managed.

Research in the WSN area, in terms of impact of ADL monitoring on health status, remains at an early stage. One of the initial technology demonstrators of remote monitoring of the free-living elderly occurred in the early 1990's in Australia as a result of research in our laboratories [27]. More recently there has been considerable work in laboratories in Japan and the USA. For example, a network of wired sensors that measured movement, door and window openings, human presence, kitchen usage, television and washing machine usage and sleep occurrence was employed on a single-subject to create a daily living template [36]. Subsequent daily patterns were compared with the normal template and any dramatic changes in behavior were identified.

In another study [37], an extensive WSN of over 100 RFID tags were deployed in a mock facility to derive fourteen different ADLs. When an activity was performed, the subject interacted by touching different RFID-tagged objects. The critical component in the decoding of these data was a probabilistic inference engine which was used to estimate the different ADLs performed.

It can perhaps be appreciated that the primary challenge of unobtrusive monitoring systems is the appropriate use of collected data to produce useful information. Continuous data streaming from numerous sensors over the network results in a situation of data overload. Data cleaning and noise reduction is critical to ensure only data with an acceptable level of noise and quality is processed to extract waveform features. Naturally the design challenge is to incorporate as much of this algorithmic intelligence within the actual sensors themselves so as to limit network traffic and to conserve sensor battery life. The additional challenge is that, unlike unsupervised monitoring where data fusion may be a very useful adjunct, in the unobtrusive environment, data fusion is essential in order to provide appropriate data context. For example, with RFID locator technology, signal interaction can be used to infer the length of time spent using kitchen facilities, time spent toileting, or the time spent dispensing medications to a particular subject.

At a higher level, the unobtrusive/undirected measurement modality of the WSN described in this section, and that of the directed self-assessment protocols described in previous sections, can be considered similar in that they constitute systems which monitor indicators of subject wellbeing over an extended period of time – be it heart rate for directed measurement, or ADLs for unobtrusive measurement. The following section discusses the significance of these longitudinal records and how one might begin to leverage the health related data contained within them.

4 Future Trends

4.1 Interpretation of Longitudinal Records

Most of the biosignal processing challenges described earlier in this chapter relate directly to the reliable acquisition and basic feature extraction from physiological measurements recorded directly by the patient. Similar, though even more complex challenges arise when using sensor network technologies to infer functional health status in the unobtrusive monitoring environment. However, leaving aside the outstanding challenges of designing the signal quality indication algorithms outlined above, future directions in telehealth will be primarily driven by the need for automated management, fusion and interpretation of multiple sources of longitudinal patient data, within the context of previous medical history and additional sources of relevant information, such as medications. Data fusion and knowledge management techniques become even more critical in the unobtrusive monitoring environment, where a multitude of sensor data sets must be analyzed to detect changes in patterns of interaction with a patient and their surrounds. Fig. 3 shows a sample of the physiological parameters of a COPD sufferer using a home telecare device, over a period of three months.

The following sections provide a brief discourse on how telehealth data is often utilized, in current state of the art systems, and some possible pathways of improving on these methods in the coming years.

4.1.1 Human Analysis

One of the current paradigms for utilizing telehealth data is by inspection, by a managing clinician. This is easily facilitated through a Website interface or portal, by which the clinician may browse their patients' records, retrieving charts of measurement parameters, or questionnaires, plotted against time (on a timescale of days, weeks or months). Viewing this information, the clinician may flag a subject based on their level of criticality for further intervention. However, several concerns with such a methodology for information delivery immediately arise: (1) Similar to patients, clinicians are often unfamiliar with such technologies and must progress through an education program to learn how to correctly interact with the technology and interpret the information on display. (2) Clinicians may be managing tens or hundreds of patients, resulting in information overload; the result of which being that critical patients are not identified. (3) Most importantly, given the newness of the telehealth data modality, there is no best-practice evidence base for interpreting longitudinal data records. The technology which helps address the three issues highlighted above, relating to the automated management and interpretation of information, is generally termed a decision support system (DSS).

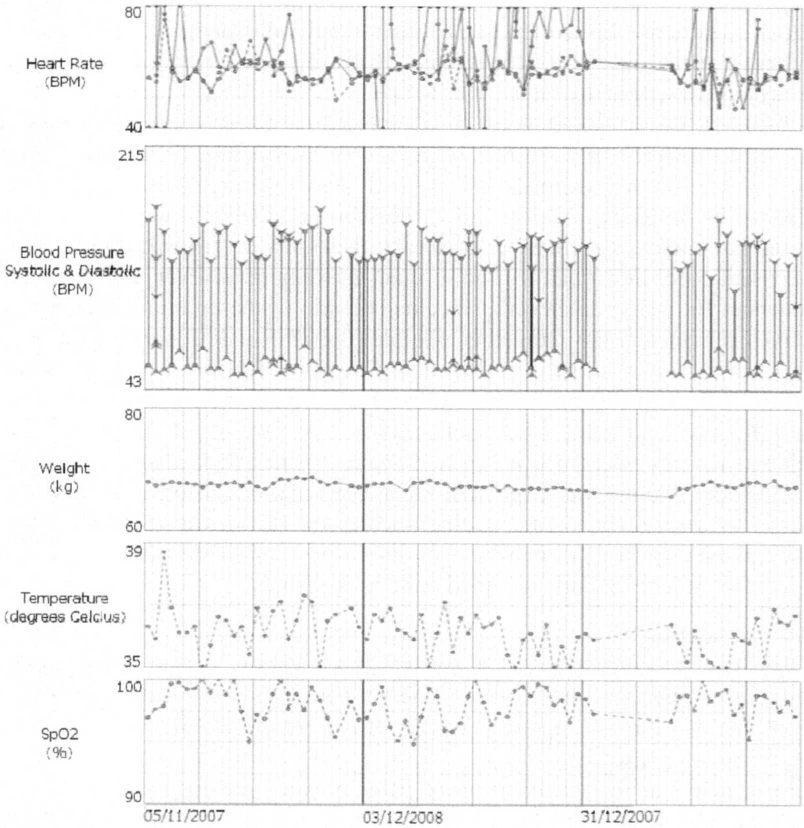

Fig. 3. Five longitudinal clinical measurement graphs, over a three month time span, are shown for a 76 year old male patient with COPD. The Heart Rate plot shows estimated heart rates from ECG, blood pressure and pulse oximetry recordings, superimposed – the heart rate estimate showing the largest variation is that drawn from the ECG, without the use of artifact detection.

Existing evidence-based clinical guidelines use simple criteria, infrequent measures, or rely on simple calculations, as they are feasible for providers to calculate during consultations. However, longitudinal telehealth records provide much more scope for advanced interpretation, in terms of examining trends in the data and projecting or predicting the anticipated health status, based on the previous data recorded. We shall demonstrate in the following sections, not only how trend information might be extracted from longitudinal health records, but also how this trend information might be employed in a structured pattern classification scheme to provide a robust evidence-based DSS.

4.1.2 Automated Analysis

While, the literature [38-40] and Internet abounds with DSSs applied in the practice of clinical medicine, very little published works exist relating to DSSs as applied to data collected using telehealth technologies; the principle reason being the relative originality of systems that collect such longitudinal data sources and the complexity and richness of the data that is generated. However, one exception is the relatively generic system described by Falas *et al.* [41].

Telehealth DSSs are built upon physiological parameters extracted through the analysis of the underlying biosignals [42]. However, as demonstrated above, standard analyses of telehealth biosignals are typically based upon algorithms which were intended for use with signals acquired in the supervised environment. As a result, the automated parameter extraction performed by these procedures will frequently fail on unsupervised recordings due to noise and movement artifact components. The consequences for a DSS are severe, since its operation is based on the assumption that the derived parameters are altogether reliable; the DSS will generate false alerts, or overlook other subjects in need of attention. One approach under investigation in our laboratory to mitigate against false alerts is to derive and attach a signal quality indicator for each extracted parameter. Such a measure would be derived from an algorithm that examined the level of noise and artifact in the underlying waveform from which the feature was extracted. Once the longitudinal data has a guaranteed reliability, as assured by signal quality indicators, automatic interpretation of the longitudinal data by the DSS may begin; a first step is often the extraction of features of interest for these irregularly sampled parameter signals.

Feature Extraction

The DSS may be considered as a pattern classification system whose task is to categorize a patient into one of several health categories, i.e. Stable or Critical, for example. One of the most important aspects of any pattern classification system is the feature extraction phase.

There is potential for stratification of patients when the long-term pattern of the patient's parameters is considered – in the form of their longitudinal data record. Longitudinal records, acquired across multiple parameters, provide the utility to learn the normal range of the subject and hence recognize any variation away from that baseline range. Given that this is a nascent technology, there is little precedence for which exact novel features extracted from the longitudinal record are of most relevance.

One particular feature extraction technique which our group has investigated is the detection of trends within the data [43]. Our method of choice involves regressing the longitudinal parameter signals onto a series of a simple piecewise linear splines. Since it is uncertain whether any newly proposed features are of clinical significance, it is suggested that human evaluation (as an intermediate step) of the usefulness of these particular trend features might be achieved more readily if each section of data is adequately described by start and end values and a slope, or rise-time. This decomposition of the data is intuitive compared to more traditional smoothing and filtering techniques, which we consider unsuitable for an intermediate interpretation by a clinician.

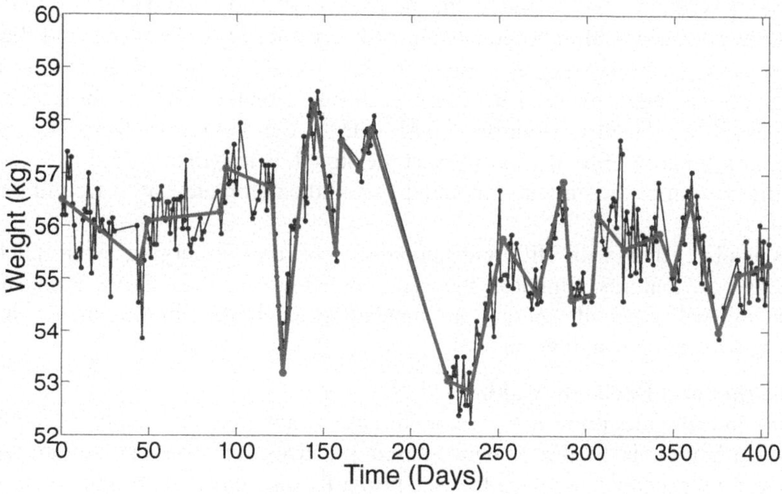

(a)

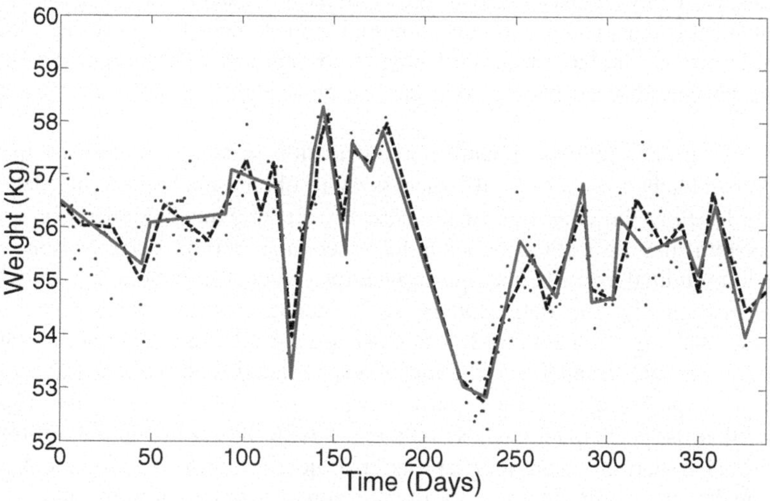

(b)

Fig. 4. (a) shows a sample illustration of the human-annotated fit to longitudinal weight data for one subject. (b) shows the automated piecewise-linear fit to the same data (dashed) superimposed on the human-annotated fit (for clarity the original data marker size has been reduced).

Validation of this proposed fitting technique was performed by comparing against a human-annotated set of simulated and real longitudinal records. During annotation, the human fits the piecewise regression to the data using a graphical user interface. Once the automated fit is applied to the data, the error between the human-annotated and automatic fit are calculated. On simulated data, the piecewise linear fitting method matches or improves upon the human performance in most cases; with the largest improvement demonstrated in more noisy data. Similarly, for the real longitudinal physiological data (consisting of heart rate, FEV1, systolic and diastolic blood pressure, arterial oxygen saturation and weight, collected from 24 home-dwelling patients, aged from 54-92 years, suffering either COPD and/or CHF, using the TeleMedCare Health Monitor - TMC-HM), the deviation from the human marking, as a fraction of total variability of the signal, is less than 0.35. Fig. 4 shows a sample illustration of the human-annotated fit and the automated fit to weight data for one subject.

We speculate that with different modalities of data emerging from telehealth monitoring, new features, similar to the trend detection described above, encapsulating patient health will gradually merge into standard clinical guidelines as their benefit in identifying worsening health is realized.

Classification and Decision-Making

Standard clinical guidelines for chronic disease draw upon a wealth of published research, with regards to the evidence-based practices for diagnosis, treatment and prevention. In this chapter we have focused primarily on the recognition of deterioration in the health status of a patient suffering chronic disease, under continuous telehealth monitoring. In this regard we notice that, for COPD in particular, the clinical guidelines do not incorporate any advanced statistical classification models. There is great scope to utilize statistical pattern recognition decision making techniques in emerging DSSs for telehealth. However, the largest hurdle facing the design and validation of the DSS for monitoring chronic disease patients is a lack of an evidence-base for training such a system.

In order to train a statistical pattern classification system, to serve as the alert generation mechanism for a DSS, it is necessary to obtain labeled training data. This amounts to labeling the condition of the patient over an extended period of time, so that the system may be trained to accurately recognize certain deterioration events, from the longitudinal records and questionnaires, when the patient was labeled as unwell. Unfortunately, the definition of such categories of wellness constitutes a considerable challenge, since these labels must be derived from information which is objective and reliable. In addition, the logistics associated with the collection of such labels is also problematic, since current trials of telehealth systems typically necessitate that the participant has less interaction with the care team, therefore, there is no medical expert at hand to perform an objective clinical assessment of the subject's health; the result is large gaps in the training data labeling, meaning that the false alarm rate for the system cannot be quantified.

Given the obstacles involved in designing a DSS for telehealth applications, a practical approach will likely involve a combination of a heuristic system design, inspired by existing guidelines and clinical expertise, which facilitates incremental improvement through user feedback and ongoing assessment of classification performance.

5 Summary

In this chapter we have provided some insight into the practical issues currently facing the successful proliferation of telehealth patient monitoring and management paradigms. The most immediate concern for unsupervised directed biosignal measurement is the need for signal quality indicators to validate the correct execution of the signal recording; although in many cases a solution to these problems is well within our grasp, as demonstrated earlier with some basic ECG signal analysis to identify motion artifact during acquisition. Additionally, we recognize that more robust signal validation techniques should perhaps involve the fusion of information from several simultaneous biosignal recordings; such as ECG in parallel with pulse oximetry.

Looking forward, the ability to identify the quality of telehealth biosignals also provides a means to assess subject performance of unsupervised measurement. If a subject is seen to frequently spoil signal recordings, through poor technique (for example, excessive movement during ECG or BP recordings), they may be targeted for further education in the use of the technology. This will also act as a fail-safe to recognize those subjects who have 'slipped through the net' and are not benefiting from telehealth care, due to an inability to correctly engage with the equipment and perform recordings; otherwise there is an unreasonable onus placed on the monitoring clinician to identify that the subject consistently spoils recordings, and that the returned data is unreliable.

Removing the requirement for traditional directed assessment, the wireless sensor network approach to unsupervised patient monitoring raises some of the most complex signal processing problems for the telehealth care model. While the hardware instantiation of the WSN is in its embryonic stages, the potential of such systems is vast. These technologies will likely have a major impact in unobtrusively facilitating the care or monitoring of elderly subjects in their own homes, or in residential care facilities. Furthermore, the economic benefit of autonomously monitoring and delegating resources, using the WSN, in the residential care environment is substantial. However, much algorithmic development and validation must be performed at the level of interpretation of the raw sensor network signals and fusion of multiple data sources, to map the signals to an intermediate meta-level, by identifying activities of daily living and simple interactions between the subject and their environment, before an overarching assessment of wellbeing is attempted.

As clinicians and carers face an increasing burden of data to assimilate, systems that reduce data to knowledge and that support clinical decision-making processes are becoming more and more essential. It is anticipated that the combination of biosignal processing algorithms, signal quality measures, longitudinal trend analysis, domain knowledge and an over-arching decision support system will allow telehealth systems of the future the ability to risk stratify patients according to disease severity, and to assist in patient care. The demand for such support tools will become increasingly apparent as telehealth systems provide increasingly more clinical measurement capabilities.

References

1. Pipberger, H.V., Stallman, F.W.: Use of computers in electrocardiogram interpretation. Am. Heart J. 64, 285–286 (1962)
2. Adam, D., Burla, E.: Blood pressure estimation by processing of echocardiography signals. Computers in Cardiology (2001)
3. Colak, S., Isik, C.: Blood pressure estimation using neural networks. In: IEEE International Conference on Computational Intelligence for Measurement Systems and Applications, Boston, MA, USA (2004)
4. Jung, J.H., et al.: Estimation of the Blood Pressure using Arterial Pressure-Volume Model. In: 6th International Special Topic Conference on Information Technology Applications in Biomedicine, Tokyo, Japan (2007)
5. Lass, J., et al.: Continuous blood pressure monitoring during exercise using pulse wave transit time measurement. In: 26th Annual International Conference of the IEEE Engineering in Medicine and Biology Society, San Francisco, USA (2004)
6. Köhler, B.U., Hennig, C., Orglmeister, R.: The principles of software QRS detection. IEEE Engineering in Medicine and Biology Magazine 21(1), 42–57 (2002)
7. Morizet-Mahoudeaux, P., et al.: Simple microprocessor-based system for on-line ECG arrhythmia analysis. Med. Biol. Eng. Comput. 19(4), 497–500 (1981)
8. Ciaccio, E.J., Dunn, S.M., Akay, M.: Biosignal pattern recognition and interpretation systems. IEEE Engineering in Medicine and Biology Magazine 12(3), 89–95 (1993)
9. Afonso, V.X., Tompkins, W.J.: Detecting ventricular fibrillation. IEEE Engineering in Medicine and Biology Magazine 14(2), 152–159 (1995)
10. La Foresta, F., Mammone, N., Morabito, F.: Artifact Cancellation from Electrocardiogram by Mixed Wavelet-ICA Filter. In: Neural Nets, pp. 78–82. Springer, Berlin (2006)
11. Kigawa, Y., Oguri, K.: Support Vector Machine Based Error Filtering for Holter Electrocardiogram Analysis. In: 27th Annual International Conference of the Engineering in Medicine and Biology Society, Shanghai, China (2005)
12. Celler, B.G., Lovell, N.H., Chan, D.: The Potential Impact of Home Telecare on Clinical Practice. Medical Journal of Australia 171, 512–521 (1999)
13. Stoecke, J.D., Lorch, S.: Why go see a doctor? - Car goes from office to home as technology divorces function from geography. Int. J. Technology Assessment in Health Care 13(4), 537–546 (1997)
14. Barlow, J., et al.: A systematic review of the benefits of home telecare for frail elderly people and those with long-term conditions. Journal of Telemedicine and Telecare 13, 172–179 (2007)
15. Celler, B.G., Lovell, N.H., Basilakis, J.: Using information technology to improve the management of chronic disease. Medical Journal of Australia 179(5), 242–246 (2003)
16. Ruggiero, C., Sacile, R., Giacomini, M.: Home Telecare. Journal of Telemedicine and Telecare 5, 11–17 (1999)
17. American Thoracic Society, 1994 Update. Standardization of Spirometry. Am. J. Respir. Crit. Care Med., 1107–1136 (1995)
18. Edelberg, R.: Local electrical response of the skin to deformation. J. Appl. Physiol. 34(3), 334–340 (1973)
19. Redmond, S.J., et al.: ECG Recording and Rhythm Analysis for Distributed Health Care Environments. In: Acharya, R.U., Tamura, T. (eds.) Distributed Diagnosis and Home Healthcare. American Scientific Publishers (2009)
20. Srikureja, W., Darbar, D., Reeder, G.S.: Tremor-Induced ECG Artifact Mimicking Ventricular Tachycardia. Circulation 102(11), 1337–1338 (2000)

21. Redmond, S.J., et al.: ECG Quality Measures in Telecare Monitoring. In: Proc. 30th Annual International Conference of the IEEE Engineering in Medicine and Biology Society, Vancouver, BC, Canada. IEEE Press, Los Alamitos (2008)
22. Barker, S.J.: Motion-Resistant Pulse Oximetry: A Comparison of New and Old Models. Anesthesia and Analgesia 95, 967–972 (2002)
23. Karantonis, D., et al.: Implementation of a real-time human movement classifier using a triaxial accelerometer for ambulatory monitoring. IEEE Trans. on Information Technology in Biomedicine 10(1), 156–167 (2006)
24. Narayanan, M.R., et al.: A Wearable Triaxial Accelerometry System for Longitudinal Assessment of Falls Risk. In: Proc. of the 30th Annual International Conference of the IEEE EMBS, Vancouver, BC, Canada. IEEE Press, Los Alamitos (2008)
25. Li, Q., Mark, R.G., Clifford, G.D.: Robust heart rate estimation from multiple asynchronous noisy sources using signal quality indices and a Kalman filter. Physiological Measurement 29, 15–32 (2008)
26. Ohta, S., et al.: A health monitoring system for elderly people living alone. Journal of Telemedicine and Telecare 8(3), 151–156 (2002)
27. Celler, B.G., et al.: Remote monitoring of the elderly at home. A multidisciplinary project on aging at the University of New South Wales. Int. J. Biomed. Comp. 40, 147–155 (1995)
28. Stanford, V.: Using Pervasive Computing to Deliver Elder Care. IEEE Pervasive Computing 1(1), 10–13 (2002)
29. Srovnal, V., Penhaker, M.: Health Maintenance Embedded Systems in Home Care Applications. In: Second International Conference on Systems (2007)
30. Voon, R., Celler, B.G., Lovell, N.H.: The use of an energy monitor in the management of diabetes: a pilot study. Diabetes Technology and Therapeutics (2008) (in press)
31. Virone, G., Noury, N., Demongeot, J.: A system for automatic measurement of circadian activity deviations in telemedicine. IEEE Transactions on Biomedical Engineering 49(12), 1463–1469 (2002)
32. Mathie, M.J., et al.: Accelerometry: providing an integrated, practical method for long-term, ambulatory monitoring of human movement. Physiol. Meas. 25(2), R1–R20 (2004)
33. Mathie, M.J., et al.: A pilot study of long-term monitoring of human movements in the home using accelerometry. J. Telemed. Telecare 10(3), 144–151 (2004)
34. Pentland, A.: Healthwear: Medical technology becomes wearable. Computer 37(5), 42–49 (2004)
35. Tamura, T., et al.: E-Healthcare at an Experimental Welfare Techno House in Japan. The Open Medical Informatics Journal 1(1), 1–7 (2008)
36. Suzuki, R., et al.: Analysis of activities of daily living in elderly people living alone: single-subject feasibility study. Telemedicine Journal and e-health: the Official Journal of the American Telemedicine Association 10(2), 260–276 (2004)
37. Philipose, M., et al.: Inferring activities from interactions with objects. IEEE Pervasive Computing 3(4), 50–57 (2004)
38. Carson, E.R., et al.: Clinical decision support, systems methodology, and telemedicine: their role in the management of chronic disease. IEEE Transactions on Information Technology in Biomedicine 2(2), 80–88 (1998)
39. Montani, S., et al.: Integrating model-based decision support in a multi-modal reasoning system for managing type 1 diabetic patients. Artificial Intelligence in Medicine 29(1-2), 131–151 (2003)
40. O'Neill, E.S., et al.: Knowledge acquisition, synthesis, and validation: a model for decision support systems. Journal of Advanced Nursing 47(2), 134–142 (2004)

41. Falas, T., Papadopoulos, G., Stafylopatis, A.: A review of decision support systems in telecare. Journal of Medical Systems 27(4), 347–356 (2003)
42. Lovell, N.H., et al.: The Application of Decision Support Systems in Home Telecare. In: Acharya, R.U., Tamura, T. (eds.) Distributed Diagnosis and Home Healthcare. American Scientific Publishers (2009)
43. Redmond, S.J., et al.: Piecewise-linear trend detection in longitudinal physiological measurements. In: Proc. 31st Annual International Conference of the IEEE Engineering in Medicine and Biology Society, Minneapolis, USA (in press, 2009)

Calibration of Automated Non Invasive Blood Pressure Measurement Devices

E. Balestrieri and S. Rapuano

Dept. of Engineering, University of Sannio, Benevento, Italy

Abstract. Starting from a brief history of blood pressure measurement, the chapter provides a critical overview of the existing standards and protocols dealing with the calibration of automated non-invasive measurement of blood pressure. Some issues about the device specifications and the test methods, including the most used clinical protocols are pointed out. The lack of a clear and unique set of definitions and objective methods for calibrating blood pressure measuring devices is highlighted and the IEEE P1721 project for a new standard on Objective Measurement of Systemic Arterial Blood Pressure in Humans is introduced in the chapter.

Keywords: Blood pressure, non-invasive measurement, calibration, standardization, IEC, BHS, ANSI/AAMI.

1 Introduction

Blood pressure (BP) measurements are used for diagnosis, determination of prognosis and for initiating, evaluating and discontinuing several medical treatments. Due to the widespread interest in and reliance on the procedure of BP measurement, an adequate training of all the personnel and the standardization of the involved equipment is necessary to minimize the major sources of uncertainty that contribute to variations in BP measurement, and may adversely influence clinical treatment decisions. The standardization should involve: the selection of quality BP measuring equipment, the proper maintenance of such equipment, the calibration, instruction and accreditation in the measurement techniques of all personnel directly involved in BP measurement [1].

For more than a century, a manual non-invasive BP measurement (NIBPM) device, the mercury sphygmomanometer, has been the "gold standard". However, the potential of mercury spillage contaminating the environment has led to the decreased use or elimination of mercury in sphygmomanometers as well as in thermometers. As a result, electronic BP measuring instruments have increased dramatically over the past several years and represent a good candidate for replacing the mercury sphygmomanometers.

Concerns regarding the accuracy of non-mercury sphygmomanometers have created new challenges for accurate BP determination. When mercury sphygmomanometers are replaced, the new equipment, including all electronic BP measurement devices used at home, must be appropriately validated and checked regularly for accuracy by adopting specific procedures [2].

S.C. Mukhopadhyay and A. Lay-Ekuakille (Eds.): Adv. in Biomedical Sensing, LNEE 55, pp. 281–304.
springerlink.com

Automated BP measurement devices, the latest technological advance in BP measurement, improve on current technology and reduce the human error factor. Finally, BP can be measured automatically several times during the same examination [3].

The majority of non-invasive automated BP measuring devices (NIBPM) currently available use the oscillometric method to calculate systolic and diastolic BP implemented by means of empirically derived algorithms, generally patent protected [4]. This means that algorithms used by different manufacturers can produce different results from device to device [5]. Moreover, although the measurement uncertainty of these devices should be assessed by a clinical trial involving an adequately sized group of patients, such evaluation is not mandatory for all Countries [6]. Therefore, only a small portion of automated devices on the market have been qualified by clinical evaluations according to generally accepted protocols of an independent institution or scientific society.

At now, specific protocols for the clinical trials are available from the British Hypertension Society (BHS), the American Association for the Advancement of Medical Instrumentation (AAMI) and the European Society of Hypertension (ESH). The available standards for the NIBPM devices include the international IEC 60601-2-30 [7], the European EN 1060 [8], the German DIN58130 [9], substituted from EN 1060-4, and the American ANSI/AAMI SP-10 [10], while the international organization of legal metrology (OIML) produced a specific recommendation on non invasive automated sphygmomanometers, the OIML R16-2 [11].

Unfortunately, even the successfully evaluated devices may not guarantee a specific uncertainty for all kinds of users [6] and still display significant inaccuracies in BP measurement [5] due to the limited reproducibility of the calibration methods provided in the above quoted standards.

Very recently, the IEEE started working on two independent standards for evaluating and calibrating cuff-less (P1708) [12] and cuff-based (P1721) [13] devices that measure BP.

The chapter starts with a short overview of the BP measurement history to introduce the currently used BP measurement techniques. Then the main sources of the BP measurement uncertainty and its consequences are introduced. The current standards and protocols devoted to the calibration of automated non-invasive BP measurement devices are critically analyzed in terms of metrological effectiveness and applicability, considering their estimated cost and time requirements. Such analysis put in evidence the need for standardized terminology and calibration procedures to establish minimum accuracy specifications, to increase confidence in the quality and reliability of automated NIBP measuring devices, as well as to facilitate the comparison of one device with the others.

2 Blood Pressure Measurement History

The ancient Greek physician Galen first proposed the existence of blood in the human body. According to Hippocrates, the body was comprised of three systems. The brain and nerves were responsible for sensation and thought. The blood and arteries filled the body with life-giving energy. He also theorised that blood travels backward and forward in unconnected veins and arteries. In 1616 William Harvey announced that Galen was wrong in his assertion that the heart constantly produces blood. Instead, he proposed that there was a finite amount of blood that circulated the body in one direction only (Fig.1). Interestingly, Harvey was neither the only nor the first to question Galen's ideas. The Egyptians knew that blood flowed through the body and used leeches to unblock what they thought were passages of blood [14].

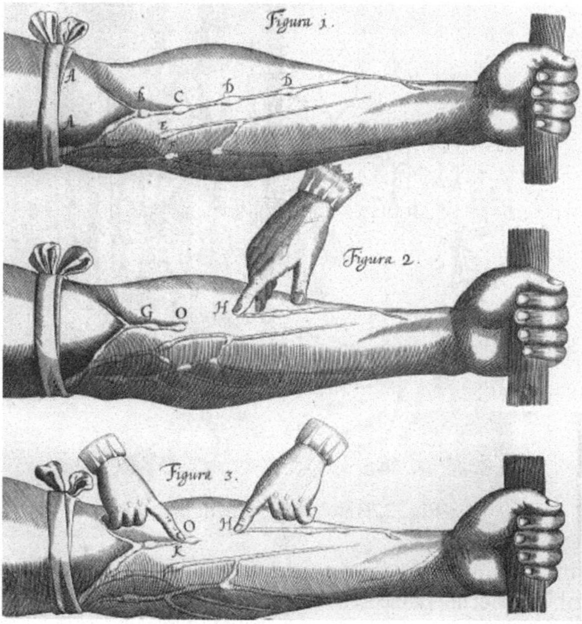

Fig. 1. Illustration from William Harvey: De motu cordis (1628). *Figura 1* shows distended veins in the forearm and position of valves (*B,C,D,E*). *Figura 2* shows that if a vein is occluded centrally and the peripheral end compressed, it does not fill until the finger is released. *Figura 3* shows that blood cannot be forced in the 'wrong' direction. [15].

The first recorded instance of the measurement of BP was in 1733 by the Reverend Stephen Hales. Reverend Hales spent many years recording the BPs of animals. He inserted one end of a brass pipe into the ligated left crural artery of a horse, and to the other end, he attached a vertically positioned glass tube, nine feet in length. On untying the ligature on the artery, blood rose in the tube to a height of 8 feet 3 inches above the left ventricle of the heart (Fig. 2). This was the first recorded estimation of BP. He also demonstrated that the pulse rate was more rapid in small animals than large animals and that BP was proportionate to the size of the animal [14].

John Leonard Marie Poisseuille introduced the BP measurement unit mmHg, which he described in his medical school thesis in 1828. Such unit is still being used today to measure BP.

The use of mercury allowed for smaller height of the column needed for measurements. Poisseuille improved upon the original BP-measuring apparatus by substituting the short tube of a mercury manometer for the inconveniently long tube used by Hales. Connection with the artery was established by means of a hollow lead tube filled with potassium carbonate, to prevent coagulation. This was Posseuille's haemodynamometer of 1828, with which he showed that BP rises and falls with expiration and inspiration [14].

Fig. 2. Hale's experiment [16]

Karl Ludwig in 1846 added a float to the mercury manometer with a connecting arm, which inscribed arterial pulse wave on a recording cylinder that gave a permanent record. Ludwig's kymograph consisted of a U-shaped manometer tube connected to a brass pipe cannula into the artery (Fig. 3). The manometer tube had an ivory float onto which a rod with a quill was attached. This quill would sketch onto a rotating drum, and hence the name 'kymograph', 'wave writer' in Greek. However BP could still only be measured by invasive means [14].

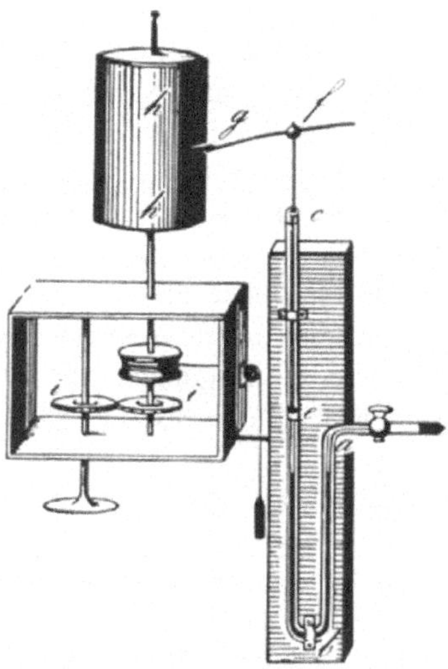

Fig. 3. Ludwig's kymograph [14]

Karl von Vierordt described in 1855 that with enough pressure, the arterial pulse could be obliterated. In 1854, he created the sphygmograph, a pulse recorder usable for routine non-invasive BP monitoring on humans (Fig. 4). Sphygmographs worked by transmitting the movement of the pulse to a long lever that traced a curve onto prepared paper. By adding weights to little pans attached to a lever, he attempted to estimate the BP. His instrument was cumbersome and his measurements incorrect, but he established the principle that the estimation of BP can be accomplished by measuring the outside pressure necessary to obliterate the pulse – a method employed even today [14].

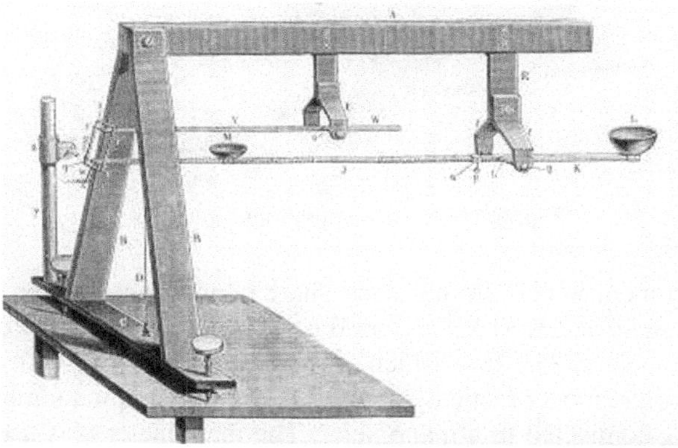

Fig. 4. Vierordt Sphygmograph [17]

Etienne Jules Marey in 1860 devised on the sphygmograph and enabled recording graphically the features of the pulse and variations in BP. His basic instrument, with modifications, is still used today. His sphygmograph could accurately measure the pulse rate but was very unreliable in determining the BP (Fig. 5). Doctors found the sphygmograph cumbersome and difficult to use accurately – as well as intimidating for the patient. However, the ability to see variations and abnormalities in the circulation of blood and the heartbeat was of enormous value to the development of experimental physiology and cardiology [14].

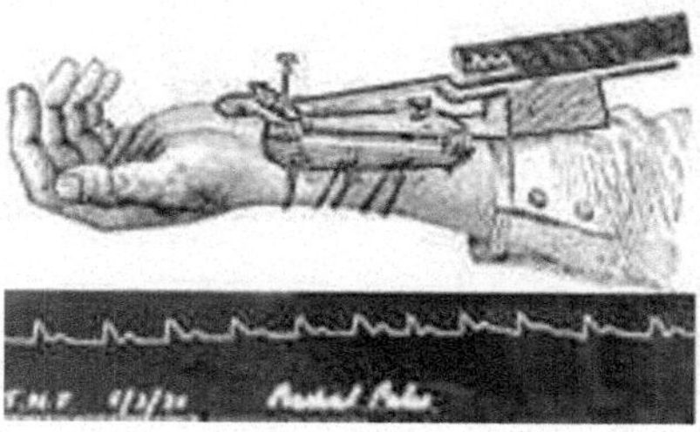

Fig. 5. Marey sphygmograph for taking a tracing of the radial pulse [18]

In 1881, Robert Ellis Dudgeon introduced a new, highly portable sphygmograph. Dudgeon's sphygmograph was strapped to the wrist (Fig. 6). The pulse at the wrist caused a metal strip to move a stylus, transmitting a record of the pulse onto smoked paper. Dudgeon's instrument quickly became popular, as it was compact and easy to use. The sphygmograph traces an undulating line, which represents a record of BP and pulse over time [14].

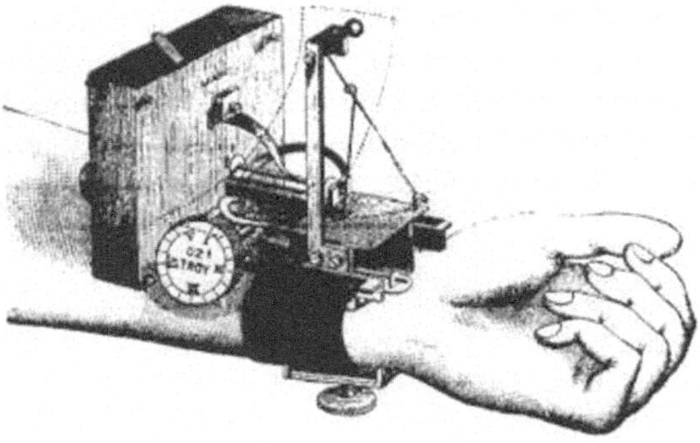

Fig. 6. Dudgeon's sphygmograph [14]

The first instrument which did not necessitate puncturing the skin was devised in 1880 by Samuel Siegfried Karl Ritter von Basch (Fig. 7). He fabricated three models of sphygmomanometers. The first which was invented in 1881, with a mercury column, proved to be the most practical and useful. His sphygmomanometer consisted of a water-filled bag connected to a manometer. The manometer was used to determine the pressure required to obliterate the arterial pulse. Direct measurement of BP by catheterisation confirmed that von Basch's design would allow a non-invasive method to measure BP. Feeling for the pulse on the skin above the artery was used to determine when the arterial pulse disappeared [14].

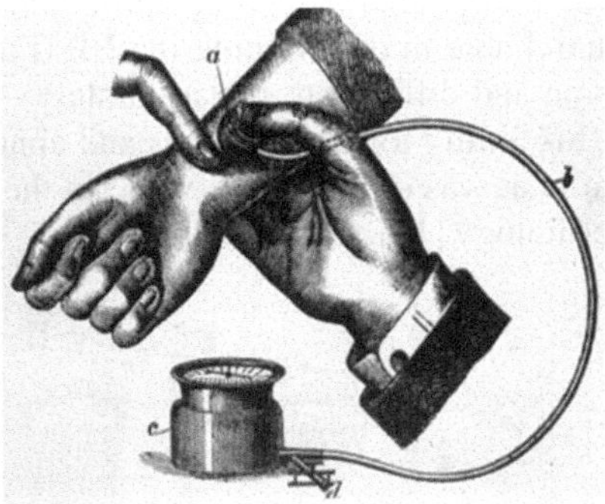

Fig. 7. A sphygmomanometer by von Basch [19]

In 1896, Scipione Riva-Rocci developed the mercury sphygmomanometer. The Riva-Rocci sphygmomanometer contained an elastic inflatable cuff that was placed over the upper arm to constrict the brachial artery, a rubber bulb to inflate the cuff and a glass manometer filled with mercury to measure the cuff pressure (Fig. 8). Riva-Rocci measured the systolic pressure by registering the cuff pressure at which the radial pulse was obliterated as determined by palpation. The palpation technique did not allow the measurement of diastolic pressure [14].

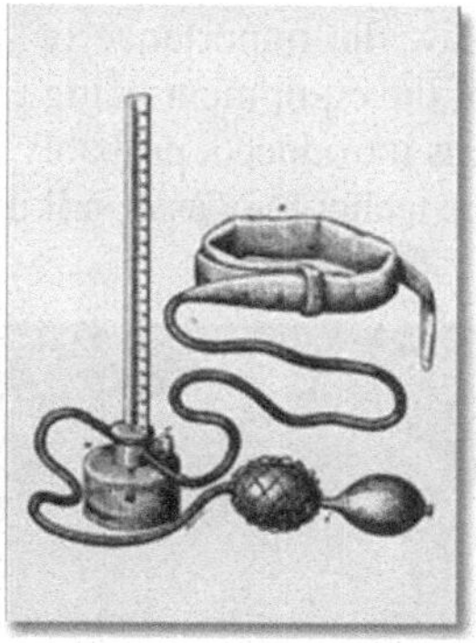

Fig. 8. Riva-Rocci's sphygmomanometer [20]

Soon after Riva-Rocci's technique was described, Hill and Barnard in 1897 reported an apparatus with an arm-encircling inflatable cuff and a needle pressure gauge that allowed measurement of the diastolic pressure by the oscillatory method. This method used the oscillations transmitted to the gauge, as the pulse wave came through the compressed artery. When the cuff pressure was reduced slowly from the suprasystolic pressure, the appearance of definitive oscillations denoted the systolic pressure, whereas the change from maximal oscillations to smaller ones denoted the diastolic pressure. In 1900, von Recklinghausen increased width of the cuff from 5 to 13 cm [14].

Korotkoff, using a stethoscope and the apparatus proposed by Riva-Rocci, established in 1905 that certain specific sounds could be heard during the decompression of the arteries. This auscultatory method proved to be more reliable than the previous palpation techniques and thus became the standard practice. This specific phenomenon, known in world literature as 'Korotkoff sounds', became the basis of the new method of BP measurement [14].

In 1909, Pachon introduced his first oscillometer that allowed users to estimate arterial rigidity and blockages by making simultaneous measurements on different parts of the body and to compare the amplitude of the oscillations (Fig. 9) [21].

Fig. 9. Pachon oscillometer [21]

In 1916, William A. Baum saw the importance of accurate blood pressure measurement and the imperfections of the equipment being used and proposed.

In 1918 the Vaquez-Laubry was introduced, probably the first aneroid blood pressure monitor that uses exactly the same technology as is still used today (Fig. 10) [21].

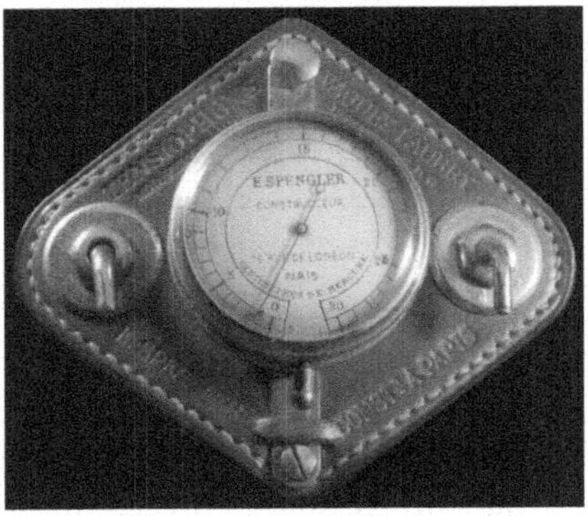

Fig. 10. The Vaquez-Laubry Sphygmo-tensiophone [21]

In 1974, Panasonic released the first digital oscillometric device. These sphygmomanometers measure the pressure imparted onto the cuff by the blood pushing through the constricted artery over a range of cuff pressures. This data is used to estimate the systolic and diastolic blood pressures [21].

The first automated oscillometric method was introduced to medicine by Ramsey in 1979. The first instrument displayed only mean pressure (cuff pressure for maximal oscillations) and heart rate. At that time, mean pressure was not measured, nor was it well understood, so Ramsey created algorithms, based on the oscillation amplitude, to identify systolic and diastolic pressures, as well as mean pressure; thus was born the Dynamap automatic noninvasive blood-pressure monitor [22].

3 Blood Pressure Measurement Methods

BP measurement techniques are generally grouped in two broad classes: direct and indirect measurements.

Indirect measurements are often called noninvasive measurements because the body is not entered in the process. The upper arm, containing the brachial artery, is the most common site for indirect measurement because of its closeness to the heart although many other sites may have been used, such as forearm or radial artery, finger, etc. Distal sites such as the wrist, although convenient to use, may give much higher systolic pressure than brachial or central sites as a result of the phenomena of impedance mismatch and reflective waves [23].

An occlusive cuff is normally placed over the upper arm and is inflated to a pressure greater than the systolic blood pressure so that the blood flow is completely stopped. The cuff is then gradually deflated, while a detector system simultaneously employed determines the point at which the blood flow is restored to the limb. The detector system does not need to be a sophisticated electronic device. It may be as simple as manual palpation of the radial pulse. The most commonly used indirect methods are auscultation and oscillometry [23] (Fig. 11).

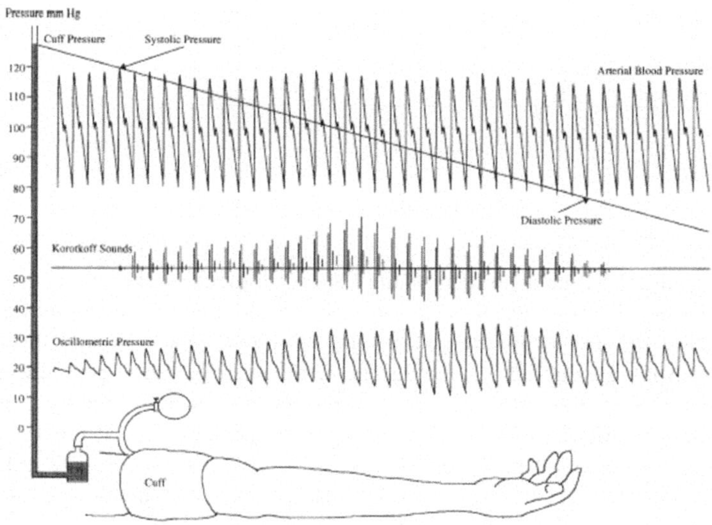

Fig. 11. Indirect blood pressure measurements: oscillometric measurement and auscultatory measurement [23]

The auscultatory method most commonly employs a mercury column or aneroid manometer, an occlusive cuff, and a stethoscope.

The stethoscope is placed over the blood vessel for auscultation of the Korotkoff sounds, which define both systolic pressure (SP) and diastolic pressure (DP). The Korotkoff sounds are mainly generated by the pulse wave propagating through the brachial artery. The Korotkoff sounds consist of five distinct phases (Fig.12). The onset of Phase I Korotkoff sounds (first appearance of clear, repetitive, tapping sounds) signifies SP and the onset of Phase V Korotkoff sounds (sounds disappear completely) often defines DP.

Observers may differ greatly in their interpretation of the Korotkoff sounds. Simple mechanical error can occur in the form of air leaks or obstruction in the cuff or coupling tubing. Mercury can leak from a column gage system. In spite of the errors inherent in such simple systems, more mechanically complex systems have come into use. The impetus for the development of more elaborate detectors has come from the advantage of reproducibility from observer to observer and the convenience of automated operation. Examples of this improved instrumentation include sensors using plethysmographic principles, pulse-wave velocity sensors, and ultrasonic microphones [23].

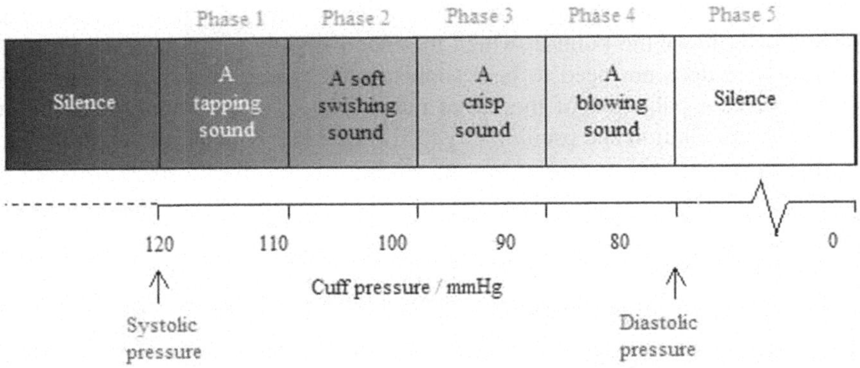

Fig. 12. Korotkoff sounds phases [24]

The principle of BP measurement using the oscillometric technique is dependent on the transmission of intra-arterial pulsation to the occluding cuff surrounding the limb. An approach using this technique could start with a cuff placed around the upper arm and rapidly inflated to about 30 mmHg above the systolic blood pressure, occluding blood flow in the brachial artery. The pressure in the cuff is measured by a sensor. The pressure is then gradually decreased, often in steps, such as 5 to 8 mmHg. The oscillometric signal is detected and processed at each step of pressure. The cuff pressure can also be deflated linearly in a similar fashion as the conventional auscultatory method.

Arterial pressure oscillations are superimposed on the cuff pressure when the blood vessel is no longer fully occluded. Separation of the superimposed oscillations from the cuff pressure is accomplished by filters that extract the corresponding signals. Signal sampling is carried out at a rate determined by the pulse or heart rate [23]. In the case of conventional oscillometric method, the pulsation measured with the cuff pressure sensor, ideally, starts at SP and ends at DP [25] (Fig. 13). However, when cuff pressure is greater than SP, it is impossible to occlude the brachial artery completely and stop arterial pulsation of the cuff even if the central part of the cuff blocks the brachial artery. Since the lateral part of the cuff has lower pressure than the center and takes arterial pulsation, it is not easy to determine the SP point. When the cuff pressure is decreased below DP, the interference of the arterial pulsation makes identification of DP hard [25].

For these reasons, the oscillation amplitudes are most often processed with an empirical algorithm to estimate SP and DP [23]. In particular, mathematical criteria are applied to the envelope curve resulting from plotting the oscillatory pulses versus the cuff base pressure. The algorithms used for detecting systolic and diastolic pressures are different from one device to another and are not divulged by the manufacturers [26]. The differences between devices have been dramatically shown by studies using simulated pressure waves, in which a SP of 120 mm Hg was registered as low as 110 and as high as 125 mmHg by different devices [26].

One advantage of the method is that no transducer need be placed over the brachial artery, so that placement of the cuff is not critical. Other potential advantages of the oscillometric method for ambulatory monitoring are that it is less susceptible to external noise (but not to low-frequency mechanical vibrations), and that the cuff can be removed and replaced by the patient [26].

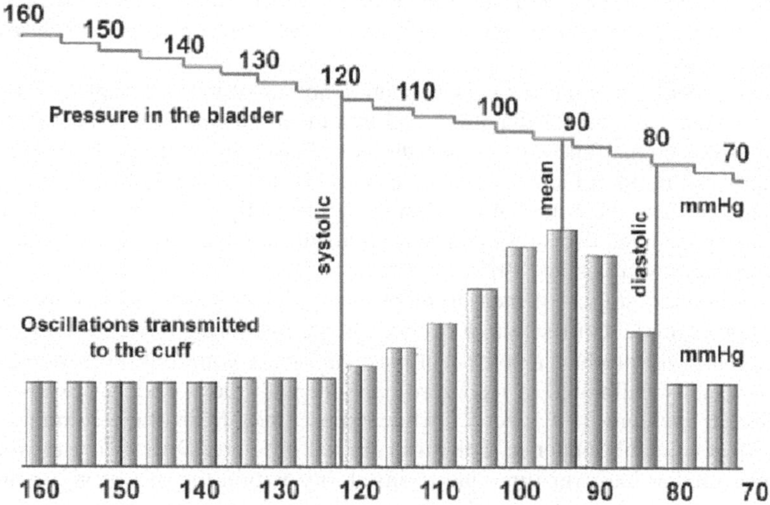

Fig. 13. Oscillometric determination of SP and DP [27]

Direct measurement are also called invasive measurements because bodily entry is made. For direct arterial BP measurement an artery is cannulated. The equipment and procedure require proper setup, calibration, operation, and maintenance [23]. Such a system yields BPs dependent upon the location of the catheter tip in the vascular system. It is particularly useful for continuous determination of pressure changes at any instant in dynamic circumstances. When massive blood loss is anticipated, powerful cardiovascular medications are suddenly administered, or a patient is induced to general anesthesia, continuous monitoring of blood pressures becomes vital. Most commonly used sites to make continuous observations are the brachial and radial arteries [23].

Invasive access to a systemic artery involves considerable handling of a patient. The longer a catheter stays in a vessel, the more likely an associated thrombus will form. In spite of well-studied potential problems, direct BP measurement is generally accepted as the gold standard of arterial pressure recording and presents the only satisfactory alternative when conventional cuff techniques are not successful. This also confers the benefit of continuous access to the artery for monitoring gas tension and blood sampling for biochemical tests. It also has the advantage of assessing cyclic variations and beat-to-beat changes of pressure continuously, and permits assessment of short-term variations [23].

4 Blood Pressure Measurement Uncertainty

Medicine has a great dependence from measurement. Most clinical measurements require a measuring device and a human operator, either of which can give rise to errors. The device may be wrongly calibrated or malfunctioning; the operator may select an inappropriate device or take the reading incorrectly [28].

Historically, medical devices were simple and it was easy for experienced practitioners to detect malfunction or inadequate performance. Modern physiological measurement systems are becoming more sophisticated and the spectrum of measurements is broadening [29].

Medical practitioners increasingly rely on quantitative measurements for the early detection of disease, and for diagnosis and treatment. However, it is becoming more difficult to detect malfunction or measurement biases in modern electronic instruments. This evolution of increasingly more complex measuring devices within clinical practice has, in fact, the potential to generate significantly biased measurements. This may lead to misguided decision-making if clinicians are unaware of the limitations of the equipment they are using [29].

The commonest source of measurement biases due to hardware is improper calibration. For example an offset error occurs when a measuring device is calibrated incorrectly. This will produce a consistent bias. An invasive arterial pressure measuring system is usually calibrated to read zero pressure with the transducer exposed to atmospheric air pressure. If it is zeroed while exposed to a hydrostatic column of fluid, then it will predictably under-read the true arterial pressure. Drift is a gradual change in the measured value over time due to instability within the measuring system and leads to increasing bias. It may be due to fluctuating temperature, and chemical changes or contamination of the sensor [28].

There is extensive literature on the sources of uncertainty encountered specifically with BP measurement. These can be divided into those that are associated with the observer, with his/her bias or inaccuracy; the physiology of the patient, coupled with the variability of BP; the manometer itself, which may be inaccurate or damaged; and the cuff which may be the wrong size. Fig. 14 shows some of the most common sources of uncertainty [3].

Concerning the observer related uncertainty, researchers shown that it could account for standard deviations ranging from 2 mmHg to 45 mmHg.

Observer Factors
Poor management technique
Digit preference
Expectation bias

Patient Factors
No rest period before measurement
Not in recommended position
Conversation with observers
White-coat effect

Equipment Factors
Aged mercury devices
Non-calibrated aneroid devices
Non-validated automated devices
Improper cuff size

Fig. 14. BP measurement uncertainty sources [3]

In addition, almost 96% of physicians did not use an average of multiple readings in a single arm with the patient in the same position.

Reasons for low reproducibility of BP measurements may include the facts that:

- auscultation requires considerable clinical expertise to obtain accurate measurements;
- detecting and recording Korotkoff signs requires good auditory acuity;
- distraction and noise from a busy clinic can negatively impact the accuracy of readings;
- up to 78% of practitioners demonstrate digit preference by rounding measurements off to the nearest 10 mmHg;
- expectation bias may be present with selected patients to avoid the institution of long-term therapy or in order to achieve a therapy reduction;
- the position of the practitioner relative to the manometer may result in an improper angle of sight for reading the measurements;
- deflation faster than 2 mmHg per heartbeat makes it impossible to record BP at the nearest 2 mmHg increments [3].

Concerning the patient related factors, it is worth to note that activities of daily living can have substantial and variable effects on BP (Fig. 15). These activities result in a variable but higher estimate of BP in the "unrested" patient relative to that obtained with a standardized technique [30].

A frequent cause for bias in BP measurement due to patient factors in the clinical environment is the white coat effect. It is defined as an office BP exceeding mean daytime ambulatory pressure by at least 20 mmHg systolic and/or 10 mmHg diastolic found in as many as 73% of treated hypertensive subjects. It may occur more frequently in women than in men and it is virtually impossible to be diagnosed on clinical examination alone [5].

	Change in blood pressure, mm Hg	
Activity	Systolic	Diastolic
Attending a meeting	+20.2	+15.0
Working	+16.0	+13.0
Commuting	+14.0	+9.2
Walking	+12.0	+5.5
Dressing	+11.5	+9.7
Doing chores	+10.7	+6.7
Talking on telephone	+9.5	+7.2
Eating	+8.8	+9.6
Talking	+6.7	+6.7
Doing desk work	+5.9	+5.3
Reading	+1.9	+2.2
Doing business (at home)	+1.6	+3.2
Watching television	+0.3	+1.1
Sleeping	−10.0	−7.6

*Changes are shown relative to blood pressure while relaxing. Adapted from reference 2.

Fig. 15. Average changes in BP associated with 14 common activities [30]

Concerning the equipment related factors, all indirect BP measuring methods tend to overestimate at low pressures and underestimate at high pressures. Instrument biases are exacerbated by the use of an incorrectly sized pneumatic cuff. Accurate measurement of BP depends on the relationship between arm circumference and the length and width of the blood pressure cuff. Typically, a cuff too small for the patient's arm circumference will overestimate BP, whereas a too large cuff will underestimate blood pressure [32]. Measured pressure alters with the site of measurement and its relationship to the level of the heart (hydrostatic pressure) [31].

Mercury sphygmomanometers, in particular, are a source of observer biases. Viewing the mercury column from different angles has been shown to generate variability in measurements. Also, given the age of many mercury devices, dirty columns, faded calibration marks and mercury oxidation have made many devices difficult to read. Of equal importance, mercury manometers and cuffs can be prone to numerous problems and need to be serviced regularly. In fact, as many as 65% of the devices considered in [3] have been shown to be out of calibration by at least 4 mmHg.

Aneroid sphygmomanometers have replaced many mercury devices. However up to 60% of the aneroid devices tested in [3] were inaccurate due to improper calibration or maintenance.

For automated NIBPM devices, shivering, patients' movement or cardiac arrhythmias such as atrial fibrillation produce unstable data leading to failure to measure or spurious readings [31].

Moreover, a significant disadvantage of the electronic technology has been a lack of validation of the available devices. Of the several hundred self-measurement devices that have been marketed to consumers, only a handful have undergone independent testing and only a few that have been tested to standardised criteria have been found to be satisfactory. Disappointingly, commercially available automated devices that have passed validating study protocols still display significant inaccuracies in BP measurement [5].

Improper calibration may lead to biases in the case of direct BP measurements. Baseline drift of the measuring system's electronics over time may occur requiring periodic re-zeroing. Damping and resonance are the second major source of error in invasive BP measuring systems. Reduction of the natural (resonant) frequency of the measuring system may result in resonant oscillations of the fluid column that add to SP and subtract from DP (overshoot), producing an erroneously widened pulse pressure. Reduction of the resonant frequency and overshoot may be caused by excessively lengthening the manometer tubing. Damping is a reduction in the energy of transmitted oscillations in a manometric system caused by friction between fluid and the walls of the tubing in which it is contained. Excessive damping results in under-reading of the SP and over-reading of the DP with an erroneous narrowing of the pulse pressure. Air bubbles and clotted blood in the system or arterial spasm are causes of excessive damping seen in clinical practice. Excessive damping also leads to a phase shift in the transduced signal. This occurs because of the increased delay in pressure waves reaching the transducer [31].

Accurate measurement of BP is crucial from a public health standpoint. Even small underestimation of BP can cost lives by failing to prevent cardiovascular disease through effective and safe therapy. Consistently underestimating the diastolic pressure by 5 mmHg could result in almost two-thirds of people with hypertension not receiving potentially life-saving treatment.

Overestimating BP can also have a significant impact. Consistently overestimating diastolic pressure by 5 mmHg could more than double the number of hypertensive patients, leading to millions of people exposed to possible inappropriate therapy.

Furthermore, use of a single measurement to define a patient's BP, while not recommended by current guidelines, still occurs in practise. These facts reinforce the importance of accurate measurement of BP to assess and manage cardiovascular risk and, conversely, avoid costly and unnecessary treatment [3]. For these reasons, all devices for BP measurement require regular calibration and maintenance. NIBPM devices have to be validated according to guidelines including commonly adopted rules to be used in all Countries.

5 Automated Non-invasive BP Measurement Device Calibration Procedures

In the years both standards and clinical trial protocols have been published concerning automated non-invasive BP measurement. A standard is defined as a "document, established by consensus and approved by a recognized body, that provides, for common and repeated use, rules, guidelines or characteristics for activities or their results, aimed at the achievement of the optimum degree of order in a given context" [33]. A clinical trial protocol, instead, is defined by the International Conference on Harmonization of Technical Requirements for Registration of Pharmaceuticals for Human Use (ICH) as: "a document that describes the objective(s), design, methodology, statistical considerations and organization of a trial" [34]. Therefore, matter of a standardization process can be a procedure or/and an object, matter of a clinical protocol is a trial. A standard describes a specific level or degree of quality or minimum requirements and is drawn up by standardization bodies which have the principal function, by virtue of their statutes, of preparing, approving or adopting, and publishing standards. Clinical trial protocols describe the set of rules, policies and procedures that the researchers and staff conducting a clinical trial (usually medical doctors and/or nurses) must follow. Clinical trial protocols can be sponsored by pharmaceutical, biotechnology or medical device companies. A protocol can become part of a standard, as happened for example in the case of DIN 58130, EN 1060-4 and ANSI/AAMI SP10. Moreover, a standard can recommend the application of an established clinical protocol, as for example in the case of the IEC 60601. Such standard, in fact, recommends (it is not mandatory) independent clinical evaluation of BP measuring devices according to the established international protocols like the British Hypertension Society (BHS) protocol or the protocols included in DIN 58130 and in ANSI/AAMI SP10 [35].

The current standards for automated BP monitors are (i) the IEC 60601-2-30 [7], particular standard of IEC 60601 for automatic cycling Non-Invasive BP (NIBP) monitoring equipment, mainly devoted to the safety requirements for such devices, ii) the ANSI/AAMI SP1O [10] for electronic or automated sphygmomanometers, initially developed as a protocol by the Association for the Advancement of Medical Instrumentation (AAMI) and then become a USA national standard, (iii) the European Union (EU) standard EN 1060 [8] that specifies the safety and basic accuracy requirements of BP measuring devices for automated non-invasive sphygmomanometers, and (iv) the DIN 58130 [9, 36] that determines test methods for the clinical investigation of noninvasive sphygmomanometers, substituted from the EN 1060 part 4 [8].

Currently, the widest acceptance as validation guidelines has been gained by the ANSI/AAMI and the BHS protocols. In fact, since their introduction, a large number of BP measuring devices have been evaluated according to one or both these protocols [26].

The international organization of legal metrology produced a specific recommendation on non invasive automated sphygmomanometers, The OIML R16-2 that "specifies general, performance, efficiency and mechanical and electrical safety requirements, including test methods for type approval, for non-invasive electronic or automated sphygmomanometers and their accessories which, by means of an inflatable cuff, are used for the non-invasive measurement of arterial blood pressure" [11]. However, concerning the assessment of the device accuracy it specifies that the ANSI/AAMI, BHS or DIN 58130 protocols should be used.

Even if the international measurement unit for pressure is the Pascal, the EN standard and the IEC standards use both Pascal and mercury millimeters to specify the device accuracy, due to the wide use of mmHg as discussed in the previous Sections.

Another main difference with the metrology literature is that the standards on BP measurement define the Device Under Test (DUT) accuracy in terms of maximum error instead of uncertainty. In the following the current standards for automated BP monitors and the AAMI, BHS and ESH protocols are discussed.

5.1 The International Standard IEC 60601-2-30

The EN 60601-2-30 is a particular standard of 60601 including specifications for the safety of automatic cycling NIBP monitoring equipment. As the standard scope states, it concerns "requirements for the safety, including essential performance" with "special attention being paid to the avoidance of hazards due to the inflation process" for such devices. In practice, the standard applies to the monitoring devices designed to work continuously attached to the patients, mainly adopted in hospital environment. The standard doesn't apply to "BPM equipment typically in which each determination needs to be initiated manually". The safety specifications are targeted to prevent the equipment from harming the patients' circulation due to anomalous operation. Therefore, the standard covers (i) alarm types and modes, (ii) maximum cuff pressure, (iii) minimum safe pressure, (iv) inflation and deflation timing, (v) safety intervention timing, (vi) immunity to temperature, humidity and power supply variations, (vii) safe conditions to be set after a manual or abnormal shut down, and (viii) EMC requirements, clearly considering the cases of electrical surgery units, or defibrillation discharges.

The standard requires compliance tests and specific measurement before the first use and after each anomalous event. However, it sends the reader to other standards concerning the test methods and the relative instrumentation. In particular, for what concerns calibration, the 60601-2-30 suggests the adoption of BHS and ANSI/AAMI protocols or the DIN 58130.

This standard is going to be replaced by the new standard IEC 80601-2-30. Specific technical changes will include: expansion of the scope to include all automated sphygmomanometers including those where the patient is the operator, identification of essential performance, new clinical accuracy requirements, additional mechanical strength requirements and prohibition of operator accessible 'Luer' connectors in the pneumatic system. Moreover, the Technical Committee ISO/TC 121 and Technical Committee IEC/TC 62 are preparing the ISO 81060-2 devoted to the clinical validation of automated measurement type and covering sphygmomanometers intended for use in all patient populations (e.g. all age and weight ranges), and all conditions of use (e.g. ambulatory blood pressure monitoring, stress testing blood pressure monitoring and blood pressure monitors for the home healthcare environment or self-measurement).

5.2 The EU Standard EN 1060

The European standard EN 1060 covers mechanical and electronic NIBP measuring systems and is divided into four parts. The EN 1060-1 contains general requirements for all kinds of non invasive sphygmomanometers. The EN 1060-2 includes supplementary requirements for mechanical sphygmomanometers. The EN 1060-3 includes supplementary requirements for electro-mechanical blood pressure measuring systems. Finally, the EN 1060-4 describes the test procedures to determine the overall system accuracy of automated noninvasive sphygmomanometers.

The overall accuracy of the NIBP devices is determined in two ways. The first test method can be find in part 1, aimed to determine "the maximum permissible errors of the cuff pressure indication". According to the test set-up described in Fig. 16 a specified pressure is applied to the DUT (2) and a reference manometer (1) by means of a pressure generator (4) and a metal vessel (3) of known volume.

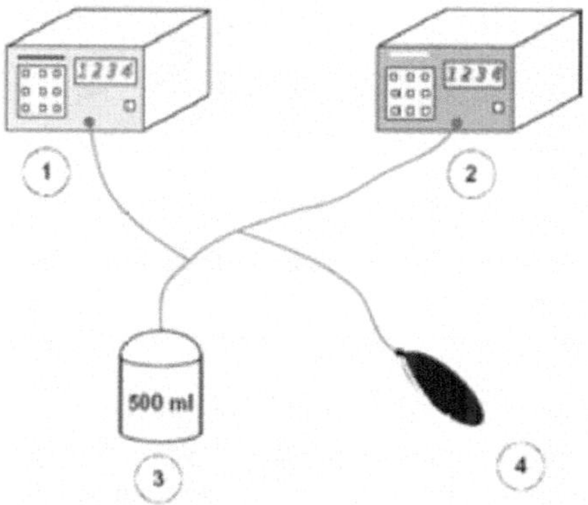

Fig. 16. Test set-up for determining the maximum permissible errors of the cuff pressure indication [8]

The reference manometer should be calibrated with an uncertainty of less than 0.1 kPa (1 mmHg), which is the required resolution for the DUT display.

The environmental conditions during the tests should be changed in the temperature range [15-25] °C and humidity range [20-85]%.

The test pressure values should range from 0 mmHg to the maximum nominal pressure of the DUT and vice versa. The maximum permissible error for the measurement of the cuff pressure at any point of the scale range is ± 0.4 kPa (± 3 mmHg) for the first time verification and ± 0.5 kPa (± 4 mmHg) for maintenance testing within the above specified environmental conditions. The standard covers specific tests for determining the effects of temperature and humidity on the DUT in part 1 and part 3. EN 1060-3 includes also the following overall accuracy specifications:

a) maximum mean error of measurement: ± 5 mmHg (± 0.7kPa);
b) maximum experimental standard deviation: 8 mmHg (1.1 kPa).

The test method is described in EN 1060-4. The measurement reference is constituted by a group of healthy subjects and a group of subjects with hypertension, that should represent the population of all possible patients. The clinical trial shall be conducted on at least 85 subjects. The standard defines as precisely as possible the characteristics of the subjects. The subjects' pressure should be measured simultaneously or successively by means of the DUT and by the auscultatory method.

The auscultatory blood pressure measurements should be carried out by two observers by means of a double stethoscope. The auscultatory reference value will then be the mean of the two values determined by the observers. Fig. 17 reports an example of a simultaneous measurement on the same arm of a subject.

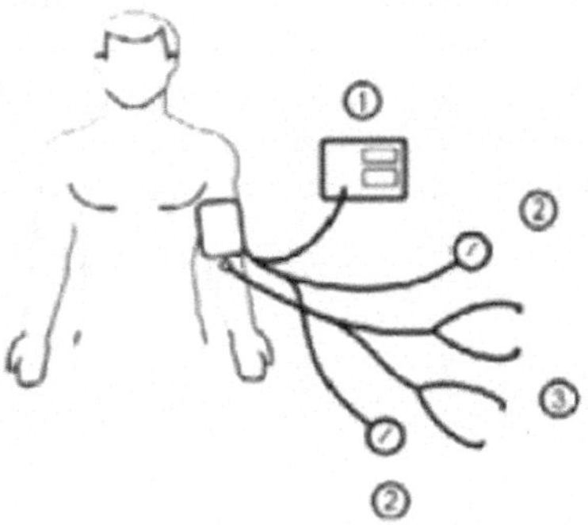

Fig. 17. Simultaneous auscultatory and automated BP measurement [8]. (1) DUT, (2) reference manometers, (3) double stethoscope.

The difference between both values shall not exceed 4 mmHg. Any measurements with observer-to-observer differences greater than 4 mmHg shall not be included in the data set. The number of discarded measurements shall not be greater than the number of the required valid measurements.

The calibrated reference manometers shall comply with the requirements of EN 1060-1 to EN 1060-3 but shall not exceed error limits of 1 mmHg (0,1 kPa) with dropping cuff pressure prior to the start of the clinical investigation.

On each subject, at least three pressure measurements should be taken. The selection of the test method is dependent on the measuring principle and the application of the sphygmomanometers to be tested. EN 1060-4 includes six non invasive (NI to N6) and two invasive (I1 and 12) test methods for the BPM devices.

The non invasive methods are differentiated on the basis of the arm where the cuff should be applied. The NI method describes the simultaneous blood pressure measurement on the same upper arm. The N2 method describes the simultaneous blood pressure measurement on opposite arms. The N3 method describes the sequential blood pressure measurement on the same upper arm. The N4 method describes the simultaneous blood pressure measurement on the same upper arm under physical load. The N5 method describes the ambulatory simultaneous blood pressure measurement on the same upper arm. The N6 method describes the ambulatory simultaneous blood pressure measurement on opposite arms.

The invasive methods, instead, are differentiated on the basis of the subject age: I1 is for testing the device on adults while I2 is for new born and baby subjects. The results of the evaluation of all measurements shall be within the limits determined in EN 1060-3.

5.3 The ANSI SP-10 Standard, the BHS and ESH Protocols

A NIBP measuring device would comply the ANSI SP-10 standard if its measurement error has a mean of no more than 5 mmHg, and a standard deviation of no more than 8 mmHg. As it can be seen, such values are exactly the same as EN 1060-1 and IEC 60601-2-30.

The BHS protocol introduces a classification of the NIBP measuring devices based on their accuracy. In particular, a grade of A device has 60% of the measures with an error within 5 mmHg, 85% of the measures with an error within 10 mmHg, and 95°% within 15 mmHg.

BHS has progressively less stringent criteria for the grades of B and C, and assigns a grade D to the worse performing. However, experience has demonstrated that the conditions demanded by the protocols are extremely difficult to fulfil.

Also in the ANSI standard and the BHS protocols, the readings of the DUT are compared with observations of 2 trained human observers using the auscultatory method as a reference one adopting the mechanical sphygmomanometers as reference instrumentation.

Moreover, except for the number of involved subjects, that should be equal to 85, the protocols do not agree each other in all circumstances [26]. For this reason, there are still many devices on the market that have never been adequately validated. More recently, an international group of experts who are members of the European Society of Hypertension (ESH) Working Group on Blood Pressure Monitoring has produced an International Protocol that could replace the 2 earlier versions being easier to perform. Briefly, it requires comparison of the device readings (4 in all) alternating with 5 mercury readings taken by 2 trained observers. Devices are recommended for approval if both systolic and diastolic readings taken are at least within 5 mmHg of each other for at least 50% of readings [26]. It is recommended that only those devices that have passed this or similar tests should be used in practice.

However, the fact that a device passed a validation test does not mean that it will provide accurate readings on all patients. There can be substantial numbers of subjects in whom the error is consistently greater than 5 mmHg with a device that has achieved a passing grade. This may be more likely to occur in elderly or diabetic patients. For this reason, it is recommended that each BP monitor should be validated on each patient before the readings are accepted. No formal protocol has yet been developed for doing this, but if sequential readings are taken with a mercury sphygmomanometer and the device, then the major inaccuracies can be detected at least [26]. In Tab. 1 the main characteristics of the three protocols are compared.

Table 1. AAMI, BHS and ESH comparison

	AAMI	BHS	ESH
Minimal number of subjects	85	85	33*
Age range (years)	Not specified	15-80	≥30
Arm circumference (cm)	<25 (10%) 25-35 (80%) >35 (10%)	No limitations	No limitations
SBP range (mmHg)	<100 (10%) 100-180 (80%) >180 (10%)	100-140 (15%) 140-180 (15%) 180-220 (15%) 220-240 (15%)	90-129 (11) 130-160 (11) 161-180 (11)
DBP range (mmHg)	<60 (10%) 60-100 (80%) >100 (10%)	60-80 (20%) 80-100 (20%) 100-120 (20%)	40-79 (11) 80-100 (11) 101-130 (11)
Measurement configuration	Simultaneous or Sequential (same arm)	Sequential (same arm)	Sequential (same arm)
Number of measurements for each subject	3	3	3
Total number of measurements	255	255	99
Postures	Supine, seated and standing in case of ambulatory measurements	No limitations	No limitations
Number of experimenters	2	2 + 1 supervisor	2 + 1 supervisor + 1 expert
Validation criteria	Differences with a mean of ±5 mmHg and standard deviation of ±8 mmHg	50% of the differences ≤5 mmHg, 75% ≤ 10 mmHg and 90% ≤15mmHg	According two phases and three validation steps (1, 2.1 e 2.2) Based on the number of measurements not exceeding differences of 5, 10 and 15 mm Hg

*15 subjects in phase I (at least 5 male and 5 female subjects) and 18 subjects in phase II (at least 10 male and 10 female subjects)

6 Considerations on Objective Calibration of Non Invasive Blood Pressure Measurement Devices

Mercury type BP meters have been in use for about 100 years and have become reliable equipment for measuring BP. However, the instrument is expensive and time-consuming to clean up the spill. Besides, exposure to mercury causes serious harms to the central nervous system of human body. Therefore, in order to reduce mercury levels in the environment and exposure to this hazardous substance, this type of BP meter is becoming unwelcome and even is banned to be commercially manufactured or sold in some countries. The replacement goes to automated sphygmomanometers, the most based on oscillometric technique, which have been used extensively nowadays. The accuracy has been a great concern to legal metrology authority since there is a need to acquire the related expertise to regulate automated sphygmomanometers [37].

Currently, the evaluation and calibration of automated BP devices is mainly carried out by taking as a reference the measurements obtained on a large number of people by two observers using stethoscope and mercury sphygmomanometer, as described above.

The methods to do that, do not always agree each other, giving questionable results. However, all of them seem to agree on two main aspects: the reference method and the reference instrumentation.

The reference method is the auscultatory one, based on the Korotkoff sounds, involving at least two trained observers and a numerous group of subjects with given characteristics. The reference instrumentation is constituted by calibrated reference manometers and double stethoscopes.

Of course, observer-based measurements are not free from uncertainty bias and prone to observer related biases. Careful training of observers can reduce but not abolish the sources of uncertainty.

Another problem is to find the number of subjects to be recruited for the validation procedure that must be representative for a wide range of blood pressures.

Most of the automated sphygmomanometers are subject to validation according to internationally accepted protocols of the AAMI, the BHS or the latest ESH. However, commercially available automated devices that have passed validating study protocols still display significant inaccuracies in BP measurement. Gerin et al. showed that BP measurements with such devices were inaccurate by at least 5 mmHg in 20 to 38% of the individuals tested [5]. Moreover, when using BP monitors that meet the AAMI and BHS validation criteria, it has been shown that more than 50% of the persons tested may have average measurements that differ by more than 5 mmHg [5]. The applicability of the ESH protocol has been questioned too, by showing that the reduced required sample size results in a reduction in statistical power from 98 to 70%, respect of the other two protocols [38].

Some questions about the comparison between oscillometric and sphygmomanometer BPs done by validation protocols have been also opened, since it cannot be excluded that oscillometry measures a different kind of physiological variable of BP than sphygmomanometers do. If oscillometrically measured BP does mark a different kind of physiological variable, this would be masked by adjusting the algorithms to mimic sphygmomanometer outcome, in order to achieve an 'A' grading in validation [39].

Therefore, different criteria are used to fulfil the existing, each of them determining the automated NIBP devices performance and accuracy in its own way. However, to make possible the comparison among different devices, a univocal definition of their performance and accuracy is required.

The need for a standard based on automated technology for BP measurement devices has been perceived by the IEEE that has begun to work on the first independent standard (IEEE P1708) for evaluating and calibrating cuff-less devices that measure BP.

However, the cuff is still present in the majority of automated devices currently available in the marketplace. Hence a unified approach for the calibration procedure, not prone to errors of interpretation, observer bias or terminal digit preference, requiring acceptable execution time and costs is still missing and required. In any way, standardized calibration procedure cannot be apart from a standardized terminology to establish a unique and unambiguous set of specifications describing automated BP device performance and accuracy.

Unfortunately, the problem of setting a standardized terminology seems to be not treated in the IEEE standard project for evaluating and calibrating cuff-less devices that measure BP.

Only by means of standardized terminology and calibration procedures, users and above all the doctors, often uneasy about trusting in automated BP devices, can have increased confidence in the quality and reliability of automated NIBP measuring devices. This will be the problem the IEEE I&M Society TC-25 Subcommittee on "Objective Blood Pressure Measurement" will deal with in the next future. The project P1721 "IEEE Standard for Objective Measurement of Systemic Arterial Blood Pressure in Humans" started with the aim of providing a comprehensive guide to assess the uncertainty of automatic BPM devices in an objective way, not depending on a group o human subjects or a specific clinical protocol.

The purpose of the standard project is to provide an objective reference for measuring systemic arterial blood pressure in humans in form of definitions and descriptions of objective techniques and procedures that are independent of specific devices, apparatus, instruments, or computing devices that may be used in blood pressure measurements.

The resulting document could be a reference for (i) capturing, recording, and communication of blood pressure measurement data, (ii) further standards for automated NIBPM devices, and (iii) further standards for devices employing systemic arterial pulses techniques.

7 Conclusion

The chapter presented a brief history of measurement of blood pressure in humans focusing on non invasive methods and presenting the different approaches to the calibration of non invasive blood pressure measurement devices. The comparative analysis of the existing standards showed some degree of harmonization but high complexity of the specified procedures leading to difficult reproducibility of a calibration. As a result it can be found that different instruments calibrated according to a given standards produce incompatible measurements. Several papers dealing with the assessment of the results of such approach to calibration have been recalled in the chapter and the IEEE projects for ensuring the compatibility of measurements have been introduced.

References

1. Utah Department of health, Blood Pressure Measurement Standardization Protocol, Heart disease and stroke prevention program (July 2006), http://www.hearthighway.org

2. Chobanian, A.V., Bakris, G.L., Black, H.R., Cushman, W.C., Green, L.A., Izzo Jr., J.L., Jones, D.W., Materson, B.J., Oparil, S., Wright Jr., J.T., Roccella, E.J.: The National High Blood Pressure Education Program Coordinating Committee, Seventh report of the joint national committee on prevention, detection, evaluation, and treatment of high blood pressure. Hypertension 42(6), 1206–1252 (2003)
3. Gelfer, M.: Addressing the Need for Accurate Blood Pressure Measurements – A Review of the Evidence. Business Briefing: global healthcare (2003)
4. Medical Device Agency, Blood pressure measurement devices - mercury and non-mercury, MDA DB2000(03) (2000)
5. Pater, C.: Beyond the evidence of the new hypertension guidelines. Blood pressure measurement- is it good enough for accurate diagnosis of hypertension? Time might be in, for a paradigm shift (I). Current Controlled Trials in Cardiovascular Medicine 6(1) (2005), `http://Hcvm.controlled-trials.com/content/6/1/6`
6. Tholl, U., Forstner, K., Anlauf, M.: Measuring blood pressure: pitfalls and recommendations. Nephrology Dialysis Transplantation 19(4), 766–770 (2004)
7. IEC 60601-2-30, Medical electrical equipment - Part 2-30: Particular requirements forthe safety, including essential performance, of automatic cycling non-invasive blood pressure monitoring equipment, Ed. 2.0 EN:2000
8. EN 1060, Non-invasive sphygmomanometers, Part 1, 2, 3, 4 (2004)
9. Deutsches Institut fur Normung: NormenausschuB feinmechanik und optik. DI-Norm58130. Pforzheim (1999)
10. ANSI/AAMI SP10, American national standard for electronic or automated sphygmomanometer (1992)
11. OIML R16-2, Non invasive automated sphygmomanometers (2002)
12. IEEE P1708, IEEE Standard for Wearable Cuffless Blood Pressure Measuring Devices
13. IEEE P1721 IEEE Standard for Objective Measurement of Systemic Arterial Blood Pressure in Humans
14. Roguin, A.: Scipione Riva-Rocci and the men behind the mercury Sphygmomanometer. International Journal of Clinical Practice 60(1), 73–79 (2006)
15. `http://www.answers.com/topic/blood-circulation`
16. `http://www.britannica.com/EBchecked/topic-art/252340/15460/Stephen-Hales-measuring-the-blood-pressure-of-a-mare-by`
17. `http://clendening.kumc.edu/dc/rm/major_19th.htm`
18. `http://www.medicine.mcgill.ca/physio/home_history1.htm`
19. `http://www.hemonctoday.com/article.aspx?rid=40807`
20. `http://pacs.unica.it/biblio/lesson8.htm`
21. O'Brien, E.: A few important milestones in the history of blood pressure, O'Brien Healthworks – The Collection, `http://www.bloodpressurehistory.com`
22. Geddes, L.A.: Counterpressure: the concept that made the indirect measurement of blood pressure possible. IEEE Engineering in Medicine and Biology Magazine 17(6), 85–87 (1998)
23. Rithalia, S., Sun, M., Jones, R.: Blood Pressure Measurement. CRC Press, Boca Raton (2000)
24. `http://www.medphys.ucl.ac.uk/teaching/undergrad/projects/2003/group_03/how.html`
25. Kim, T.K., Chee, Y.J., Lee, J.S., Nam, S.W., Kim, I.Y.: A New Blood Pressure Measurement Using Dual-Cuffs. Computers in Cardiology 35, 165–168 (2008)

26. Pickering, T.G., Hall, J.E., Appel, L.J., Falkner, B.E., Graves, J., Hill, M.N., Jones, D.W., Kurtz, T., Sheps, S.G., Roccella, E.J.: Recommendations for blood pressure measurement in humans and experimental animals. Part 1: Blood pressure measurement in humans. Hypertension 45, 142–161 (2005)
27. Department of ECE Institute of Engineering & Technology, Bhaddal, The heart and the cardiovascular system, `http://www.scribd.com/doc/6704587/Cardiovascular-System`
28. Dyer, I., Williams, D.J.: Common errors in clinical measurement. Anaesthesia and Intensive Care Medicine 6(12), 405–407 (2005)
29. Turner, M.J., Kam, P.C., Baker, A.B.: Metrology in Medicine, `http://www-personal.usyd.edu.au/~mjturner`
30. Campbell, N.R.C., McKay, D.W.: Accurate blood pressure measurement: Why does it matter? Canadian Medical Association Journal 161(3) (1999)
31. Stoker, M.R.: Common errors in clinical measurement. Anaesthesia & Intensive Care Medicine 9(12), 553–558 (2008)
32. Jones, D.W., Appel, L.J., Sheps, S.G., Roccella, E.J., Lenfant, C.: Measuring Blood Pressure Accurately: New and Persistent Challenges. Journal of the American Medical Association 289(8), 1027–1030 (2003)
33. `http://www.iec.ch`
34. ICH Guideline for Good Clinical Practice: Consolidated Guidance (1996)
35. O'Brien, E., Asmar, R., Beilin, L., Imai, Y., Mancia, G., Mengden, T., et al.: On behalf of the European Society of Hypertension working group on blood pressure monitoring, European Society of Hypertension recommendations for conventional, ambulatory and home blood pressure measurement. Hypertension (21), 821–848 (2003)
36. CEN/TC 205/WG 10 N 110 Non-invasive sphygmomanometers, Draft for Enquiry (2000), `http://www.hersmedical.com.tw/DIN58130engl.pdf`
37. APEC/APLMF Training Courses in Legal Metrology, Handbook on Trining course on Automated Sphygmomanometers (2008)
38. Friedman, B.A.: Assessment of the validation of blood pressure monitors: a statistical reappraisal. Blood Pressure Monitoring 13(4), 187–191 (2008)
39. Kiers, H.D., Hofstra, J.M., Wetzels, J.F.M.: Oscillometric blood pressure measurements: differences between measured and calculated mean arterial pressure. The Netherlands Journal of Medicine 66(11) (2008)

Augmented Reality in Minimally Invasive Surgery

Lucio Tommaso De Paolis and Giovanni Aloisio

Department of Innovation Engineering, Salento University, Lecce, Italy

Abstract. In the last 15 years Minimally Invasive Surgery, with techniques such as laparoscopy or endoscopy, has become very important and research in this field is increasing since these techniques provide the surgeons with less invasive means of reaching the patient's internal anatomy and allow for entire procedures to be performed with only minimal trauma to the patient. The advantages of the use of this surgical method are evident for patients because the possible trauma is reduced, postoperative recovery is generally faster and there is less scarring. Despite the improvement in outcomes, indirect access to the operation area causes restricted vision, difficulty in hand-eye coordination, limited mobility handling instruments, two-dimensional imagery with a lack of detailed information and a limited visual field during the whole operation. The use of the emerging Augmented Reality technology shows the way forward by bringing the advantages of direct visualization (which you have in open surgery) back to minimally invasive surgery and increasing the physician's view of his surroundings with information gathered from patient medical images. Augmented Reality can avoid some drawbacks of Minimally Invasive Surgery and can provide opportunities for new medical treatments. After two decades of research into medical Augmented Reality, this technology is now advanced enough to meet the basic requirements for a large number of medical applications and it is feasible that medical AR applications will be accepted by physicians in order to evaluate their use and integration into the clinical workflow. Before seeing the systematic use of these technologies as support for minimally invasive surgery some improvements are still necessary in order to fully satisfy the requirements of operating physicians.

Keywords: Augmented Reality, biomedical images, Minimally Invasive Surgery.

1 Introduction

In recent years the latest technological developments in medical imaging acquisition and computer systems have permitted physicians to perform more sophisticated as well as less invasive treatments of patients.

One trend in surgery is the transition from open procedures to minimally invasive laparoscopic interventions, where visual feedback to the surgeon is only available through the laparoscope camera and direct palpation of organs is not possible. To successfully perform such sophisticated interventions, the provision of additional intra-operative feedback can be of great help to the surgeon.

These techniques mean a reduction in the amount of unnecessary damage to the patient, by enabling the physician to visualize aspects of the patient's anatomy and physiology without disrupting the intervening tissues. In particular, imaging methods

S.C. Mukhopadhyay and A. Lay-Ekuakille (Eds.): Adv. in Biomedical Sensing, LNEE 55, pp. 305–320.
springerlink.com

such as CT, MRI, and ultrasound scan make the safe guidance of instruments through the body possible without direct sight by the physician.

In addition, the availability of high-speed graphic workstations and medical virtual reality techniques has expanded the possibilities of medicine in the area of diagnosis, treatment, and education.

In traditional open surgery, surgeons often have to cut through many layers of healthy tissue to reach the target of interest, thereby inflicting significant damage on the tissue. This is very traumatic for the patient.

In the last 15 years Minimally Invasive Surgery (MIS), such as laparoscopy or endoscopy, has become very important and the research in this field is ever more widely accepted because these techniques provide surgeons with less invasive means of reaching the patient's internal anatomy and allow entire procedures to be performed with only minimal trauma to the patient [1].

The diseased area is reached by means of a small incision in the body, called ports, and specific instruments are used to gain access to the operation area. The surgical instruments are inserted through the ports using trocars and a camera is also inserted. During the operation a monitor shows what is happening inside the body. This is very different to what happens in open surgery, where there is full visual and touch access to the organ.

The idea of Minimally Invasive Surgery is to reduce the trauma for the patient by minimizing the incisions and the tissue retraction. Since the incision is kept as small as possible, the surgeon does not have direct vision and is thus guided by camera images. As a promising technique, the practice of MIS is becoming more and more widespread and is being adopted as an alternative to the classical procedure.

The advantages of the use of this surgical method are evident for the patients because the possible trauma is reduced, the postoperative recovery is nearly always faster and scarring is reduced.

Despite the improvement in outcomes, these techniques have their limitations and come at a cost to the surgeons. The view of the patient's organs is not as clear and the ability to manipulate the instruments is diminished in comparison with traditional open surgery. The indirect access to the operation area causes restricted vision, difficulty in hand-eye coordination, limited mobility in handling instruments and two-dimensional imagery with a lack of detailed information and a limited field of view during the whole operation. In particular, the lack of depth in perception and the difficulty in estimating the distance of the specific structures in laparoscopic surgery can impose limits on delicate dissection or suturing.

This situation, where eye-hand co-ordination is not based on direct vision, but more predominantly on image guidance via endoscopes, requires a different approach to conventional surgical procedures.

In Fig. 1 a cholecystectomy carried out in laparoscopic and open surgery is shown.

On the other hand, the quality of medical images and the speed with which they can be obtained, the increasing ability to produce 3-dimensional models and the advanced developments in Virtual Reality technology make it possible to localize the pathology accurately, to see the anatomic relationships like never before and to practice new methods such as surgical navigation or image-guided surgery.

Given that a great deal of the difficulties involved in MIS are related to perceptual disadvantages many research groups are now focusing on the development of surgical assistance systems, motivated by the benefits MIS can bring to patients. Advances in technology are making it more and more possible to develop systems which can help surgeons to perform their tasks in ways which are both faster and safer.

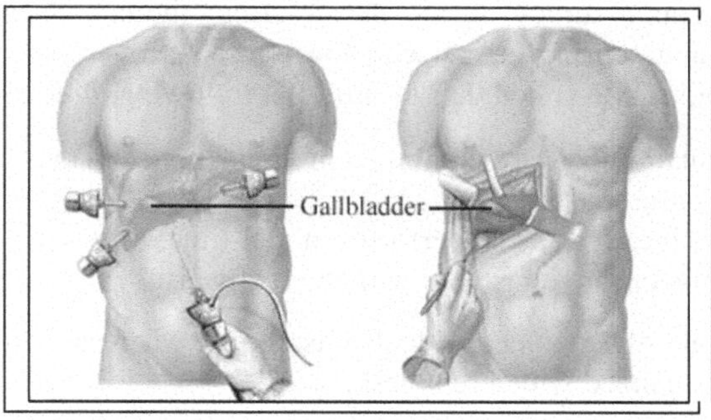

Fig. 1. Cholecystectomy in laparoscopic and open surgery

Appropriate visualization tools and techniques are playing an important role in providing detailed information regarding human organs and pathologies and realistic 3D models of the organs for the specific patient. The utilization of this visual information in combination with the operation techniques can help the surgeon during the surgical procedure and provide a possible solution to the problems associated with the practice of minimally invasive surgery.

In addition, the integration with Virtual Reality technology can change surgical preparation and surgeons may be able to practice and perform a surgical procedure before the patient arrives in the operating room; not only can complications be reduced, but individual components of the surgery can also be honed to precision.

The use of the emerging Augmented Reality technology shows the way forward in bringing the direct visualization advantage of open surgery back to minimally invasive surgery and can increase the physician's view of his/her surroundings with information gathered from patients' medical images.

Augmented Reality can avoid some drawbacks of MIS and can provide opportunities for new medical treatments.

2 Augmented Reality Systems and Technologies

2.1 Introduction to AR

Augmented Reality (AR) research aims to develop technologies that allow the real-time fusion of computer-generated digital content with the real world. With the help of Augmented Reality, a user can see hidden objects and, for this reason, AR enhances the users' perception and improves their interaction with the real world.

The virtual objects, displaying information that they cannot directly detect with their own senses, help them to perform real-world tasks better.

In contrast with Virtual Reality technology which completely immerses a user inside a synthetic environment and where he cannot see the real world around him,

Augmented Reality technology allows the user to see 3-dimensional virtual objects superimposed upon the real world.

Therefore, AR supplements reality, rather than completely replacing it. The user is under the impression that the virtual and real objects coexist in the same space.

Azuma [2] presents a survey of AR and describes the characteristics of AR systems, registration and sensing errors with the efforts to overcome these. Using the Azuma's definition, an AR system has to fulfil the following three characteristics:

- real and virtual objects are combined in a real environment, they appear to co-exist in the same space;
- the system is interactive and performs in real-time;
- the virtual objects are registered with the real world.

In Fig. 2 an example of Augmented Reality is shown where a virtual lamp and two virtual chairs are visualized with a real desk.

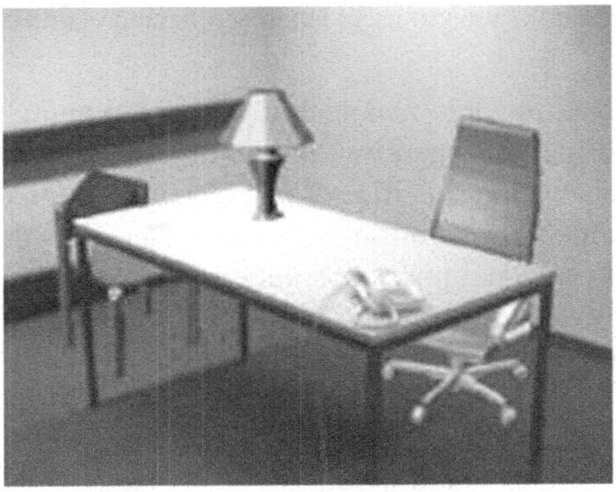

Fig. 2. Real desk with virtual lamp and two virtual chairs

Milgram and Kishino defined Mixed Reality as an environment "in which real world and virtual world objects are presented together within a single display, that is, anywhere between the extrema of the virtuality continuum" [3].

The Virtuality Continuum extends from the completely real through to the completely virtual environment with Augmented Reality and Augmented Virtuality ranging between.

Thus Augmented Reality is a mixture of reality and virtual reality and includes elements of both, virtual objects and real-world elements, where the surrounding environment is real.

In Fig. 3 Milgram's reality–virtuality continuum is shown.

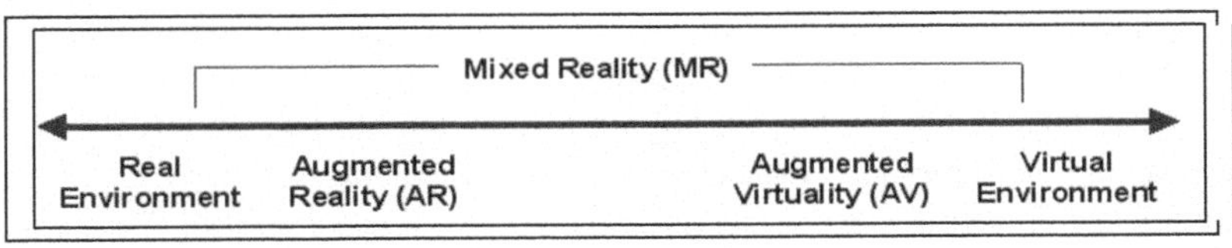

Fig. 3. Milgram's reality–virtuality continuum

Several research studies carried out have shown that Augmented Reality technology can be applied in a wide range of areas including education, medicine, engineering, military and entertainment.

It is possible to identify several research directions in:

- *tracking techniques*: how to achieve robust and accurate overlay of virtual imagery on the real world;
- *visualization technologies*: head mounted displays, handheld devices and projectors for AR;
- *interaction techniques*: methods for interaction with AR content;
- *novel AR applications* in fields which are not jet analyzed.

2.2 Tracking Systems

One of the most important tasks in developing Augmented Reality applications is to continuously determine the position and orientation of surgical instruments with regard to the patient's virtual organs and to estimate the physician's viewpoint.

For this reason tracking systems are integrated in the scene and attached to surgical instruments and to the patient's body.

AR applications require accurate knowledge of the relative positions of the camera and the scene; when either of them moves, it is necessary to keep track in real-time of all six degrees of freedom that define the camera position and orientation relative to the scene and the 3D displacements of the objects relative to the camera [4].

Many technologies have tried to achieve this goal. Typical tracking devices used in medical applications are mechanical, optical and electromagnetic systems.

Mechanical trackers are quite accurate, but the accuracy degrades with the length of the mechanical link; however, the mechanical link can be obstructive and the tracking volume is limited to the length of the mechanical linkage. Magnetic trackers are vulnerable to distortions by metal in the environment and limit the range of displacements. The optical trackers track both wired active tools with infra-red light-emitting diodes and wireless passive tools with reflective markers; the position sensor receives light from marker reflections or emissions and the system provides precise, real-time spatial measurements of the location and orientation of an object or tool within a defined coordinate system.

Computer vision technology has the potential to yield non-invasive and accurate solutions [5]. It is desirable to rely on naturally present features, such as edges, corners or texture, but this approach makes tracking much more challenging. In some case it requires the addition of fiducials, as special markers, to the scene or target objects to aid the registration task. This means that one or more fiducials are visible at all times and, if the markers are tracked, the virtual object will be blended in the real scene. However, in some applications it is not possible to place fiducials.

Planar square fiducials are used in ARToolKit, a video tracking library that calculates the real camera position and orientation relative to physical markers in real time [6]. ARToolKit software has become popular because it yields a robust, low-cost solution for real-time 3D tracking and it is publicly available.

In Fig. 4 an application in medicine based on ARToolkit is shown. The virtual environment is built using the real patient's CT images of the abdominal area and markers are used to overlap the virtual organs onto the real scene and to provide visual information which is not visible by means of normal senses [7].

Tracking technology has already entered operating rooms for medical navigation and provides the surgeon with important help to further enhance performance during the real surgical procedure.

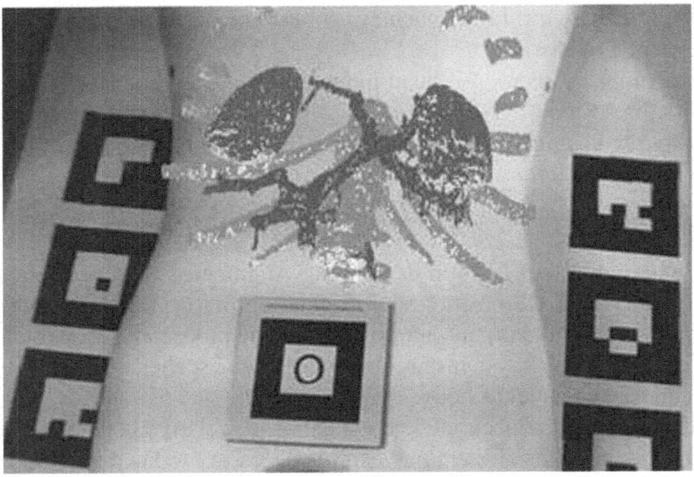

Fig. 4. An ARToolKit application in medicine (from [7])

The prevailing method in medical procedures is currently optical tracking using infrared light. The advantages of this kind of tracking are high accuracy and reliability; the use of infrared light can easily be explained by the fact that light conditions can be controlled for optimal measurements without disturbing human vision. Accuracy degradation is mainly caused by line-of-sight issues, which can be detected easily during measurement.

2.3 Visualization Devices and Modalities

Medical Augmented Reality takes its main motivation from the need of visualizing medical data and the patient within the same physical space. This would require real-time visualization of co-registered heterogeneous data and was probably the goal of many medical augmented reality solutions proposed in literature.

Augmented Reality systems often involve the use of a Head Mounted Display (HMD). A high resolution HMD is preferred for a dexterous manipulation task and is crucial in medical 3D visualization; stereoscopic view is also important for accurate operations.

There are mainly two types of see-through approaches in AR: optical and video [4].

With an optical see-through display, real and synthetic imagery is combined with a partial transmissive and reflective optical device and the synthetic imagery is overlaid on the real image. Advantages of optical see-through HMDs include a natural, instantaneous view of the real scene and simple and lightweight structures.

With a video see-through display, the real world imagery is first captured by a video camera then the captured image and the synthetic imagery are combined electronically and presented to the user.

Advantages of video see-through HMDs over optical see-through HMDs include pictorial consistency between the real and the synthetic view and the availability of a variety of image processing techniques. With appropriate vision-based tracking and synchronous processing of the captured and the rendered images, geometric and temporal consistencies can be accomplished.

Fig. 5 shows typical configurations of an optical and a video see-through display.

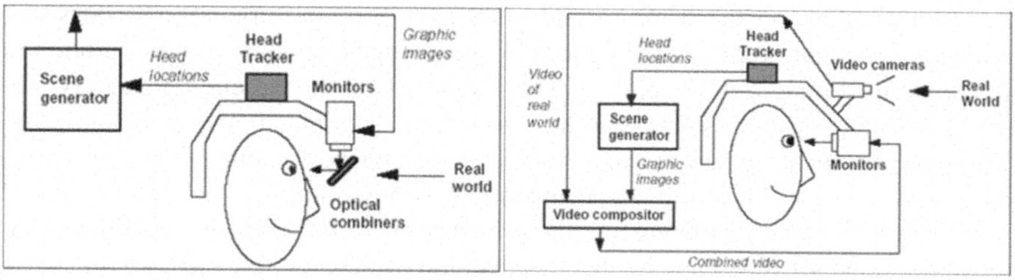

Fig. 5. Configurations of an optical and a video see-through display

Operating microscopes and operating binoculars can be augmented by inserting a semi-transparent mirror into the optics. The mirror reflects the virtual image into the optical path of the real images [8].

A drawback of augmented optics in comparison with other augmented technology is the process of merging real and computed images. As virtual images can only be added and may not entirely cover real ones, certain graphic effects cannot be realized.

It is possible to augment video images on ordinary monitors using an additional tracked video camera. As an advantage of augmented monitors, users need not wear an HMD or glasses.

Since endoscopy has been successfully introduced into many surgical disciplines, the use of augmented endoscopes is very interesting [8]. These devices require a tracking system for augmentation but, since the endoscopic setup already contains a camera, the integration of AR techniques does not necessarily introduce additional hardware into the workflow of navigated interventions.

Several research groups have investigated in order to find appropriate solutions, but for a helpful endoscopic augmentation the issues of calibration, tracking and visualization have not been completely solved.

In some applications the projection of the virtual images directly onto the patient can be used. These systems provide augmented vision without looking through additional devices such as glasses, HMDs, etc.

The simplicity of the system introduces certain limitations as a compromise, but this modality presents a beneficial feature when visualization on the skin rather than beneath it is required [8].

3 Building an AR Application for Surgery

The aim of an AR application for minimally invasive surgery is the development of a system that can help a surgeon to see in a non-invasive way the patient's internal anatomy during a minimally invasive surgical procedure.

To develop such a system, different technologies must be integrated:

- generation of the 3D model;
- calibration of the camera;
- registration;
- stereoscopic visualization and depth perception.

In order to obtain an AR environment which is as realistic as possible and, therefore, to provide information on the location and visualization of the organs, it is possible to visualize the internal organs of the patient by means of 3D models of the anatomy built from the medical images of the patient. The 3D models of the patient's organ have to be overlaid on the real patient's body and have to coincide with the real organs.

Through non-invasive sensors like Magnetic Resonance Imaging (MRI), Computed Tomography scans (CT) or ultrasound imaging it is possible to collect 3D datasets of a patient and an efficient 3D reconstruction of his/her anatomy can be provided in order to improve the standard slice view by the visualization of 3D models of the organs.

The geometric models are reconstructed by means of specific segmentation and classification algorithms in order to obtain information about the size and the shape of the human organs [9]. The grey levels in medical images are replaced by colours allocated to the different organs.

There are different software toolkits currently available for use in medicine for the visualization and analysis of scientific images and the 3D modelling of human organs; among these tools Mimics [10], 3D Slicer [11], ParaView [12] and OsiriX [13] play an important role.

Fig. 6 shows an example of the segmentation and classification results applied to CT images of the abdominal region of a human body.

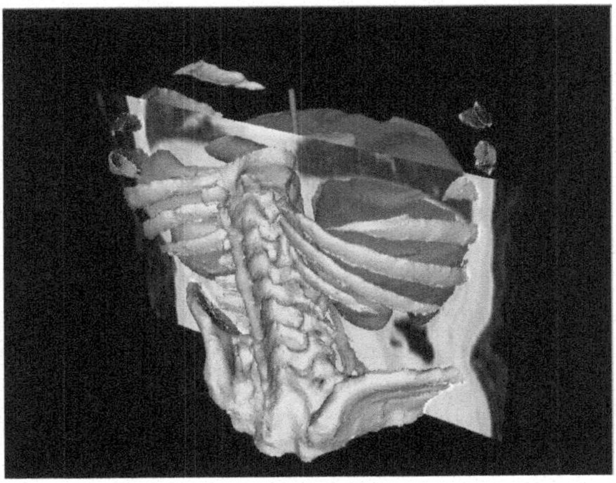

Fig. 6. A 3D model of the abdominal region obtained from CT images

In order to have an effective AR application, the real and computer generated organs must be accurately positioned relative to each other. For this reason it is necessary to carry out an accurate registration phase, which provides, as a result, the correct overlapping of the 3D model of the virtual organs on the real patient [14], [15], [16].

In medical applications it is very important to have correct detection and overlapping of the fiducial points because even a very slight error could have very serious consequences for the patient.

The integration of the registration algorithm into the surgical workflow requires a trade-off between complexity, accuracy and invasiveness. The process of registration can be computed using tracking data after an initial calibration step that provides the registration of a certain pose. For the pose determination of the real view, optical (infrared) tracking systems are currently the best choice; these devices are already in use in the modern operating rooms.

For the registration of patient data with the AR system it is possible to have a point-based registration approach where specific fiducials can be used that are fixed on the skin or implanted. These fiducials are touched with a tracked pointer and their positions have to match with the corresponding positions of fiducials placed during the patient scanning and segmented in the 3D model. Point-based registration is known to be a reliable solution if the set of fiducials is carefully chosen. The accuracy depends on the number of fiducials, the quality of measurement, and the spatial fiducial arrangement [8].

The simple augmentation of the real scene is not realistic enough because, although the organ positions are computed correctly, the relative position in depth of real and virtual images may not be perceived.

Indeed, in AR applications, although virtual objects have been correctly positioned in the scene, visually they are overlapped with all real objects, creating a situation which is not sufficiently realistic. This situation is shown in Fig. 7.

In particular, this effect is not acceptable for surgical AR applications and it is necessary, in addition to a proper positioning of the organs in the virtual scene, to ensure correct visualization.

Some solutions have been proposed [17], but the issue of a correct depth visualization remains partially unsolved.

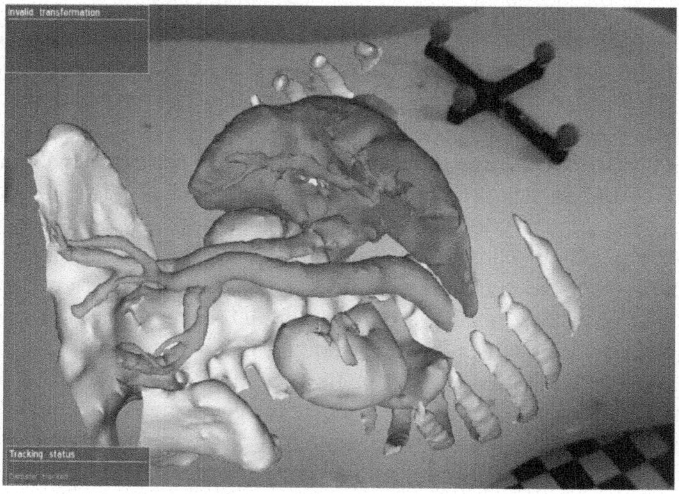

Fig. 7. Issue of the correct depth visualization

4 AR Applications in Minimally Invasive Surgery

Augmented Reality provides an intuitive human-computer interface and in surgery this technology makes it possible to overlay virtual medical images onto the patient, allowing surgeons to have a sort of "X-ray vision" of the body and providing a view of the patient's anatomy.

Augmented Reality technology has the potential to bring the visual advantages of open surgery back to minimally invasive surgery; increasing the physician's visual knowledge with information gathered from patients' medical images.

The patient becomes transparent and this virtual transparency will therefore make it possible to find tumours or vessels not by locating them thanks to touch, but simply by visualizing them thanks to augmented reality.

The virtual information could be directly displayed on the patient's body or visualized on an AR surgical interface, showing where the operation should be performed.

For instance, a physician might also be able to see the exact location of a lesion on a patient's liver or where to drill a hole into the skull for brain surgery or where to perform a needle biopsy of a tiny tumour.

To successfully perform minimally invasive interventions, highly trained and experienced specialists are required.

In general, AR technology in minimally invasive surgery may be used for:

- training purposes;
- pre-operative planning;
- advanced visualization during the real procedure.

Several research groups are exploring the use of AR in surgery and many image-guided surgery systems have been developed.

Devernay et al.[18] propose the use of an endoscopic AR system for robotically assisted minimally invasive cardiac surgery. One of the problems closely linked to endoscopic surgery is the fact that, because of the narrow field of view, it is sometimes quite difficult to locate the objects that can be seen through the endoscope. This is especially true in cardiac surgery, where it is difficult not to confuse two coronary arteries on a beating heart. The narrow field of view of the endoscope may lead to misidentifying the coronary or the position of the stenosis on the coronary. The information coming from the 3D anatomical model of the patient extracted from MRI or CT-scan and the position of the endoscope with respect to the patient are not sufficient since the organs (in particular the lungs and the heart) are displaced by the inflated gas. They propose a methodology to achieve coronary localization by Augmented Reality on a robotized stereoscopic endoscope adding "cartographic" information on the endoscopic view, by indicating the position of the coronaries with respect to the field of view. The proposed method involves five steps: making a time-variant 3D model of the beating heart using coronarography and CT-scan or MRI, calibrating the stereoscopic endoscope, reconstructing the 3D operating field, registering the operating field surface with the 3D heart model and adding information on the endoscopic images using Augmented Reality.

Samset et al. [21] present tools based on novel concepts in visualization, robotics and haptics providing tailored solutions for a range of clinical applications. Examples from radio-frequency ablation of liver-tumours, laparoscopic liver surgery and minimally invasive cardiac surgery will be presented.

Demonstrators were developed with the aim of providing a seamless workflow for the clinical user conducting image-guided therapy. The presented solutions are the results of the multidisciplinary ARIS*ER project.

Bichlmeier et al. [19] focus on handling the problem of misleading perception of depth and spatial layout in medical AR and present a new method for medical in-situ visualization that allows for improved perception of 3D medical imaging data and navigated surgical instruments relative to the patient's anatomy.

They describe a technique to modify the transparency of video images recorded by the colour cameras of a video see-through HMD. The transparency of the video images depends on the topology of the skin surface of the patient and the viewing geometry of the observer and the modified video image of the real scene is then blended with the previously rendered virtual anatomy. The presented method allows for an intuitive view on the deep-seated anatomy of the patient providing visual cues to correctly perceive absolute and relative distances of objects within an AR scene.

In addition, they describe a method for integrating surgical tools into the medical AR scene resulting in improved navigation. The effectiveness has been demonstrated in a series of experiments at the Chirurgische Klinik in Munich, Germany with a cadaver study and a thorax phantom, both visualizing the anatomical region around the spinal column, and an in-vivo study visualizing the head.

The results can be applied for designing medical AR training and educational applications. Fig. 8 shows an application of the developed method. The medical AR scene is presented to the observer using an "AR window" [20].

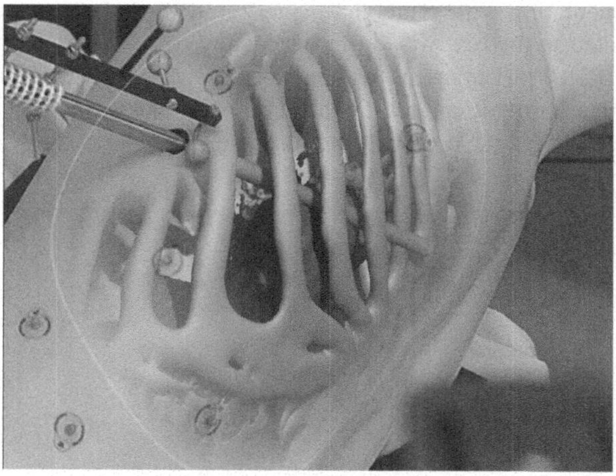

Fig. 8. The medical AR scene using an "AR window" (from [20])

Navab et al. [22] introduce an interaction and 3D visualization paradigm, which presents a new solution for using 3D virtual data in many AR medical applications. The problem becomes more evident when a single camera is used in augmented laparoscopic surgery. When augmenting a monoscopic laparoscope, which is the usual case, the 3D volume is projected onto the laparoscope's image plane, so one dimension is totally lost, leading to even more limited perception of 3D shape and depth during superimposition. However, particularly for interventions targeting the inside of organs, shape information is crucial, for instance for identifying blood vessels to be clipped during liver resection.

To recover this lost shape information they introduce the concept of a laparoscopic virtual mirror: a virtual reflection plane within the live laparoscopic video, which is able to visualize a reflected side view of the organ and its interior. The Laparoscopic Virtual Mirror is able to virtually reflect the 3D volume as well as the laparoscope or any other modelled and tracked instruments. This enables the surgeon to observe the 3D structure of, for example, blood vessels by moving the virtual mirror within the augmented monocular view of the laparoscope.

By combining this visualization paradigm with a registration-free augmentation system for laparoscopic surgery, a powerful medical augmented reality system becomes possible, which could make such minimally invasive surgeries easier and safer to perform.

To demonstrate the full advantage of this new AR interaction paradigm, the system is integrated into a medical application, which was desperately in need of such interactive visualization. Fig. 9 shows the Laparoscopic Virtual Mirror used in an experimental setup.

A clinical evaluation investigating the perceptive advantage of a virtual mirror integrated into a laparoscopic AR scenario has been carried out [23].

Kalkofen et al. [24] carefully overlay synthetic data on top of the real world imagery by taking into account the information that is about to be occluded by augmentations as well as the visual complexity of the computer-generated augmentations added to the view.

Careless augmentation with synthetic imagery may occlude extremely relevant information presented in the real world imagery. They solve the problem of augmentations occluding useful real imagery, with edges extracted from the real video stream. The extracted edges provide an additional depth cue and since they come from the real imagery, they are also able to preserve important landmarks.

Fig. 9. Laparoscopic Virtual Mirror in an experimental setup (from [23])

De Paolis et al. [25] present an Augmented Reality system that can guide the surgeon in the operating phase in order to prevent erroneous disruption of some organs during surgical procedures.

Since the simple augmentation of the real scene cannot provide information on the depth, a sliding window is provided in order to allow the occlusion of part of the organs and to obtain a more realistic impression that the virtual organs are inside the patient's body. It is possible to slide the visualization window and to locate it in a precise position which provides a view of the organs of interest; only through this window can the internal organs be seen.

In addition, distance information is provided to the surgeon and an informative box is shown in the screen in order to visualize the distance between the surgical instrument and the organ concerned. When the distance between the surgical instrument and some specified organs is under a safety threshold, a video feedback as well as an audio feedback in the form of an impulse, the frequency of which increases as the distance decreases between the surgical instrument and the organ concerned, are provided.

Fig. 10 shows the visualization of the organs with the box reporting the distance information.

In minimally Invasive Surgery a new original technique, called Natural Orifice Transluminal Endoscopic Surgery (NOTES), could replace traditional laparoscopic surgery for a large set of procedures. By replacing the rigid optic that is introduced through the skin by a flexible optic that is introduced through a natural orifice such as stomach, vagina or colon, this technique should eliminate all visible incisions.

On the other hand, such minimally invasive techniques present new difficulties for surgeons, such as a loss of their gesture capacity due to the length of surgical instruments and the gesture complexity due to the loss of orientation and inversion of movement because of the flexibility of the endoscope.

Such difficulties can be solved thanks to AR technology combined with instrument tracking that can provide information about the location and internal orientation of surgical instruments.

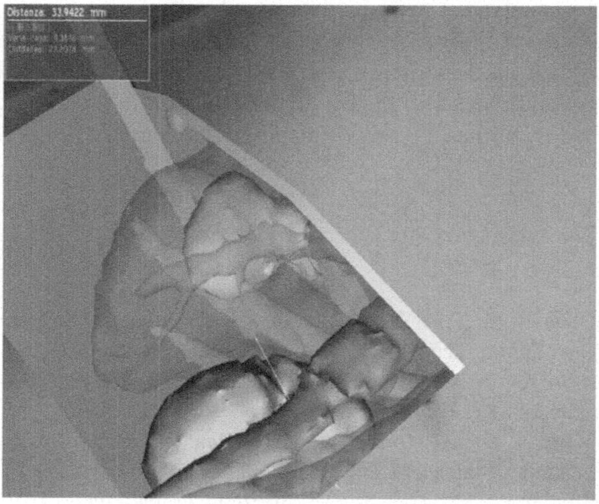

Fig. 10. Visualization of the organs with the distance information box (from [25])

Soler et al. [26] present the results of their research into the application of AR technology in laparoscopic and NOTES procedures. They have developed two kinds of AR software tools (Interactive Augmented Reality and Fully Automatic Augmented Reality) taking into account a predictive deformation of organs and tissues during the breathing cycle of the patient. A preclinical validation has been performed on pigs and results are very encouraging and represent the first phase for surgical gesture automation that will make it possible to reduce surgical mistakes.

5 Conclusions and Future Work

Minimally Invasive Therapy (MIT) is a major step forward in interventionist therapy, capable of offering quality of life to patients and decreasing costs for health care systems, the two most important considerations as well as the future of modern medicine.

This new approach, however, also brings limitations to surgeons that can only be compensated for by means of the massive use of innovative technologies in order to have MIT widespread and optimally implemented.

The challenge is related to the actions necessary to bridge the gap between the new surgical methods which are already available and the new emerging technologies, like Virtual Reality and Augmented Reality, which can provide improvements and benefits in the practice of these advanced medical treatments.

After two decades of research on medical Augmented Reality, these enabling technologies are now advanced enough to meet the basic requirements for a large number of medical applications. It is feasible that medical AR applications could be accepted by physicians in order to evaluate their use and integration into the clinical workflow.

Of course, before AR technologies can be used systematically as support for minimally invasive surgery, some improvements are still necessary in order to fully satisfy the requirements of operating physicians. For instance, a perfect medical AR user interface would be integrated in such a way that the user would not notice its existence while taking full advantage of additional in situ information it provides. Also the visualization systems still need hardware and software improvement in order to allow the surgeons to take full advantage of the augmented virtual data.

It seems likely that the superimposition of data acquired by other emerging intraoperative imaging modalities will have a great impact on future surgical interventions and will be applied to further advanced image-guided surgery in the operating rooms of the future.

References

1. Harrell, A.G., Heniford, T.B.: Minimally Invasive Abdominal Surgery: Lux et Veritas Past, Present, and Future. The American Journal of Surgery 190, 239–243 (2005)
2. Azuma, R.: A Survey of Augmented Reality. Presence: Tele-operators and Virtual Environments 4(6), 355–385 (1997)
3. Milgram, P., Kishino, F.: A Taxonomy of Mixed Reality Visual Displays. IEICE Transactions on Information Systems E77-D(12), 1321–1329 (1994)
4. Haller, M., Billinghurst, M., Thomas, B.: Emerging Technologies of Augmented Reality: Interfaces and Design. Idea Group Publishing (2007)

5. Rolland, J.P., Davis, L., Baillot, Y.: A Survey of Tracking Technology for Virtual Environments. In: Barfield, M., Caudell, T. (eds.) Fundamentals of Wearable Computers and Augmented Reality, pp. 67–112. Lawrence Erlbaum, Mahwah (2001)
6. Kato, H., Billinghurst, M., Poupyrev, I., Imamoto, K., Tachibana, K.: Virtual Object Manipulation on a Table-Top AR Environment. In: Proc. International Symposium on Augmented Reality (ISAR 2000), Munich, Germany, pp. 111–119 (2000)
7. De Paolis, L.T., Pulimeno, M., Aloisio, G.: An Augmented Reality Application for Minimally Invasive Surgery. In: 14th Nordic–Baltic Conference on Biomedical Engineering and Medical Physics (NBC-14), Riga, Latvia, pp. 489–492. Springer, Heidelberg (2008)
8. Sielhorst, T., Feuerstein, M., Navab, N.: Advanced Medical Displays: A Literature Review of Augmented Reality. IEEE/OSA Journal of Display Technology, Special Issue on Medical Displays 4(4), 451–467 (2008)
9. Laugier, C., D'Aulignac, D., Boux de Casson, F.: Modeling Human Tissues for Medical Simulators. In: IEEE International Conference on Intelligent Robots and Systems (IROS), Japan (2000)
10. Mimics Medical Imaging Software, Materialise Group, `http://www.materialise.com/materialise/view/en/92458-Mimics.html`
11. 3D Slicer, `http://www.slicer.org`
12. Ahrens, J., Geveci, B., Law, C.: ParaView: an End-User Tool for Large Data Visualization. In: Hansen, C.D., Johnson, C.R. (eds.) Visualization Handbook. Elsevier, Amsterdam (2005)
13. Faha, O.: Osirix: an Open Source Platform for Advanced Multimodality Medical Imaging. In: 4th International Conference on Information & Communications Technology, Cairo, Egypt, pp. 1–2 (2006)
14. Maintz, J.B.A., Viergever, M.A.: A survey of medical image registration. Medical Image Analysis 2, 1–36 (1998)
15. Sauer, F.: Image Registration: Enabling Technology for Image Guided Surgery and Therapy. In: 2005 IEEE Engineering in Medicine and Biology, Shanghai, China (2005)
16. Feuerstein, M., Wildhirt, S.M., Bauernschmitt, R., Navab, N.: Automatic Patient Registration for Port Placement in Minimally Invasixe Endoscopic Surgery. In: Duncan, J.S., Gerig, G. (eds.) MICCAI 2005. LNCS, vol. 3750, pp. 287–294. Springer, Heidelberg (2005)
17. Bichlmeier, C., Navab, N.: Virtual Window for Improved Depth Perception in Medical AR. In: International Workshop on Augmented Reality environments for Medical Imaging and Computer-aided Surgery (AMI-ARCS), Copenhagen, Denmark (2006)
18. Devernay, F., Mourgues, F., Coste-Manière, E.: Towards Endoscopic Augmented Reality for Robotically Assisted Minimally Invasive Cardiac Surgery. In: IEEE International Workshop on Medical Imaging and Augmented Reality, pp. 16–20 (2001)
19. Bichlmeier, C., Wimmer, F., Michael, H.S., Nassir, N.: Contextual Anatomic Mimesis: Hybrid In-Situ Visualization Method for Improving Multi-Sensory Depth Perception in Medical Augmented Reality. In: Sixth IEEE and ACM International Symposium on Mixed and Augmented Reality (ISMAR 2007), pp. 129–138 (2007)
20. Bichlmeier, C., Navab, N.: Virtual Window for Improved Depth Perception in Medical AR. In: International Workshop on Augmented Reality environments for Medical Imaging and Computer-aided Surgery (AMI-ARCS 2006), Copenhagen, Denmark (2006)
21. Samset, E., Schmalstieg, D., Vander Sloten, J., Freudenthal, A., Declerck, J., Casciaro, S., Rideng, Ø., Gersak, B.: Augmented Reality in Surgical Procedures. In: SPIE Human Vision and Electronic Imaging XIII, vol. 6806, pp. 68060K.1-68060K.12 (2008)

22. Navab, N., Feuerstein, M., Bichlmeier, C.: Laparoscopic Virtual Mirror - New Interaction Paradigm for Monitor Based Augmented Reality. In: IEEE Virtual Reality Conference 2007 (VR 2007), Charlotte, North Carolina, USA, pp. 10–14 (2007)
23. Bichlmeier, C., Heining, S.M., Rustaee, M., Navab, N.: Laparoscopic Virtual Mirror for Understanding Vessel Structure: Evaluation Study by Twelve Surgeons. In: 6th IEEE International Symposium on Mixed and Augmented Reality (ISMAR 2007), Nara, Japan (2007)
24. Kalkofen, D., Mendez, E., Schmalstieg, D.: Interactive Focus and Context Visualization in Augmented Reality. In: 6th IEEE International Symposium on Mixed and Augmented Reality (ISMAR 2007), Nara, Japan, pp. 191–200 (2007)
25. De Paolis, L.T., Pulimeno, M., Lapresa, M., Perrone, A., Aloisio, G.: Advanced Visualization System Based on Distance Measurement for an Accurate Laparoscopy Surgery. In: Joint Virtual Reality Conference of EGVE - ICAT - EuroVR, Lyon, France (submitted, 2009)
26. Soler, L., Nicolau, S., Fasquel, J.-B., Agnus, V., Charnoz, A., Hostettler, A., Moreau, J., Forest, C., Mutter, D., Marescaux, J.: Virtual Reality and Augmented Reality Applied to Laparoscopic and NOTES Procedures. In: IEEE 5th International Symposium on Biomedical Imaging: from Nano to Macro, pp. 1399–1402 (2008)

Advances in EEG Signal Processing for Epilepsy Detection

Aimé Lay-Ekuakille[1], Amerigo Trotta[2], Antonio Trabacca[3], and Marta De Rinaldis[3]

[1] D.I.I.-University of Salento, Italy
[2] D.E.E.-Polytechnic of Bari, Italy
[3] Scientific Institute for Research, Hospitalization and Health Care "E. Medea", Ostuni, Italy

Abstract. Epilepsy is one of the most common neurological disorders, affecting around 1 in 200 of the population. However, identifying epilepsy can be difficult because seizures tend to be relatively infrequent events and an electroencephalogram (EEG) does not always show abnormalities. The aim of this project is to develop a new methods that could improve the diagnosis of epilepsy, leading to earlier treatment and to a better quality of life for epileptic patients. The above methods must be composed with a flexible hardware development in order to discriminate noise and bad signals from correct EEG, MEG (Magnetoencephalogram), Eye Image recognition, Somnography and DTI (Diffusion Tensor Imaging). Even if there are EEG signal classifiers, it is suitable to perform a correct signal processing according to particular clinical reference, that is, it is difficult to have a classifier for all circumstances but it is possible to adapt EEG processing on current patient. Preliminary results are described for processing biomedical signals, namely EEG signals, in order to train the adaptive filtering in recognizing and choosing correct frequencies at which it is possible to reduce noise.

Keywords: Epilepsy detection, signal processing, EEG, beamforming.

1 Introduction

Epilepsy is defined as the recurrent paroxysmal transient disturbance of brain function due to disturbance of electrical activity in the brain, where the disturbance is unrelated to infection or acute cerebral insult. The disturbances may be manifested as episodic impairment or loss of consciousness, abnormal motor phenomena, psychic or sensory disturbances, or perturbation of the autonomic nervous system. Causes: In about 60% of cases (in developed countries) there is no known cause. Of the remaining 40%, the following are the most frequent: Head trauma, especially from automobile accidents, gunshot wounds, sports accidents, falls and blows at work or in the home; the more severe the injury, the greater the risk of developing epilepsy - Brain tumor and stroke - Poisoning, such as lead poisoning. Many people each year are reported to suffer from seizures caused by alcoholism and pollution - Infection, such as meningitis, viral encephalitis, lupus erythmeatosus and, less frequently, mumps, measles, diptheria and

S.C. Mukhopadhyay and A. Lay-Ekuakille (Eds.): Adv. in Biomedical Sensing, LNEE 55, pp. 321–334.
springerlink.com

others - Maternal injury, infection or systemic illness affecting the developing brain of the fetus during pregnancy. Patient behaviour can be studied by means of electric potential measurements, blood flux in a specific cerebral region (rCBF–regional Cerebral Blood Flux) and metabolism. The human electroencephalography (EEG) can commonly help to diagnose conditions such as brain tumors, brain injury, cerebral palsy, stroke, liver, epilepsy, and can also facilitate neurologists to discover medical issues concerning headaches, weakness, blackout, or dizziness [1] The vast majority of EEG acquisition [2], for clinical studies, is made from electrodes glued onto standard locations on the scalp, as depicted in fig.1. These electrodes average the action potentials from large numbers of cells, and therefore do not provide action potentials from single cells [3].

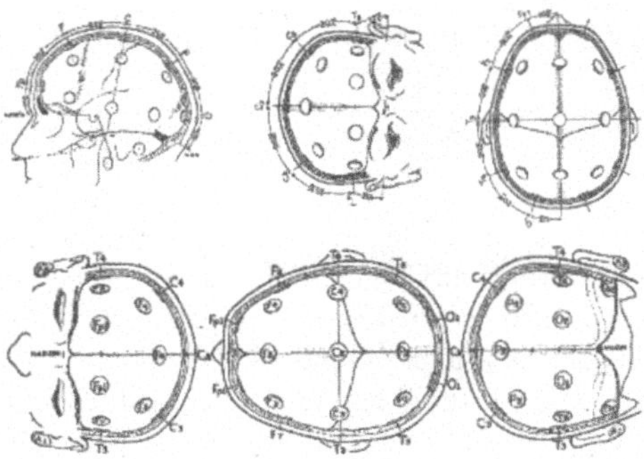

Fig. 1. The 10-20 electrode system for measuring the EEG

EEG signal includes many different components and especially four principal ones as indicated in fig.2. α wave is the most significant having the frequency in the range of 8-12 Hz. Lower frequency components, less than 8 Hz, are called δ (under 4 Hz) and θ (4-8 Hz) respectively. Upper frequency components, greater than 12 Hz, are named α. Fig. 3 illustrates different kinds of epilepsy according to appropriate representation of wave amplitudes.

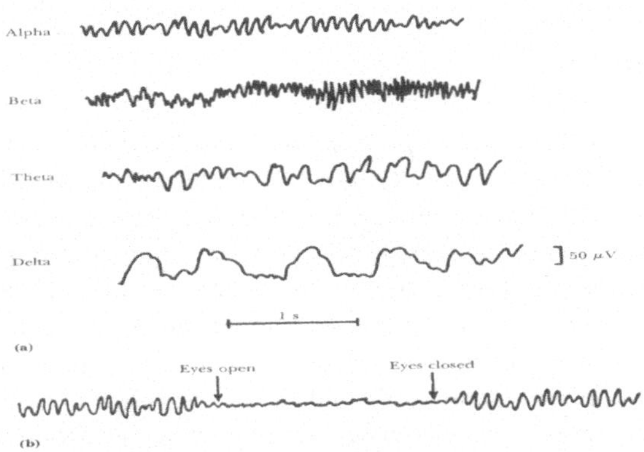

Fig. 2. (a) Different types of normal EEG waves (b) Replacement of alpha rhythm by an asynchronous discharge when patient opens eye

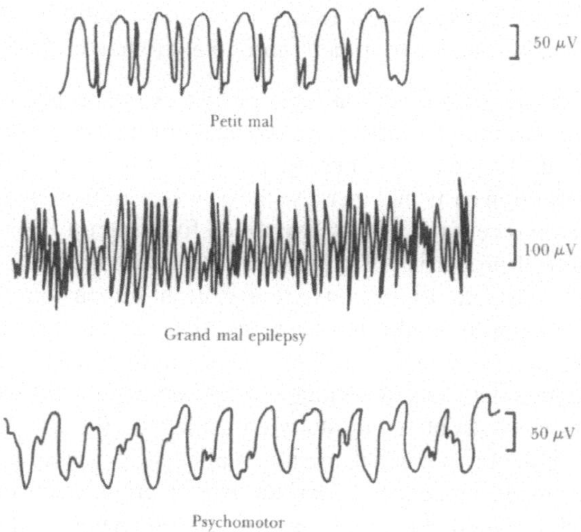

Fig. 3. Representative abnormal EEG waveforms in different types of epilepsy

2 Epilepsy Detection

Epilepsy is defined, as aforementioned, as the recurrent paroxysmal transient disturbance of brain function due to disturbance of electrical activity in the brain, where the disturbance is unrelated to infection or acute cerebral insult. The disturbances may be manifested as episodic impairment or loss of consciousness, abnormal motor phenomena, psychic or sensory disturbances, or perturbation of the autonomic nervous system. Causes: In about 60% of cases (in developed countries) there is no known cause. Of the remaining 40%, the following are the most frequent: Head trauma, especially from automobile accidents, gunshot wounds, sports accidents, falls and blows at work or in the home; the more severe the injury, the greater the risk of developing epilepsy - Brain tumor and stroke - Poisoning, such as lead poisoning. Many people each year are reported to suffer seizures caused by alcoholism and pollution - Infection, such as meningitis, viral encephalitis, lupus erythmeatosus and, less frequently, mumps, measles, diphtheria and others - Maternal injury, infection or systemic illness affecting the developing brain of the fetus during pregnancy. In facing epilepsy seizures, a basic approach regards its detection in neonates [4-7]. Some authors [8] have done pioneering work in video electroencephalographic (EEG) monitoring of neonatal seizures, which has contributed to the improved ability to detect seizures early in the course of disease and differentiate these subtle epileptic attacks in neonates from other pathology. It is particularly important to predict the onset of a seizure disorder in neonates and to anticipate the recurrence by applying new techniques to a predefined population. Methods currently available for routine seizure detection include:

- Automated seizure detection, such as video-EEG
- Standardized training for clinical observation involving the recognition of clinical seizure activity in the neonate

– Analysis of the background EEG, which can be helpful since neonates with abnormal background activity are more likely to have seizures.

These approaches, however, are very labor intensive. An important discovery made by Dr. Mizrahi's group was that the increasing sharpness of the sharp wave was a predictor for the future risk of seizures. Advantages for using this methodology would be to predict a seizure condition early and increase the accuracy of that prediction.

Different and even recent researches have been focusing on EEG signal processing techniques. Dynamical properties of brain electrical activities [9], using a non linear prediction and an estimate of effective correlation dimension together with the method of iterative amplitude adjusted surrogate date, are analyzed for different sets of EEG signals in diverse conditions. Preseizure prediction [10] can be traced out from prolonged human data sets to depict several behaviours with temporal lobe epilepsy. It allows to open frontiers in studying epilepsy. To complete this scientific promenade about epileptic seizure detection, statistical processing [11] plays a particular role in putting in evidence univariate and bivariate measurements, and their performance in order to predict the existence of preictal changes. These changes could help, whether a mathematical correlation [12] is made with video and/or EEG evaluation, neurologists in detecting seizures in patients suffering from the unpredictability of epileptic disorders.

In last years, higher order statistics (HOS) have been finding a wide applicability in many different fields, e.g. biomedicine, harmonic retrieval and adaptive filtering. In power spectrum estimation, the signal under consideration is processed in such a way that the distribution of power among its frequency is estimated and phase relations between the frequency components are suppressed. For EEG applications, as faced in this research, where it is necessary to preserve information whatever it located, higher order statistics and their associated Fourier transforms reveal not only amplitude information about a signal, but also phase information. If a non-Gaussian signal is received along with additive Gaussian noise, a transformation to higher order cumulant domain removes the noise. These are some methods for estimation of signal components, based on HOS.

Higher Order Statistics

The definitions emphasize the 2nd-, 3rd- and 4th order statistics and their respective Fourier transforms: power spectrum, bispectrum, and trispectrum.

Let $\{X(k)\}$, k=0, ±1, ±2, ±3,... be a real stationary discrete-time signal and its moments up to order n exist, hence

$$m_n^x(\tau_1,\tau_2,....\tau_{n-1}) \triangleq E\{X(k)\,X(k+\tau_1)...X(k+\tau_{n-1})\} \tag{1}$$

represents the nth-order moment function of the stationary signal, which depends only on the time differences $\tau 1, \tau 2, \ldots \tau n\text{-}1$, $\tau i = 0, \pm 1, \ldots$ for all i. Obviously, the 2nd-order moment function, $m_2^x(\tau_1)$ is the autocorrelation of $\{X(k)\}$ whereas $m_3^x(\tau_1,\tau_2)$ and $m_4^x(\tau_1,\tau_2,\tau_3)$ are the 3rd- and the 4th-order moments, respectively. $E\{\bullet\}$ denotes statistical expectation. The nth-order cumulant function of a non-Gaussian stationary random signal X(k) can be written as (for n=3,4 only):

$$c_n^x(\tau_1,\tau_2,....\tau_{n-1}) = m_n^x(\tau_1,\tau_2,....\tau_{n-1}) - m_n^G(\tau_1,\tau_2,....\tau_{n-1}) \quad (2)$$

where $m_n^x(\tau_1,\tau_2,....\tau_{n-1})$ is the nth-order moment function of $\{X(k)\}$ and $m_n^G(\tau_1,\tau_2,....\tau_{n-1})$ is the nth-order moment function of an equivalent Gaussian signal that has the same mean value and autocorrelation sequence as $\{X(k)\}$. Clearly, if is Gaussian $m_n^x(\tau_1,\tau_2,....\tau_{n-1}) - m_n^G(\tau_1,\tau_2,....\tau_{n-1})$ and hence $c_n^x(\tau_1,\tau_2,....\tau_{n-1}) = 0$. Note, however, that although Eq.2 is only true for orders n=3 and 4, it is equal to zero for all n if $\{X(k)\}$ is Gaussian. Since the dimension of the moments increases with its order, some cumulants for zero-mean-processes are defined by expanding Eq. 2. The power spectrum of the signal is defined by discrete Fourier transform (DFT):

$$P_2^X(f) = DFT\left(c_2^X(m)\right) = \sum_{m=-\infty}^{\infty} c_2^X(m)e^{-j2\pi mf} \quad (3)$$

Consequently, the spectrum of the 3rd order cumulant, the bispectrum, is:

$$B^X(f_1,f_2) = \sum_{m=-\infty}^{\infty}\sum_{n=-\infty}^{\infty} c_3^X(m,n)\, e^{-j2\pi(mf_1+nf_2)} \quad (4)$$

A further important parameter is the bicoherence, which is given by

$$Bic(f_1,f_2) = \frac{B(f_1,f_2)}{\sqrt{P(f_1)P(f_2)P(f_1+f_2)}} \quad (5)$$

HOS analysis offers diverse advantages, among them it is suitable to find the extra information provided the analysis in order to better estimate parameters and sheds light on nonlinearities in the source of the signal.

A brief description of HOS for EEG signals is provided in this paragraph by recalling a specific research [13]. The authors used the results of HOS analysis to feed a Gaussian mixture model (GMM) in order to investigate the features. An EEG database from Bonn University [14] has been used for the purposes of this research as well as a database from local hospital. In another previous work [15], it was possible to use EEG signals without GMM. For every procedure using HOS, it is necessary to calculate diverse indicators as follows after having computed the bispectrum Ω, that is, the main domain included in the area depicted in fig.4 from which all parameters are calculated.

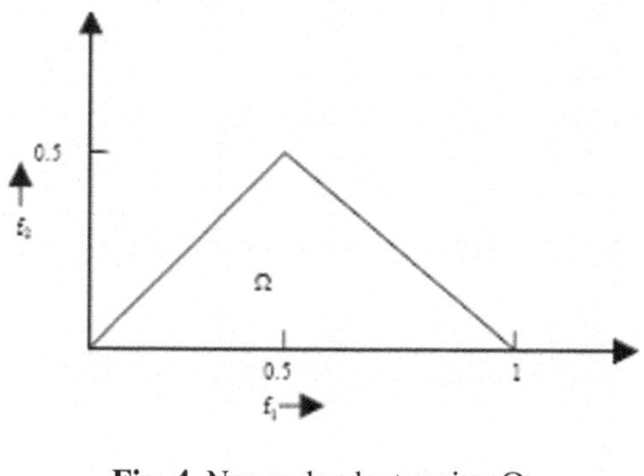

Fig. 4. Non-redundant region Ω

Spectral magnitude mean for PSD is given by:

$$M_{ave} = \frac{1}{N}\sum_{k=0}^{N-1}|X_k| \tag{6}$$

while the spectral magnitude for HOS is explained the relationship below

$$M_{ave} = \frac{1}{L}\sum_{\Omega}|B(f_1, f_2)| \tag{7}$$

In the Eq.6 and Eq.7, X_k stands for the input signal DFT and B is the signal bispectrum. Other interesting parameters to be used to represent the features are based on the entropy of the signal, that are,

Entropy 1

$$P_1 = -\sum_k p_k \log p_k \tag{8}$$

in which for PSD, $p_k = \frac{|X_k|}{\sum_{k=1}^{N=1}|X_k|}$ and for HOS, $P_k = \frac{|B(f_1, f_2)|}{\sum_{\Omega}|B(f_1, f_2)|}$

likewise, entropy 2

$$P_2 = -\sum_i q_i \log q_i \tag{9}$$

in which for PSD, $q_i = \frac{|X_k|^2}{\sum_k |X_k|^2}$ and for HOS, $q_i = \frac{|B(f_1, f_2)|^2}{\sum_{\Omega}|B(f_1, f_2)|^2}$

Beamforming Approach

Consider an array of sensors or 'Smart Antenna' collecting spatial samples of propagating wave fields. Signals may be present from any given direction and always in the presence of noise. The desired signal is from a particular spatial location. From a signal processing perspective, the main objective is to detect the signal arriving [16] from a particular look direction and cancel out any interfering signals and noise. A sensor array receives incoming signal information; a beamformer processes the spatial samples collected to provide the required spatial filtering. The beamformer output is formed by applying a complex weights vector, *w*, to the *N* incoming signals received from the antenna array and summing the result, as shown in fig 5. The beamformer output, *y(t)* to a signal approaching from direction [17] is expressed as

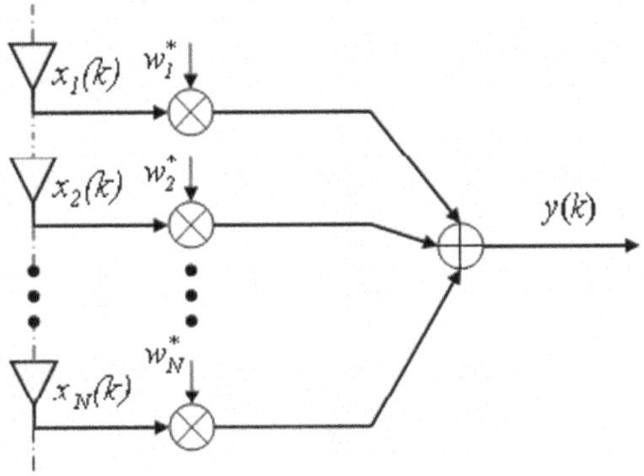

Fig. 5. Beamforming basics

$$y(t) = w^H x(t) \tag{10}$$

where

$$w = [w_1\ w_2\ \cdots\ w_N]^T$$
$$x(t) = [x_1(t)\ x_2(t)\ \cdots\ x_N(t)]^T$$

where *x(t)* is defined as a function of the received signal and the steering vector $f(\theta)$. The power response, *F* of the beamformer is defined as

$$F = |y(t)|^2 \tag{11}$$

The use of linear constraints in beamforming is to provide main-lobe protection in the case of a main-lobe interferer and to avoid performance degradation [18] in the presence of DOA (direction of arrival) mismatch and/or array perturbations. The array gain is a key issue for this research. The relationships for the array gain of LCMV (Linear Constrained Minimum Variance Beamforming) and LCMP (Linear Constrained Minimum Power)can be easily derived. For the LCMP beamformer [19], the weight *w* is given by

$$w_{lcmp}^H = g^H \left[C^H S_x^{-1} C\right]^{-1} C^H S_x^{-1} \tag{12}$$

where *g* is the gain vector, *C* is the subspace of constraints, *S* is the spectral matrix [20]. The output signal power is

$$P_s = \sigma_s^2 \left|w^H v_s\right|^2 \tag{13}$$

in which v_s is a steering vector; instead, the output noise power is

$$P_n = \sigma_n^2 \left|w^H \rho_n w\right|^2 \tag{14}$$

where σ^2 is the variance and ρ_n is the normalized spatial spectral matrix and includes both the white noise and any additional interference. Hence the gain is

$$A_0 = \frac{\left|w^H v_s\right|^2}{\left|w^H \rho_n w\right|} \tag{15}$$

If a distortionless constraint is included, then the numerator of the previous equation is unity and the array gain is

$$A_0 = \left|w^H \rho_n w\right|^{-1} \tag{16}$$

The output SNR_0 is

$$SNR_{output} = \left(\frac{\sigma_s^2}{\sigma_n^2}\right) A_{optimum} \tag{17}$$

The array gain for the LCMV beamformer, according to common literature is

$$A_{lcmv} = \frac{1}{g^H \left[C^H \rho_n^{-1} C\right]^{-1} g} \tag{18}$$

while to derive the array gain for LCMP it is suitable to use the optimum processor expression defined by Frost in [18] and put it in Eq.18 in order to get the following expression:

$$A_{lcmp} = \left\{g^H \left[C^H S_x^{-1} C\right]^{-1} C^H S_x^{-1} \rho_n S_x^{-1} C \left[C^H S_x^{-1} C\right]^{-1} g\right\}^{-1} \tag{19}$$

The array gain for the LCMP case is equal to the array gain for the LCMV case when the distortionless constraint is imposed and the signal is perfectly matched. Thus,

$$A_{lcmp} = \frac{1}{g^H \left[C^H \rho_n^{-1} C\right]^{-1} g} \tag{20}$$

One interpretation of the optimum processor is shown in fig.6. The beamformer first forms a set of M_c constraint beams and then combines them to form *Y(ω)*. One can note the processor can be viewed as operating in a M_c-dimensional constraint subspace [21].

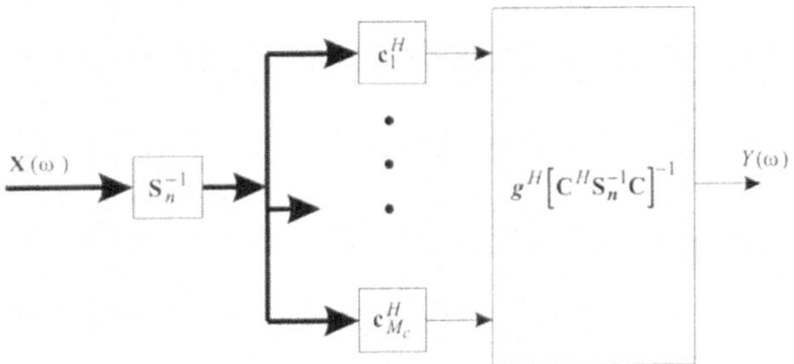

Fig. 6. Optimum constrained receiver

Diffusion Tensor Imaging

Diffusion tensor imaging (DTI) is a new imaging technique that can be used to noninvasively assess the molecular and biochemical environment of cerebral tissue [22-23]. DTI can aid in characterizing and measuring the diffusive transport of water molecules by means of an effective diffusion tensor, D. These symmetric tensor measurements contain useful information about the tissue microstructure and architecture. Of the several indices used to characterize diffusion tensor, those most commonly used are the trace of the tensor, which measures mean diffusivity (trace D), and fractional anisotropy (FA) [24-25]. These characteristic measurements may represent the changes in cerebral structure that occur in various neurologic conditions. DTI has been shown to be useful in the study of diseases, such as cerebral ischemia [26], acute stroke [27], multiple sclerosis [28] and schizophrenia [29]. Noninvasive MR imaging techniques are becoming increasingly important in lateralizing and localizing the seizure focus in a noninvasive manner. Only a few studies have addressed the utility of DTI in epilepsy [30-32]. In general, these studies have demonstrated increased diffusivity and decreased FA in cerebral tissue, which corresponded to the seizure focus in the entire group of the patients evaluated. However, these studies included a mixture of temporal epilepsy and extratemporal epilepsy with or without lesions. In addition, these studies failed to demonstrate a high yield of abnormal DTI measurements in individualized patients. A recent study evaluated DTI of white matter in temporal lobe epilepsy (TLE) [33]. In that study, significantly lower diffusion anisotropy and higher diffusivity in directions perpendicular to the axons were detected in several white matter structures in the patients, when compared with control subjects. However, none of these measured structures were in the temporal lobes. Another recent study [34] was conducted to evaluate apparent diffusion coefficient (ADC) values in patients with unilateral hippocampal sclerosis. The investigators found abnormal values on the side with hippocampal sclerosis in all patients, as compared with healthy volunteers. That study, however, was limited to unilateral hippocampal sclerosis findings on MR images, and it involved large regions of interest (ROIs) on axial sections, which are subject to partial-volume effects. An example of DTI, illustrated in fig. 7, is related to a flumazenil-positron emission tomography (FMZ PET) of a 7.2-year-old girl with intractable epilepsy projected on a 3-dimensional brain surface. Areas of >10% decreased FMZ binding are seen in black. Seizure onset was noted in the right inferior temporal cortex (yellow diamond) and areas of frequent (>10/min) interictal spiking (orange circle) were noted in the right temporal and frontal cortex. Both the seizure onset zone and the area of rapid seizure spread (circle with cross) were overlapping and/or adjacent to the areas of decreased FMZ binding. Scalp ictal EEG showed an anterior temporal focus but did not disclose epileptiform activity in the frontal region [35].

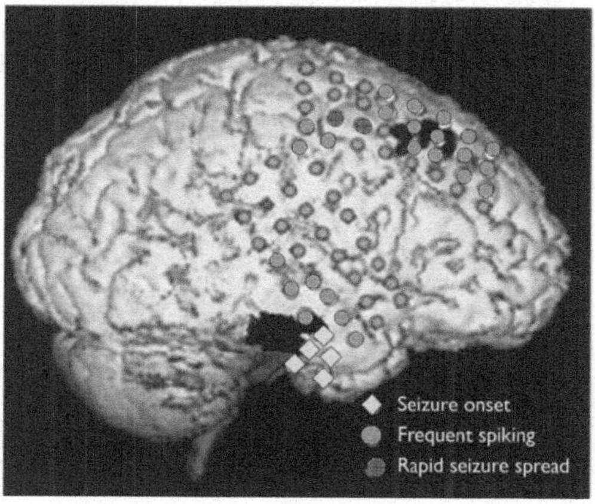

Fig. 7. FMZ PET

3 Applications

Using beamforming techniques, as illustrated in fig.8, it is possible to perform a qualitative characterization of EEG signals according to the amplitude using the initial signal (fig.9) by determining the amplitude levels as illustrated in fig.10 in order to discriminate petit mal from grand mal. Afterwards, the quantitative approach allows to overcome the LCMV beamformer limitations using the GSC (generalized sidelobe cancellers) as depicted in fig. 11. The GSC permits to control and to cancel the noise where it is necessary to remove by increasing the accuracy of the technique. These beamformers rely on the a priori information on the direction of arrival (DOA) of the desired signal, which can be adaptively determined via a source localization routine. The LCMV beamformer attempts to minimize the output power of the beamformer subject to the constraint that signals from the DOA (direction of arrival) do not get attenuated. The GSC achieves the same tasks but instead turns this constrained optimization problem into an unconstrained optimization problem by breaking up the LCMV optimal filter into two orthogonal subspaces.

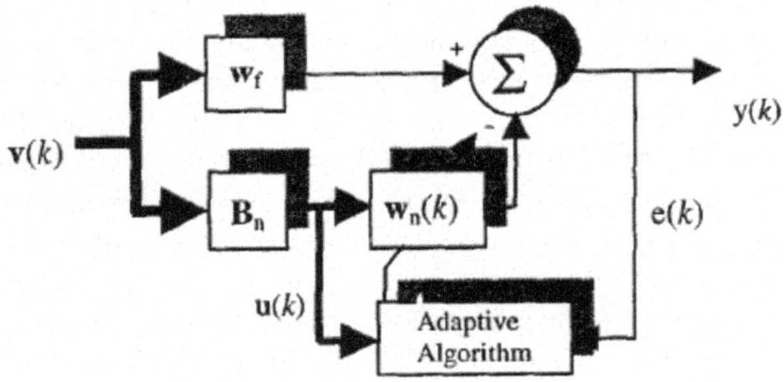

Fig. 8. A beamforming algorithm

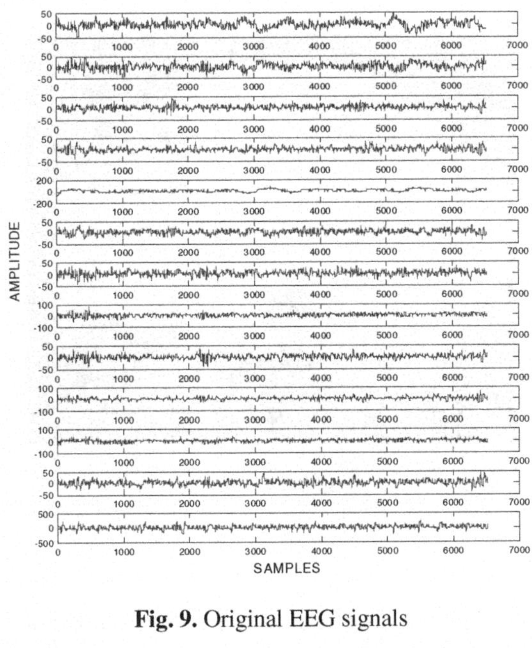

Fig. 9. Original EEG signals

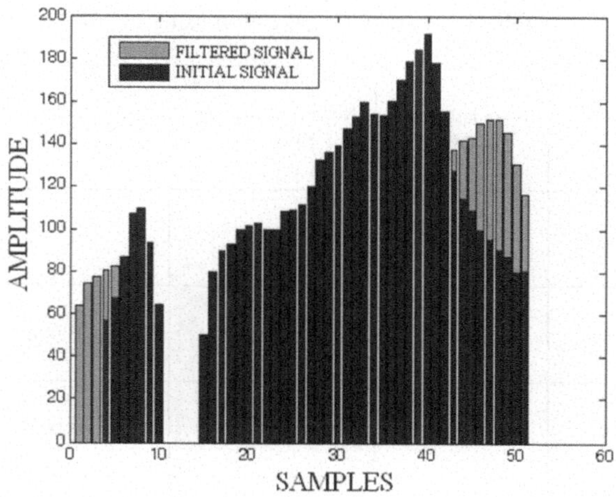

Fig. 10. Amplitudes

The complete approach is to use HOS to be combined with the beamforming for increasing the detection of epilepsy [36] as indicated in Table 1. This combination increases the accuracy of the method because it considers the EEG electrodes as an array of sensors. So, the classification is excellent.

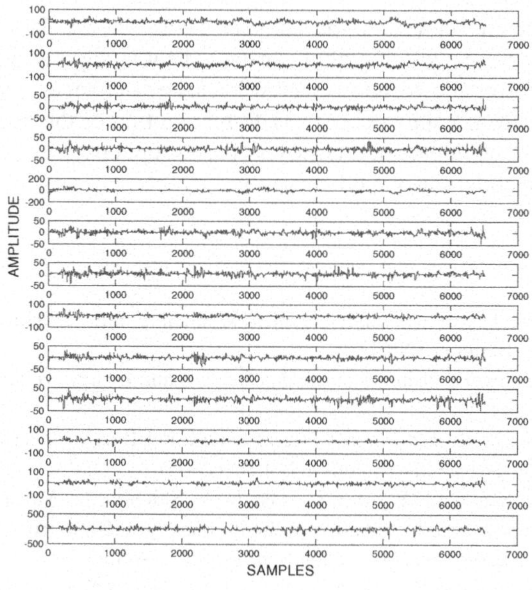

Fig. 11. EEG-based GSC processing

Table 1. Accuracy classification using LCMP

Classes	No of testing records	No of correctly classified data	Ratio of correct classification
Normal	300	275	91.66%
Pre-ictal	300	281	93.66%
Ictal	300	294	98.00%
Average			94.44%

4 Final Words

For many interesting cases faced in this research, beamfoming algorithm has delivered good and encouraging results, especially where it was necessary to increase accuracy in epilepsy detection about the state of patients. One limitation could be seen in terms of gain constant, if LMS (least mean square) algorithm is performed instead of LCMV: it is difficult to preserve the same gain constant for all electrodes involved. Electrodes 5 and 13 need a specific gain constant. That is important in terms of choosing the kind of information to be preserved because it changes according to patient age range: children from young people. With recognition of the potential role of DTI in the localization of epileptogenic cortex in partial epilepsy, its potential use in the identification of the epileptogenic tuber is useful.

References

1. WFUBMC, Wake Forest University Baptist Medical Center. Diagnostic Neurology Department. EEG Laboratory (2007)
2. Webster, J.G.: Medical Instrumentation. Application and Design, II edn. Wiley, Chichester (1977)
3. Lay-Ekuakille, A., et al.: Power Line Interference Cancelling in EEG Inspection. In: Proc. IMEKO TC-4 Congress, Gdynia-Jurata, Poland (2005)
4. Abeyratne, U.R., Kinouchi, Y., Oki, H., Okada, I., Shichijo, F., Matsumoto, K.: Artificial neural networks for source localization in the human brain. Brain Topogr 4(I), 3–21 (1991)
5. Abeyratne, U.R., Zhang, G., Saratchandran, P.: EEG source localization: a comparative study of classical and neural network rnethods. Int. J. Neural Syst. ll (4), 349–360 (2001)
6. Ahlfors, S.P., Simpson, G.V., Dale, A.M., Belliveau, J.W., Liu, A.K., Korvenoja, A., Virtanen, J., Huotilainen, M., Tootell, R.B., Aronen, H.J., Ilmoniemi, R.J.: Spatiotemporal activity of a cortical network for processing visual motion revealed by MEG and fMRI. Journal of Neurophysiology 82(5), 2545–2555 (1999)

7. Algaver, T., Smith, T., Vijai, F.: The use of artificial neural networks in biomedical technologies: an introduction. Biomed. Instrum. Technol. 28, 315–322 (1994)
8. Mizrahi, E.M.: Clinical, electroencephalographic, and quantitative predictors of neonatal seizures: the ACNs Presidential Address. In: Program and abstracts of the 55th Annual Meeting of the American Epilepsy Society, Philadelphia, Pennsylvania. Epilepsia, 42(suppl. 7), p.3 (2001)
9. Andrzejak, R.G.: Indications of nonlinear deterministic and finite-dimensional structures in time series of brain electrical activity: Dependence on recording region and brain state. Physical Review E 64, 061907 (2001)
10. Litt, B., Lehnertz, K.: Seizure prediction and the preseizure period. Current Opinion in Neurology 15, 173–177 (2002)
11. Mormann, F., et al.: On the predictability of epileptic seizures. Clinical Neurophysiology 116, 567–587 (2005)
12. Navarro, V., et al.: Seizure anticipation: Do mathematical measures correlate with video-EEG evaluation? Epilepsia 46(3), 385–396 (2005)
13. Chua, K.C., Chandran, V., Acharya, R., Lim, C.M.: Automatic identification of Epilepsy by HOS and power spectrum parameters using EEG signals: a comparative study. In: 30th Annual International IEEE EMBS Conference, Vancouver, BC, Canada, August 20-24 (2008)
14. EEG time series database, `http://www.meb.unibonn.de/epileptologie/science/physik/eegdata`
15. Chua, K.C., Chandran, V., Acharya, R., Lim, C.M.: Higher Order Spectral (HOS) Analysis of Epileptic EEG Signals. In: 29th Annual International IEEE EMBS Conference, Lyon, France, August 23-26 (2007)
16. Bell, K.L., Ephraim, Y., Van Trees, H.L.: Robust Adaptive Beamforming under Uncertainty in Source Direction-of-Arrival. In: Proc. IEEE Signal Processing Workshop on Statistical Signal and Array Processing, USA (1996)
17. Bell, K.L., Van Trees, H.L.: Adaptive and Non-Adaptive Beampattern Control Using Quadratic Beampattern Constraints. In: Proc. Conference Record of the Thirty-Third Asilomar on Signals, Systems, and Computers, USA (1999)
18. Bell, K.L., Van Trees, H.L.: Adaptive Beamforming for Spatially Spread Sources. In: Proc. Ninth IEEE SP Workshop on Statistical Signal and Array Processing, USA (1998)
19. Van Veen, B., Roberts, R.: A Framework for Beamforming Structures. IEEE Transactions on Acoustics, Speech, and Signal Processing 35(4), 584–586 (1987)
20. Van Veen, B., Roberts, R.: Partially Adaptive beamformer design via output power minimization. IEEE Transactions on Acoustics, Speech, and Signal Processing 35, 1524–1532 (1987)
21. Lay-Ekuakille, A., Vendramin, G., Trotta, A.: Acoustic Sensing for Safety Automotive Applications. In: Proc The 2nd International Conference on Sensing Technology, New Zealand (2007)
22. Le Bihan, D., Mangin, J.F., Poupon, C., Clark, C., Pappata, S., Molko, N., Chabriat, H.: Diffusion tensor imaging: concepts and applications. J. Magn. Reson. Imag. 13, 534–546 (2001)
23. Alsop, D.C., Connelly, A., Duncan, J.S., Hufnagel, A., Pierpaoli, C., Rugg-Gunn, F.J.: Diffusion and perfusion MRI in epilepsy. Epilepsia 43(suppl. 1), 69–77 (2002)
24. Pierpaoli, C., Jezzard, P., Basser, P.J., Barnett, A., Di Chiro, G.: Diffusion tensor MR imaging of the human brain. Radiology 201, 637–648 (1996)

25. Basser, P.J., Pierpaoli, C.: Microstructural and physiological features of tissues elucidated by quantitative-diffusion-tensor MRI. J. Magn. Reson. B 111, 209–219 (1996)
26. Lythgoe, M.F., Busza, A.L., Calamante, F., et al.: Effects of diffusion anisotropy on lesion delineation in a rat model of cerebral ischemia. Magn. Reson. Med. 38, 662–668 (1997)
27. Van Gelderen, P., de Vleeschouwer, M.H.M., DesPres, D., Pekar, J., Van Zijl, P.C.M., Moonen, C.T.W.: Water diffusion and acute stroke. Magn. Reson. Med. 31, 154–163 (1994)
28. Werring, D.J., Clark, C.A., Barker, G.J., Thompson, A.J., Miller, D.H.: Diffusion tensor imaging of lesions and normal-appearing white matter in multiple sclerosis. Neurology 52, 1626–1632 (1999)
29. Lim, K.O., Hedehus, M., Moseley, M., de Crespigny, A., Sullivan, E.V., Pfefferbaum, A.: Compromised white matter tract integrity in schizophrenia inferred from diffusion tensor imaging. Arch. Gen. Psychiatry 56, 367–374 (1999)
30. Wieshmann, U.C., Clark, C.A., Symms, M.R., Barker, G.J., Birnie, K.D., Shorvon, S.D.: Water diffusion in the human hippocampus in epilepsy. Magn. Reson. Imaging 17, 29–36 (1999)
31. Rugg-Gunn, F.J., Eriksson, S.H., Symms, M.R., Barker, G.J., Duncan, J.S.: Diffusion tensor imaging of cryptogenic and acquired partial epilepsies. Brain 124, 627–636 (2001)
32. Eriksson, S.H., Rugg-Gunn, F.J., Symms, M.R., Barker, G.J., Duncan, J.S.: Diffusion tensor imaging in patients with epilepsy and malformation of cortical development. Brain 124, 617–626 (2001)
33. Arfanakis, K., Hermann, B.P., Rogers, B.P., Carew, J.D., Seidenberg, M., Meyerand, M.E.: Diffusion tensor MRI in temporal lobe epilepsy. Magn. Reson. Imag 20, 511–519 (2002)
34. Yoo, S.Y., Chang, K.H., Song, I.C., et al.: Apparent diffusion coefficient value of the hippocampus in patients with hippocampal sclerosis and in healthy volunteers. AJNR Am. J. Neuroradiol. 23, 809–812 (2002)
35. Luat, A.F., Chugani, H.T.: Molecular and diffusion tensor imaging of epileptic networks. Epilepsia 49(suppl. 3), 15–22 (2008)
36. Lay-Ekuakille, A., Vendramin, G., Trotta, A.: Beamforming-aided processing of EEG signals for analyzing epileptic seizures. International Journal of Advanced and Communications 3(I/2), 110–125 (2009)

A Novel Portable Device for Laryngeal Pathologies Analysis and Classification

A. Palumbo[1,2], B. Calabrese[1], P. Vizza[1], N. Lombardo[3], A. Garozzo[3], M. Cannataro[1], F. Amato[1,*], and P. Veltri[1]

[1] School of Computer and Biomedical Engineering, University of Catanzaro Magna Græcia, Viale Europa, Campus di Germaneto "Salvatore Venuta", 88100 Catanzaro, Italy
amato@unicz.it
[2] Department of Electronics, Computer Science and Systems, University of Calabria, 87036, Rende, Italy
[3] School of Otorhinolaryngology, University of Catanzaro Magna Græcia, Viale Europa, Campus di Germaneto "Salvatore Venuta", 88100 Catanzaro, Italy

Abstract. Voice production is a process that involves the whole pneumophonoarticulatory apparatus. Nowadays, voice diseases are increasing quickly, affecting the quality of the human voice. So, an early detection and diagnosis of these diseases are very important for rehabilitation purposes and/or prevention purposes. In this chapter, a description of anatomy and physiology of voice phonoarticulary apparatus is presented. Moreover, a spectrographic analysis for pathological vocal signals is described. Finally, an example of portable device for acquisition and elaboration of vocal analysis is proposed.

Keywords: vocal tract pathologies, spectroacustic analysis, portable device.

1 Introduction

Voice is the result of a complex mechanism involving different organs of the pneumophonoarticulatory apparatus. In particular, it is the result of the vibration of the upper part of the mucosa covering the vocal cords. Such vibration determines the production of a sound, the larynx-fundamental tone, that is enriched by a set of harmonicas, generated by the resonance cavities in the upper part of the larynx. Any modification of this system may cause a qualitative and/or quantitative alteration of the voice, defined as dysphonia.

Dysphonia can be due to both organic factors (organic dysphonia) and other factors (dysfunctional dysphonia). Dysphonia is one of the major symptoms of benign laryngeal diseases, such as polyps or nodules, but it is often the first symptom of neoplastic diseases such as laryngeal cancer as well. Spectral "noise" is strictly linked to air flow turbulences in the vocal tract, mainly due to irregular vocal folds vibration and/or closure, causing dysphonia. Such symptom requires a set of endoscopic analysis (by using videolaryngoscope, VLS) for accurate analysis.

* Corresponding author.

S.C. Mukhopadhyay and A. Lay-Ekuakille (Eds.): Adv. in Biomedical Sensing, LNEE 55, pp. 335–352.
springerlink.com

However, clinical experience has pointed out that dysphonia is often underestimated by patients and, sometimes, even by family doctors. Hence, some delay in diagnosis is often found in case of neoplastic laryngeal diseases. The clinical experience proved that a dysphonia not early detected implies larynx cancer at advanced stadium. As widely reported in literature [1], an early detected glottis tumour (T1, T2 stadium) can be solved in 100% of cases with surgical intervention. Thus, the screening of voice alteration is extremely important in larynx diseases. Nowadays, the techniques of partial over-cricoid laryngectomy are widespread thus considerably reducing the number of cases of complete removal of the larynx, that cause serious personal problems for patients. Hence, early detection of dysphonia is of basic importance for pathology recovering.

Several experiences of using algorithmic approaches for the automatic analysis of signals exist [2, 3, 4, 5]. Software tools (commercial and freely available) allow manipulating voice components in an efficient way and permit specialists to manipulate and analyze voice signals. Automatic systems for the detection of illness related to abnormalities of the vocal signal have been developed and are mainly based on signal processing or on machine learning and data mining techniques. The problem is that most of them are usable only locally.

Home medical monitoring improves the quality of health care delivered at home and reduces costs associated with health care for health institutions. The use of portable devices can satisfy the need for a monitoring of the status of the vocal cords for early disease detection with no restrictions on accessibility and logistic. Furthermore, it is appropriate in case of particular monitoring requirements as for instance in rehabilitation of patients after surgery or medical treatment. Pre-surgical and post-surgical parameters evaluation allows the physician objectively quantifying the surgical effectiveness, and possibly provide an effective recovering.

The chapter begins with a brief description of anatomy and physiology of voice phonoarticulary apparatus. The following paragraphs introduce the fundament of clinical acoustic voice analysis. The description has focused on spectrographic analysis for pathological vocal signals. In the final paragraph an example of portable device for voice monitoring is presented.

1.1 Anatomy of Voice Production

The apparatus of articulation and resonance consists of structures and cavities that extend from the vocal cords (excluded) to the lips with the insertion of the nose. In Figure 1 is depicted this apparatus. It is called also supraglottic vocal tract and it is made up of laryngeal ventricle (or Morgagni's ventricle), the false vocal cords, the laryngeal vestibule, the pharynx, the oral cavity, the nasal cavity and the paranasal sinuses.

The larynx is placed on the anterior neck, slightly below the point where the pharynx divides and gives rise to the separate respiratory and digestive tracts. Because of its location, the larynx plays a critical role in normal breathing, swallowing and speaking.

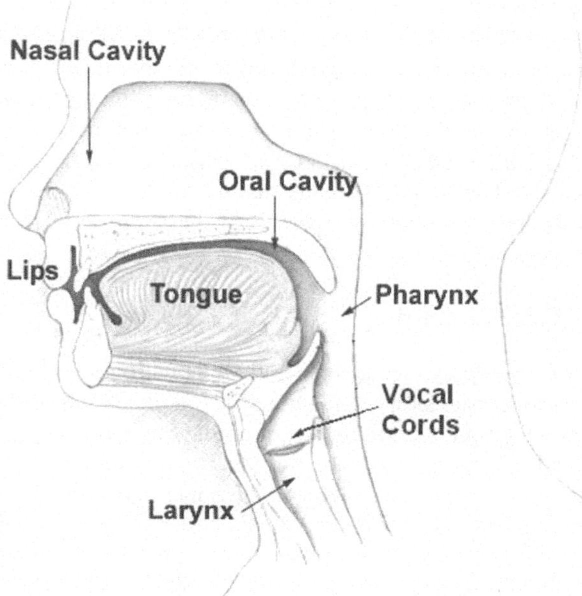

Fig. 1. Apparatus of articulation and resonance

The framework of the larynx is comprised mainly of the following cartilages: the upper thyroid cartilage (whose anterior prominence is oftentimes felt as the "Adam's apple"), the lower and smaller cricoid cartilage, the epiglottis, that lies superiorly, and the arytenoids cartilages, to which the vocal cords are attached. This structure protects the larynx during swallowing and prevents aspiration of food.

Inside the larynx are the vocal cords. The upper (false) vocal cords have protective function and they are not used in the phonation process. The vocal cords (true) delimit a virtual space, called the glottis, through which the passage of air into the trachea occurs.

The pharynx is a hollow tube that starts behind the nose and ends at the top of the trachea and esophagus. It can be divided into three parts: the upper pharynx or rino-pharynx, the oral pharynx or oropharynx and the lower pharynx or ipopharynx.

The nasal cavity is a large air-filled space lying between the floor of the cranium and the roof of the mouth and extending from the face to the pharynx.

Paranasal sinuses are air-filled spaces, communicating with the nasal cavity, within the bones of the skull and face. They are divided into subgroups (maxillary, frontal, ethmoid and sphenoid sinuses) that are named according to the bones within which the sinuses lie.

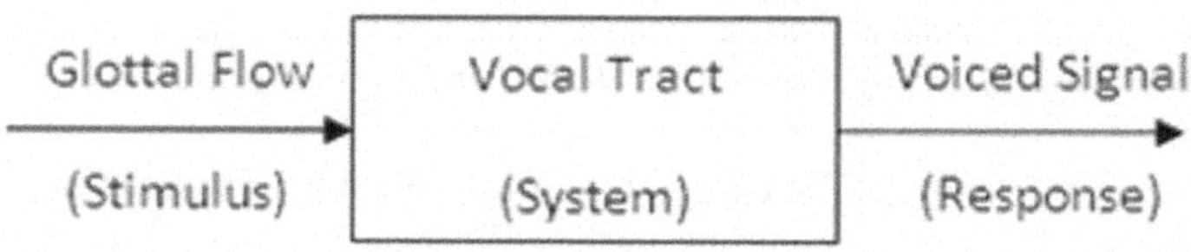

Fig. 2. Voice production model

The mouth, called also oral cavity, in addition to its primary role as the beginning of the digestive system, also plays a significant role in communication in humans. While primary aspects of the voice are produced in the throat, the tongue, lips, and jaw are also needed to produce the range of sounds included in human language.

The tongue is a muscle on the floor of the mouth that manipulates food for chewing and swallowing (deglutition). A secondary function of the tongue is speech, in which the organ assists. The tongue is made mainly of skeletal muscle. The tongue extends much further than is commonly perceived, past the posterior border of the mouth and into the oropharynx.

1.2 Physiology of Voice Production

Voice is the result of a complex mechanism, called phonation, involving different organs of the pneumophonoarticulatory apparatus [6]. It is the result of the vibration of the upper part of the mucosa covering the vocal cords. Such vibration determines the production of a sound, the larynx-fundamental tone, that is enriched by a set of harmonicas, generated by the resonance cavities in the upper part of the larynx.

During the phonation process, the aerodynamic energy generated by the respiratory system is transformed into a laryngeal acoustic energy; the larynx through the rhythmic glottic opening and closing behaves as a transducer of energy. The continuous air flow from the trachea is modulated. The glottal flow is then modified from the supraglottic structures.

The glottal flow is composed of glottal pulses. The period of a glottal pulse is the pitch period. The reciprocal of the pitch period is the fundamental frequency, also known as pitch. The vocal tract works as a time-varying filter to the glottal flow. The characteristics of the vocal tract include the frequency response, which depends on the position of organs. The peak frequencies in the frequency response of the vocal tract are formants, also known as formant frequencies.

In signal processing, a voice signal is a convolution of a time-varying stimulus and a time-varying filter. In particular, the time-varying stimulus is the glottal flow, whereas the time-varying filter is the vocal tract.

Figure 2 shows the voice production model used in signal processing.

1.3 Fluctuations and Perturbations of the Voice

Any modification of the apparatus of articulation and resonance may cause a qualitative and/or quantitative alteration of the voice, defined as dysphonia.

Dysphonia can be due to both organic factors (organic dysphonia) and other factors (dysfunctional dysphonia).

Dysphonia is one of the major symptoms of benign laryngeal diseases, such as polyps or nodules, but it is often the first symptom of neoplastic diseases such as laryngeal cancer as well. Spectral "noise" is strictly linked to air flow turbulences in the vocal tract, mainly due to irregular vocal folds vibration and/or closure, causing dysphonia.

Alteration of the voice can be caused by various phenomena occurring at all points in the production of speech or singing. In the following, the main causes of voice diseases are listed and briefly described:

- Acoustical Causes: the shaping of the vocal tract for various consonants, such as [b], [d] and [g], causes the vocal tract to be occluded for brief periods of time. This can halt phonation or cause register changes.
- Aerodynamic Causes: instability in the flow of air through the glottis is another source of fluctuations. This airflow can become turbulent and unstable creating a breathy-sounding voice.
- Biomechanical Causes: A first biomechanical source of fluctuations in vocal output is connected to the nature of the larynx itself. It presents various irregularities and variations in its composition, which inevitably cause the vocal folds to never vibrate exactly the same way twice. A second biomechanical cause is the irregular flow of blood through the vocal folds. Lastly, the various vocal articulators, such as the tongue, soft palate, and jaw, are continually moving during speech and singing, in order to create vowels and consonants. These articulators are connected to the larynx in various ways, and thus can affect vocal fold vibration.
- Neurological Causes: muscular contractions of various sorts are needed for voice production; as with all muscular activity, a certain amount of natural shaking, or tremor, is present. Various diseases, such as Parkinson's disease, can also cause large abnormal tremors in various muscles, and, in turn, in the voice.

2 Vocal Signal Analysis for Diagnostics

Specialists use different procedures to assess and diagnose the (eventual) anomalies of vocal fold; e.g. the flexible and rigid fiberscopic laryngoscopy, consisting in invasive examination with a fiber-optic instrument, the video stroboscopy, with a strobe illumination of the larynx that allows the visualization of movements, the electromyography that is an indirect observation of the functional state of the larynx, the videofluoroscopy that is a radiographic technique with radio-opaque substance to assess the swallowing function. All these methods are invasive and require a visit by a specialist.

Acoustic voice analysis is an effective and non-invasive method for evaluation and detection of laryngeal pathologies and is thus the first analysis performed on patients in the otorhinolaryngoiatric laboratories.

Up to now a substantial amount of research has been devoted to the determination of the influence of pathological changes of the larynx upon the voice signal [7, 8, 9]. The clinical research results put in evidence that laryngeal pathologies can cause large variations in voice fundamental frequency (pitch, F0) and voice amplitude (pitch and amplitude perturbation); presence of a loud turbulent (additive) noise; presence of sub-harmonic and non-harmonic components; interruptions in the pitch period generation. These changes are not always observed simultaneously. Only part of them could be present, depending on the disease and its stage.

The most widely used objective parameters in pathological voice analysis in the clinical context are the following:

1. Mean fundamental frequency (F0);
2. Standard deviation of fundamental frequency;

3. Pitch perturbation (jitter);
4. Relative average perturbation (RAP);
5. Amplitude perturbation (shimmer);
6. Signal to noise ratio (SNR);
7. Noise to harmonic ratio (NHR);
8. Normalized noise energy (NNE).

The systems, used for acoustic analysis and decision support in the otorhinolaryngoiatric laboratories, are: MDVP (Multi-Dimensional Voice Program) [10]; CSL (Computerized Speech Laboratory) [11]; Dr. Speech [12].

2.1 Vocal Parameters

In this subsection, the parameters used in the acoustic analysis and listed in the previous section, are defined.

Fundamental frequency. In clinical setting, the extraction of fundamental frequency and its time variations are useful for diagnostics and rehabilitation. The average numerical value may be part or not in the normal range of an adult male subject, an adult female or a child. In Table 1 the normal ranges of fundamental frequencies for different subjects are reported.

Table 1. Normal ranges of fundamental frequency parameter

Age [years]	Female Frequency [Hz]	Male Frequency [Hz]
Infant	440-590	440-590
3	255-360	255-360
8	215-300	210-295
12	200-280	195-275
15	185-260	135-205
Adults	175-245	105-160

The quasi-periodic pulses generated by the larynx, and subsequent changes made by the resonance of the vocal tract can determine some difficulties in extracting the fundamental frequency of the signals [13]. Further, correct F0 extraction is actually only reliable in quasi-periodic voices. In pathological situations, calculation of F0 in pathological voice analysis is complicated by the presence of strong amplitude and fundamental frequency perturbations of the vocal signal due to disturbed laryngeal functions.

The methods for the determination of F0 can be classified in the following three main groups:

- methods based on the signal in the time domain;
- methods based on the signal in the frequency domain;
- methods based on the signal in the cepstral domain.

Formant. The formants are the resonance frequencies of the vocal tract. The study of their characteristics allows an assessment of how the subject uses its glottis cavity.

Jitter. Jitter is the relative evaluation of the period-to-period (very short term) variability of the pitch within the analyzed voice sample. Voice break areas are excluded. The Jitter's equation is the following

$$\text{Jitt} = \frac{\frac{1}{N-1}\sum_{i=1}^{N-1}\left|To^{(i)} - To^{(i+1)}\right|}{\frac{1}{N}\sum_{i=1}^{N} To^{(i)}}$$

where To(i), i =1, 2 ... N is the extracted pitch period data and N = PER is the number of extracted pitch periods.

Relative Average Perturbation. RAP (Relative Average Perturbation) is the relative evaluation of the period-to-period variability of the pitch within the analyzed voice sample with smoothing factor of 3 periods. Voice break areas are excluded. RAP is defined as

$$\text{RAP} = \frac{\frac{1}{N-2}\sum_{i=2}^{N-1}\left|\frac{To^{(i-1)} + To^{(i)} + To^{(i+1)}}{3} - To^{(i)}\right|}{\frac{1}{N}\sum_{i=1}^{N} To^{(i)}}$$

where To(i), i =1, 2 ... N is the extracted pitch period data and N = PER is the number of extracted pitch periods.

Shimmer. Shimmer Percent /%/ is relative evaluation of the period-to period (very short term) variability of the peak-to-peak amplitude within the analyzed voice sample. Voice break areas are excluded. The Shimmer is defined as

$$\text{Shim} = \frac{\frac{1}{N-1}\sum_{i=1}^{N-1}\left|A^{(i)} - A^{(i+1)}\right|}{\frac{1}{N}\sum_{i=1}^{N} A^{(i)}}$$

where A(i) , i =1, 2...N is the extracted peak-to-peak amplitude data and N = PER is the number of extracted impulses.

Signal to Noise Ratio. SNR (signal to noise ratio) is defined as

$$\text{SNR} = 10\log_{10}\frac{\sum_{n=1}^{M} y^2(n)}{\sum_{n=1}^{M} w(n)^2}$$

where y(n) is the vocal signal sample at time n and w(n) is the noise signal. In the case of comparison of pre and post-surgical vocal signal, SNR can be defined as

$$\mathrm{SNR} = 10\log_{10}\frac{\sum_{n=1}^{M} y_{pre}^{2}(n)}{\sum_{n=1}^{M}\left(y_{pre}(n) - y_{post}(n)\right)^{2}}$$

where ypre(n)= pre-surgical signal sample at time n, ypost(n)=post-surgical signal sample at time n. SNR is, thus, the ratio between the noisy signal energy and that of removed noise. Negative SNR values correspond to voice quality enhancement.

Noise to Harmonic Ratio. NHR (Noise-to-Harmonic Ratio) is the ratio of the harmonics related to the noise energy present in the vocal signal. Both values are measured in dB. This is a general evaluation of noise present in the analyzed signal.

Normalized Noise Energy. NNE (Normalized Noise Energy) is a measure of the dysphonic component of the voice spectrum related to the total signal energy. It is defined as the ratio between the energy of noise and the total energy of the signal. Both values are measured in dB.

3 Spectrographic Analysis for Vocal Signals

Clinical experience has pointed out that dysphonia is often underestimated by patients and, sometimes, even by family doctors. A dysphonia not early detected implies larynx cancer at advanced stadium. Hence, early detection of dysphonia is of basic importance for pathology recovering.

Recently, a web-based system for the acquisition and remote automatic analysis of vocal signals with an online screening has been developed [14]; vocal signals are submitted through a web-interface and then they are analyzed by using signal processing technique, providing first-level information on possible voice alterations. The system is not able to satisfy the need of a monitoring status of the vocal cords and it is inappropriate in case of particular monitoring requirements, with restrictions on accessibility and logistic.

So, to improve the screening in a greater number of patients, it is necessary the use of portable devices, that can be used by patients directly from home. The home monitoring is important not only in the screening phase but also in rehabilitation for monitoring the larynx progress and support the recovery of vocal functions.

A system that performs real time monitoring of progress of laryngeal dysfunctions and allows a pre-surgical and post-surgical comparison of the main voice characteristics exhibits the following architecture (Fig. 3).

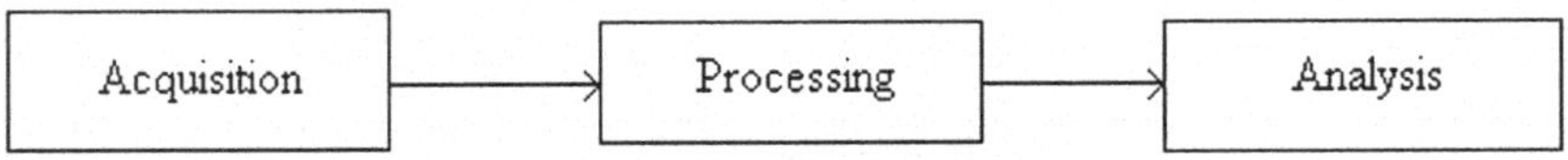

Fig. 3. Architecture of the system

The first step consists of vocal signals acquisition through a microphone connected to an audio circuit. The signal is suitably amplified and conditioned with a specific pass-band filter. Then the signal is processed and the elaboration results are analyzed.

In the literature, different methods for extracting acoustical and noise parameters (described above) are reported [15, 16]. Considering both the spectral distribution, the intensity of the a-periodic component (noise) and alteration of harmonics, Yanagihara [17] proposed a spectrographic classification of severity of dysphonia. The study interested 167 patients who had perceptually mild, moderate and serious dysphonia:

- Type I: regular harmonic components are mixed to the component of noise in the formant of vowels [a], [i] [u], [o] and [e] (under 3000 Hz): dysphonia light.
- Type II: the noise components in the second formant of [i] and [e] predominates the harmonic components, and slight noise appears even at frequencies above 3000 Hz, again in the same voice ([i] and [e]): dysphonia moderate.
- Type III: the second formant of [i] and [e] is completely replaced by noise, which further increases above the 3000 Hz: dysphonia serious.
- Type IV: the second formant of [a], [i] and [e] are replaced by noise, the first formants of all vowels lose their periodic component and the noise at high frequency increases in intensity: dysphonia very serious.

In particular, a classical method to extract the information about the fundamental frequency and harmonic content of a given signal is the Fast Fourier Transform (FFT) and Short Time Fourier Transform (STFT) analysis.

Figures 4 and 5 show the results relative to FFT elaborations of normal and pathological vocal signals. The power spectrum of the healthy subject reveals a well-defined value of the fundamental frequency in a normal range and the energy of harmonics (or formants) is lower than the fundamental.

The spectrum in Figure 5 indicates a pathological condition of the vocal tract. The fundamental frequency has a very low energy value, similar to that one of other frequency components located at higher frequencies in a pathological range. In addition, noise contaminates the entire spectrum.

The pathological vocal signals are not stationary signals because the frequency contents change over time. Because the basic functions used in the classical Fourier analysis do not associate with any particular time instant, the resulting measurements, Fourier transforms, do not explicitly reflect a signal's time-varying nature.

The Short-Time Fourier Transform (STFT) is a known method used to analyze a time-varying signal or non-stationary signal whose frequency components varies over time. To compute the STFT of the whole signal, a sliding window is used to divide the signal into several blocks and then the fast Fourier transform (FFT) is applied to each data block to obtain the frequency contents.

The following equation to compute the STFT is used

$$\mathrm{STFT}[\mathrm{u},\omega] = \int f(t)g(t-u)e^{-i\omega t}dt$$

where f(t) is the signal and g is the window function. The STFT algorithm aligns the center of the first sliding window with the first sample of the signal and extends the signal at the beginning with the zeroes or the signal itself.

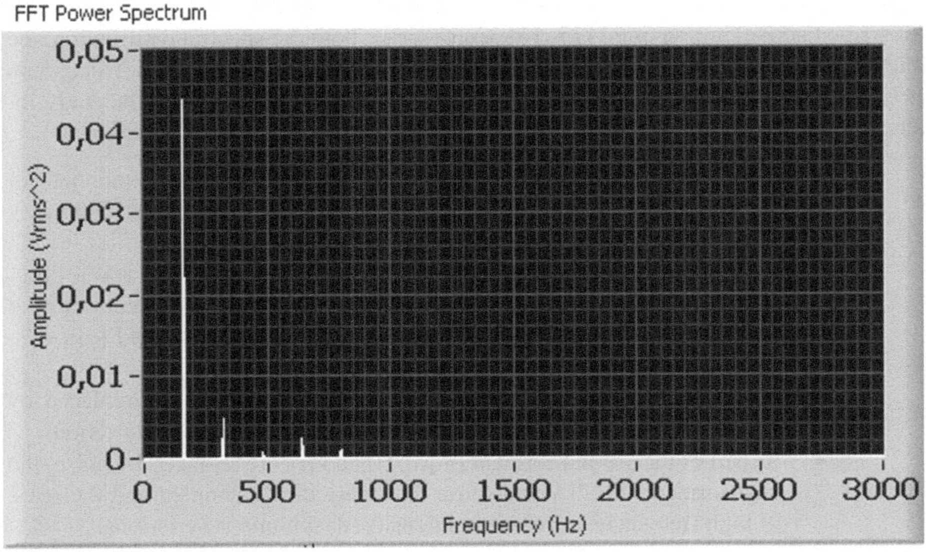

Fig. 4. Power spectrum of a normophonic subject

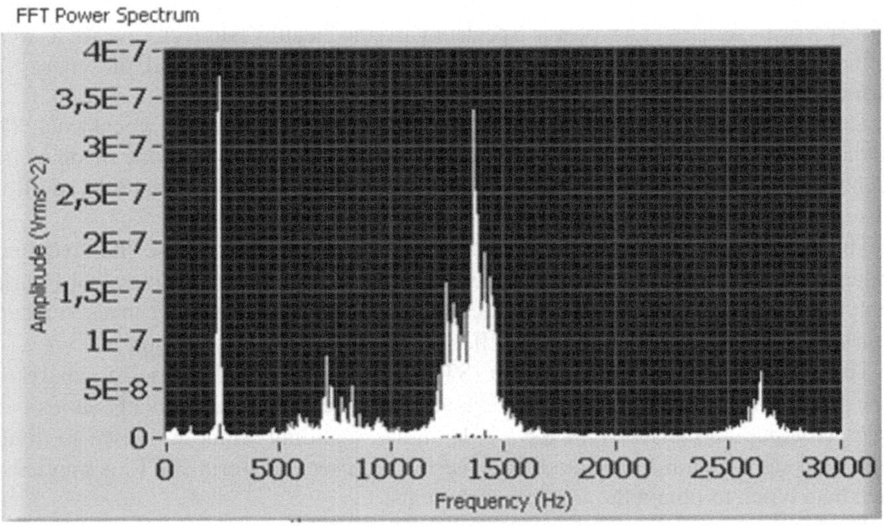

Fig. 5. Power spectrum of a pathological subject

The window function determines the joint time-frequency resolution of the STFT. The longer the window length, the better the frequency resolution and the worse the time resolution.

The spectrogram represents the temporal variation of the spectral content of the voice; it is a three dimensional graphic. The spectrograms shown in Figure 6 and

Figure 7 are related to a normal and pathological vocal signals, respectively. In the case of normophonic subject, the value of fundamental frequency remains almost constant during the acquisition.

The main characteristic of irregular spectrograms is given by the inability of discriminating fundamental frequency and different harmonic components. Indeed pathological spectrograms are characterized by:

- lack of energy in correspondence of the fundamental frequency and/or harmonics;
- presence of noise;
- higher energy density at the harmonics;
- higher energy density at the sub-harmonics or non-harmonic components in the pathological frequency range (higher frequencies than the normal).

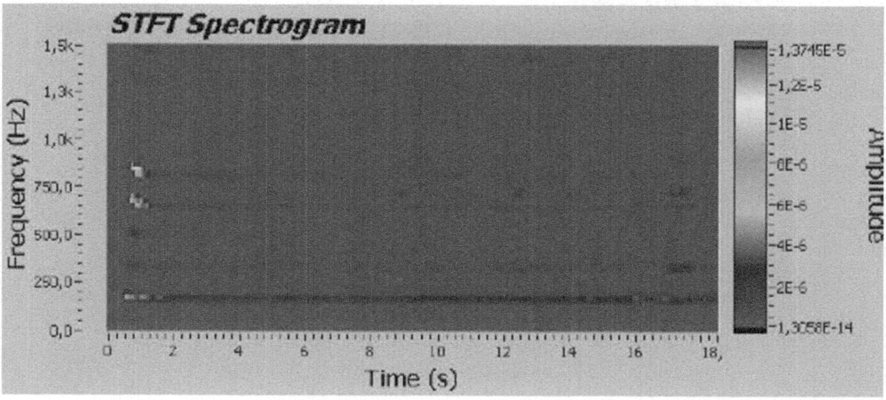

Fig. 6. Spectrogram of vocal signal acquired from a normophonic subject

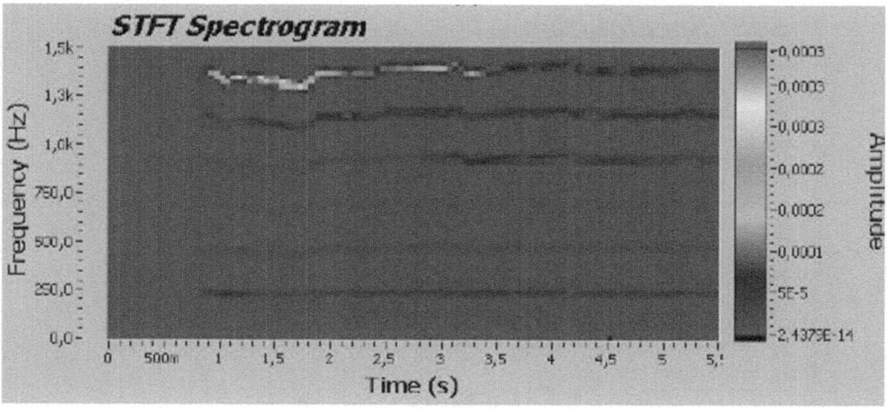

Fig. 7. Spectrogram of vocal signal acquired from a patient affected by dysphonia

In particular spectrogram analysis puts in evidence that the frequency content of a vocal signal shows strong variation over the entire acquisition time caused by vocal fold pathologies. The spectrogram in Figure 7 reveals some of these characteristics, such as the presence of higher energy density of the frequency components in the pathological range than the fundamental frequency and the variability of these frequencies values.

The plots reported in Figure 8 and Figure 9 are relative to a pre- and post-surgical monitoring of a patient affected by polyps. The pre-surgical spectrogram analysis reveals the presence of several harmonics and sub-harmonics components characterized by higher energy than the fundamental (Fig. 8). After medical treatment, the spectrogram shows that the fundamental frequency is the highest energy component, whereas there are not evident sub-harmonic components (Fig. 9). This analysis allows to confirm the effectiveness of medical treatment and the recovery of the vocal functions.

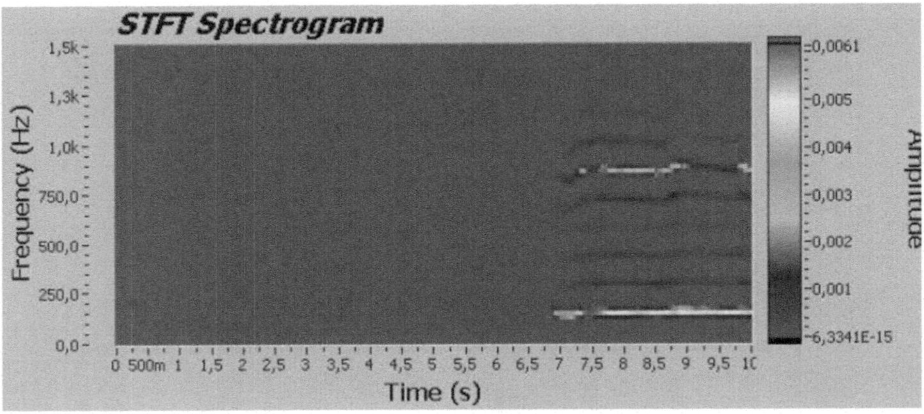

Fig. 8. Spectrogram of pathological vocal signal acquired from a patient before surgery

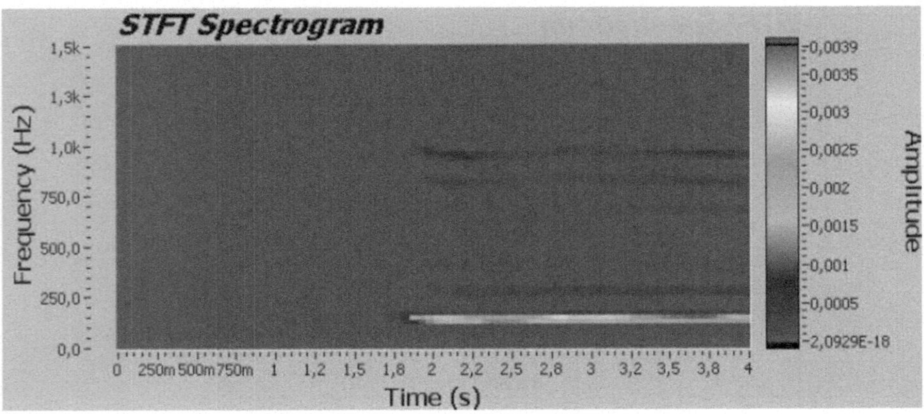

Fig. 9. Spectrogram of vocal signal acquired from the same patient after surgery

4 Development of a Portable Device for Pathological Voice Analysis

4.1 The Reference Hardware

A novel portable device for vocal signal analysis and classification is described in [18]. The implemented system allows to realize a quantitative and non invasive home-monitoring of vocal system diseases. The reference architecture is based on a digital signal processor DSP. The proposed acquisition system requires portability and usability with minimum weight and size. For these reasons, a low voltage DSP was used equipped with internal analog to digital converter (ADC) and internal memory for the user code. To develop the application a starter kit, developed from a joint venture between Analog Devices and National Instruments, has been used (Fig. 10). The DSP is the ADSP-BF537 BLACKfin Processor. The ADSP-BF537 processor is a member of the Blackfin family of products, incorporating the Analog Devices/Intel Micro Signal Architecture (MSA) [19]. Blackfin Processors are a new class of devices that combine the characteristics of the SIMD (Single Instruction Multiple Data) processor with elements such as Memory Management Unit (MMU), watchdog timer, UART and SPI ports. Typically these features are available in microcontrollers and microprocessors.

The architecture of the processor is the Micro Signal Architecture (MSA) developed by Analog Devices and Intel. It is a modified Harvard architecture combined with a hierarchical memory structure characterized by a L1 memory (that operates at the same speed of the processor) and an external memory.

The Blackfin processors contain a rich set of peripherals connected to the core via several high bandwidth buses, providing flexibility in system configuration as well as excellent overall system performance (Fig. 11). The processor contains dedicated network communication modules and high speed serial and parallel ports, an interrupt controller for flexible management of interrupts from the on-chip peripherals or external sources, and power management control functions. The Blackfin processor core contains two 16-bit multipliers, two 40-bit accumulators, two 40-bit ALUs, four video ALUs, and a 40-bit shifter. The core clock frequency reaches up to 600 MHz.

The Blakfin processors have up to 132K bytes of on-chip memory. Blackfin processors are produced with a low power and low voltage design methodology and feature on-chip dynamic power management, which is the ability to vary both the voltage and frequency of operation to significantly decrease overall power consumption. This allows longer battery life for our portable appliance. The board provides four push buttons and six leds for general-purpose I/O.

The application was developed by using LabVIEW Embedded for ADI Blackfin Processors 2.0 [20] for programming DSP Blackfin, and Analog Devices VisualDSP++4.5 for translating the LabVIEW code and transferring it on DSP. LabVIEW Embedded Module for Blackfin represents a comprehensive graphical development approach for embedded design, jointly developed by Analog Devices and National Instruments. Seamlessly it integrates LabVIEW and Visual DSP++ to deliver an easy to use programming toolset for quicker time to application development.

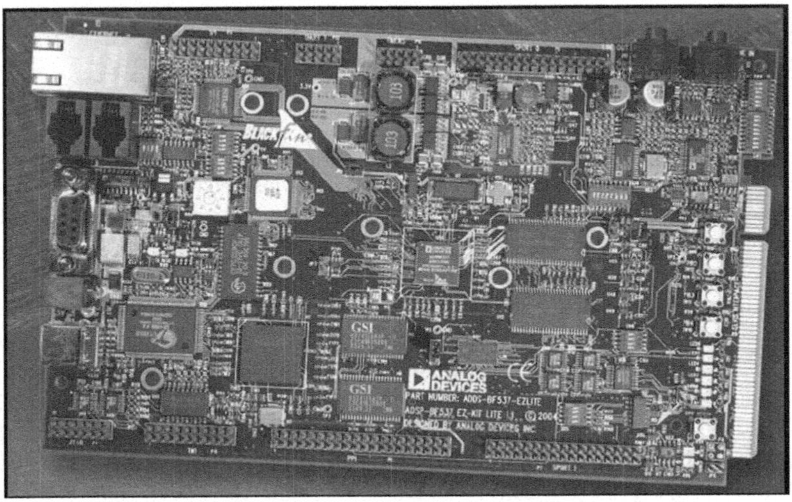

Fig. 10. The DSP Board

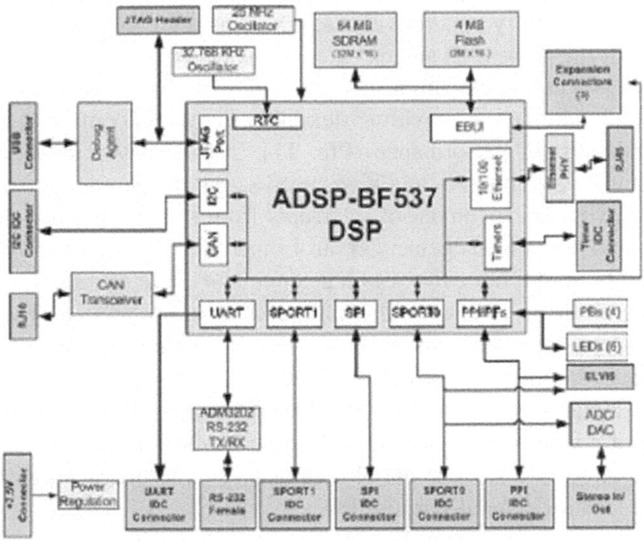

Fig. 11. DSP Board Architecture

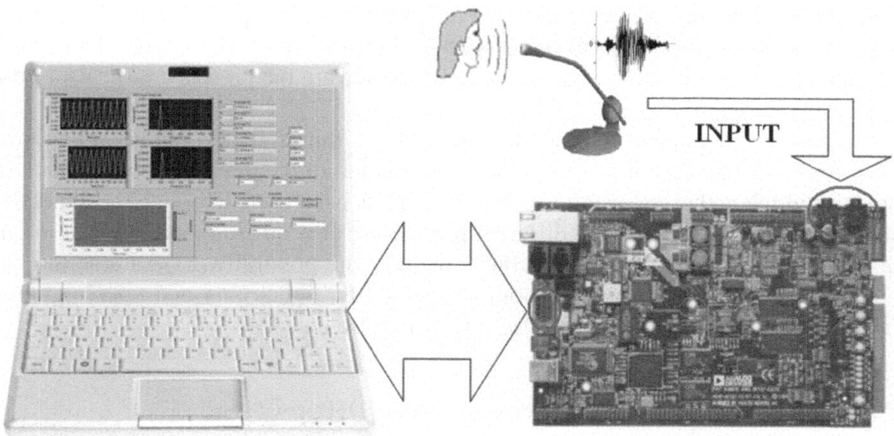

Fig. 12. Setup Configuration

4.2 Architectural Description of the Proposed System

The proposed system can work independently or connected to a laptop or PC (Fig. 12). The vocal signals are acquired through an audio circuit, available on board. The audio circuit consists of an AD1871 analog-to-digital converter (ADC) and a AD1854 digital-to-analog converter (DAC). The audio circuit provides one channel of stereo input and one channel of stereo output via 3.5 mm stereo jacks. The audio interface samples data at a 48 kHz sample rate. The AD1871 is a stereo audio ADC and it features two 24-bit conversion channels each with programmable gain amplifier (PGA), multi-bit sigma-delta modulator and decimation filters. Each channel provides 97 dB of THD+N and 107 dB of Dynamic Range.

A 4th order Butterworth pass-band filter was used to filter vocal signals in the range 50-400 Hz. The filter frequencies can be modified through four push button available on board.

Fast Fourier Transform (FFT) and Short Time Fourier Transform (STFT) analysis have been implemented on DSP to identify and classify vocal anomalies (Yanagihara classification [17]).

The application searches the fundamental frequency (F0) in the range between 100Hz and 300Hz (RangeF0), the second harmonic (or first formant, F1) in the frequency range 2x (RangeF0) and the third harmonic (or second formant, F2) in the range 3x (RangeF0), by evaluating the average power over time of coefficients extracted from STFT matrix.

After the search of the fundamental frequency and the first two harmonic components, the system looks for significant non-harmonic or sub-harmonic components. If these components have amplitude values greater than defined thresholds, they are not negligible.

Then, the system classifies vocal signals between normal or pathological. In particular a normal signal has a well-defined fundamental frequency, eventual first and second harmonic and no components in the pathological range. A pathological signal presents a fundamental frequency and harmonic components amplitude values below thresholds and the most relevant peaks in the pathological range. In stand-alone mode the six leds give information to patients about voice quality. Three leds identify the 1st, 2nd and 3rd harmonics, whereas other three indicate the presence of peaks in the pathological range.

The device is equipped with a software tool offering a configuration and control web-based user interface that allows the configuration of the device to be furnished to patients and the gathering of information after it has been used. Through the panel the user can dynamically configure different parameters of acquisition and elaboration in the debug mode. In particular, the clinicians have the possibility to examine the acquired vocal signals in the time domain and power spectrums and spectrograms. Furthermore, the system evaluates also different noise indexes.

Table 2. Elaboration results of DSP-based system

Patients	Medical Diagnosis	Evaluated Parameters				
		F0	SNR	SINAD	THD	Spectrogram Classification
Patient A	Normophonic voice	214,156	-10,966	2,525	-Inf	Regular
Patient B	Normophonic voice	143,659	-13,658	0,922	-12,851	Regular
Patient C	Normophonic voice	176,334	-16,025	0,325	-7,282	Regular
Patient D	Pathological voice	123,324	-21,019	0,229	-16,951	Irregular
Patient E	Pathological voice	143,657	-17,071	0,722	-16,951	Irregular
Patient F	Pathological voice	156,015	-11,424	2,016	-32,088	Irregular

Table 2 reports some of the elaboration results of the presented system. During the voice acquisitions, the subjects spell a vowel for a few seconds. In the second column medical diagnosis are reported. The other columns show, instead, the elaboration results of the DSP-based system:

- Fundamental Frquency (F0);
- Signal to Noise Ratio (SNR);
- Signal to Noise and Distortion (SINAD);
- Total Harmonic Distortion (THD);
- Spectrogram Classification.

The system is able to discriminate between normophonic and pathological voice signals through spectrogram analysis based on STFT elaboration of acquired vocal signals. As shown by the results in Table 2, the evaluation of only one type of parameters does not permit a correct classification of a given vocal signal. For example, fundamental frequencies for pathological voices fall, in the reported cases, into the normal frequency range. Similarly, the evaluated noise parameters do not have values that allow to discriminate between the two voice categories, whereas spectrogram analysis succeeds in classifying the different vocal signals.

However, the system is aimed to be a support for doctors and is not intended to provide a complete diagnosis. The DSP-based system alerts the subject of any vocal tract anomalies and/or gives information about the recovery of the phonatory abilities during the rehabilitation process. In the presence of anomalies, patients should contact a specialist.

5 Conclusions

Dysphonia is the major symptom of different laryngeal pathologies. An early detection of dysphonia is necessary for the diagnosis of neoplastic laryngeal diseases. Different techniques are used in clinical practice to assess the vocal tract conditions. Among these, acoustic analysis is a non invasive and reliable technique. Different parameters, used in the clinical acoustic analysis for quantifying dysphonia severity, are defined in this chapter. In particular, spectrographic analysis allows to obtain more interesting results. The use of time-frequency techniques permits to improve the accuracy of discriminating between normal and pathological voices, because pathological vocal signals are non-stationary.

The use of portable devices can increase the rate of laryngeal pathologies detection. Home monitoring is appropriate, also, in the rehabilitation process saving time and costs for health institutions. In this context, a proposal of portable DSP-based device has been illustrated; the description of the architecture and the system functionalities has also been provided. A first set of interesting results are reported to demonstrate the efficiency of the proposed approach.

References

1. Stemple, J.C., Glaze, L.E., Gerdermann, B.K.: Clinical Voice Pathology: Theory and Management. Thomson Delmar Learning (2000)
2. Moran, R.J., Reilly, R.B., Chazal, P., Lacy, P.D.: Telephony–based voice pathology assessment using automated speech analysis. IEEE Transaction on Biomedical Engineering 53(3), 468–477 (2006)
3. Hadjitodorov, S., Mitev, P.: A computer system for acoustic analysis of pathological voices and laryngeal diseases screening. Medical Engineering & Physics 24(6), 419–429 (2002)
4. COST Action 2103, Advanced Voice Function Assessment, Monitoring Progress Report (2008)
5. Umapathy, K., Krishnan, S., Parsa, V., Jamieson, D.: Discrimination of pathological voices using a time-frequency approach. IEEE Transaction on Biomedical Engineering 52(3), 421–430 (2005)
6. Guyton, A.C., Hall, J.E.: Medical Physiology, 10th edn. W. B. Saunders Company, Philadelphia
7. Boyanov, B., Hadjitodorov, S.: Acoustic Analysis of Pathological Voice. IEEE Engineering in Medicine and Biology Magazine 16(4), 74–82 (1997)
8. Martinez, C.E., Rufiner, H.L.: Acoustic Analysis of Speech for Detection of Laryngeal Pathologies. In: Proc. 22nd Annual International Conference of the IEEE Engineering in Medicine and Biology, vol. 3, pp. 2369–2372 (2000)
9. Dibazar, A.A., Berger, T.W., Narayanan, S.S.: Pathological Voice Assessment. In: Proc. 28th IEEE EMBS Annual International Conference of the IEEE Engineering in Medicine and Biology, pp. 1669–1673 (2006)

10. Multi-Dimensional Voice Program (MDVP), KayPENTAX, http://www.kayelemetrics.com/Product%20Info/CSL%20Options/5105/5105.htm
11. Computerized Speech Lab, KayPENTAX, http://www.kayelemetrics.com/Product%20Info/CSL%20Family/4500/4500.htm
12. Dr. Speech, Tiger Electronics, Inc., http://www.drspeech.com/
13. Mitev, P., Hadjitodorov, S.: Fundamental frequency estimation of voice of patients with laringeal disorders. Information Sciences 156, 3–19 (2003)
14. Amato, F., Cannataro, M., Cosentino, C., Garozzo, A., Lombardo, N., Manfredi, C., Montefusco, F., Tradigo, G., Veltri, P.: Early Detection of Voice Diseases via a Web-based System. Biomedical Signal Processing and Control 4(4), 206–211 (2009)
15. Manfredi, C., Peretti, G.: A New Insight Into Postsurgical Objective Voice Quality Evaluation: Application to Thyroplastic Medialization. IEEE Transaction on Biomedical Engineering 53(3), 442–451 (2006)
16. Costa, S.C., Aguiar Neto, B.G., Fechine, J.M.: Pathological Voice Discrimination using Cepstral Analysis, Vector Quantization and Hidden Markov Models. In: Proc. 8th International IEEE Conference on Bioinformatics and Bioengineering, pp. 1–5 (2008)
17. Yanagihara, N.: Significant of harmonic changes and noise components in hoarseness. J. Speech Hear 10, 513–541 (1967)
18. Palumbo, A., Amato, F., Calabrese, B., Cannataro, M., Veltri, P., Garozzo, A., Lombardo, N.: A Novel Portable Device for Pathological Voice Analysis. In: Proc. International Workshop on Medical Measurements and Applications, Cetraro, Italy, May 29-30 (2009)
19. ADSP-BF537 Blackfin® Processors Hardware Reference, Analog Devices, Revision 3.4 (April 2009), http://www.analog.com
20. Anderson, G.: LabVIEW Embedded for Blackfin, Analog Devices (2007), http://www.ni.com/labview/blackfin

Author Index

The manufacturer's authorised representative in the EU is Springer Nature Customer Service Centre GmbH, Europaplatz 3, 69115 Heidelberg, Germany. If you have any concerns regarding our products, please contact ProductSafety@springernature.com

Printed and bound by CPI Group (UK) Ltd, Croydon, CR0 4YY
25/09/2026
02222891-0012